Python
开发实例大全 上卷

张善香　田蕴琦　张晓博◎编著

人民邮电出版社
北　京

图书在版编目（CIP）数据

Python开发实例大全. 上卷 / 张善香，田蕴琦，张晓博编著. -- 北京：人民邮电出版社，2023.5
　ISBN 978-7-115-52855-1

Ⅰ. ①P… Ⅱ. ①张… ②田… ③张… Ⅲ. ①软件工具－程序设计 Ⅳ. ①TP311.561

中国国家版本馆CIP数据核字(2023)第061100号

内 容 提 要

本书内容齐全，通过实例循序渐进地讲解了开发 Python 应用程序的知识。本书主要内容包括：搭建开发环境、基础语法、标准库函数、进程通信和联网操作、结构化标记处理、数据持久化操作、特殊文本格式处理、图形化界面开发等知识。

本书既适合零基础的 Python 读者或已经了解了 Python 语言基础语法的读者阅读，也适合希望进一步提高自己 Python 开发水平的读者阅读，还可作为程序员的参考书。

◆ 编　著　张善香　田蕴琦　张晓博
　　责任编辑　张　涛
　　责任印制　王　郁　焦志炜
◆ 人民邮电出版社出版发行　北京市丰台区成寿寺路 11 号
　　邮编 100164　电子邮件 315@ptpress.com.cn
　　网址 https://www.ptpress.com.cn
　　固安县铭成印刷有限公司印刷
◆ 开本：787×1092　1/16
　　印张：26　　　　　　　　　　　2023 年 5 月第 1 版
　　字数：685 千字　　　　　　　　2023 年 5 月河北第 1 次印刷

定价：109.80 元

读者服务热线：(010)81055410　印装质量热线：(010)81055316
反盗版热线：(010)81055315
广告经营许可证：京东市监广登字 20170147 号

前　　言

本书通过实例讲解 Python 开发程序的知识。书中的每一个实例都是作者精心规划并挑选出的，目的是希望能够帮助程序员掌握每一个知识点，并提高程序员的开发效率。实践才是学习 Python 编程的最佳方法之一。为了帮助更多读者成为 Python 的开发者，笔者特意编写了本书。

本书的特色

1. 内容全面，实例多

本书的实例几乎涵盖了常用的 Python 应用领域，笔者通过讲解众多的实例，帮助读者高效地学习 Python 的强大功能，逐步步入 Python 开发高手之列。

2. 贴心提示和注意事项提醒

本书根据需要在内文里安排了"注意"等小板块，帮助读者理解相关知识点及概念，更快地掌握 Python 的应用技巧。

3. 通过 QQ 群提供答疑服务

为了方便给读者答疑，本书特提供了 QQ 群作为售后支持。读者无论是读书时有疑惑，还是学习中有问题，都可以随时通过 QQ 群与作者互动。

本书的 QQ 群号是：169526387。

本书的内容

本书实例齐全，几乎涵盖了主流的 Python 开发应用领域，并且以极简的文字介绍了复杂的案例，易于阅读，是学习 Python 开发的完美教程。

本书的读者对象

Python 开发初学者；

软件工程师；

测试和自动化框架开发人员。

致谢

笔者在编写本书的过程中，得到了人民邮电出版社多位编辑的大力支持。正是多位编辑的求实、耐心和高效才使得本书能够在短时间内出版。另外，也十分感谢我的家人给予的大力支持。本人水平有限，书中难免会有错误或不妥当的地方，敬请读者指正，以便修订并使之更臻完善。编辑联系邮箱：zhangtao@ptpress.com.cn。

感谢您购买本书，希望本书能成为您编程路上的领航者，祝您阅读快乐！

笔者

目 录

第1章 搭建开发环境实战 ··1
1.1 安装Python环境 ··2
实例01-01：在Windows操作系统中下载并安装Python ·······································2
实例01-02：在macOS操作系统中下载并安装Python ··3
实例01-03：在Linux操作系统中下载并安装Python ··4
1.2 安装开发工具 ··4
实例01-04：使用Python自带工具IDLE ···4
实例01-05：安装PyCharm ···5
实例01-06：安装Eclipse ···8
1.3 编写并运行Python程序 ··12
实例01-07：使用IDLE编写并运行Python程序 ··12
实例01-08：使用命令行运行Python程序 ··12
实例01-09：交互式运行Python程序 ··13
实例01-10：使用PyCharm编写并运行Python程序 ···13
实例01-11：使用Eclipse编写并运行Python程序 ···16

第2章 基础语法实战 ··20
2.1 使用列表 ···21
实例02-01：创建数字列表 ··21
实例02-02：访问列表中的值 ··21
实例02-03：删除列表中的重复元素并保持顺序不变 ··22
实例02-04：找出列表中出现次数最多的元素 ···23
实例02-05：对类的实例进行排序 ··23
实例02-06：使用列表推导式 ··24
实例02-07：命名切片操作 ··25
2.2 使用元组 ···26
实例02-08：创建并访问元组 ··26
实例02-09：修改元组的值 ··26
实例02-10：删除元组 ··27
实例02-11：使用内置方法操作元组 ··27
实例02-12：将序列分解为单独的变量 ··28
实例02-13：将序列的最后几项作为历史记录 ···29
实例02-14：实现优先级队列 ··29
2.3 使用字典 ···31

目　录

实例02-15：创建并访问字典 ·· 31
实例02-16：向字典中添加数据 ·· 32
实例02-17：修改字典中的数据 ·· 32
实例02-18：删除字典中的元素 ·· 32
实例02-19：创建一键多值字典 ·· 33
实例02-20：使用OrderedDict类创建和修改有序字典 ······································ 34
实例02-21：获取字典中的最大值和最小值 ··· 35
实例02-22：获取两个字典中的相同键值对 ··· 35
实例02-23：使用函数itemgetter()对字典进行排序 ··· 36
实例02-24：使用字典推导式 ·· 37
实例02-25：根据字段进行分组 ·· 38
实例02-26：转换并换算数据 ·· 39
实例02-27：将多个映射合并为单个映射 ··· 40

2.4　变量 ·· 41
实例02-28：实现变量赋值 ·· 41
实例02-29：使用局部变量 ·· 42
实例02-30：使用全局变量 ·· 42
实例02-31：体验关键字global的作用 ·· 43
实例02-32：定义常量功能类_const ··· 44

2.5　条件语句和循环流程控制 ·· 45
实例02-33：使用条件语句判断年龄 ··· 45
实例02-34：使用for循环遍历单词"Python"中的各个字母 ···························· 46
实例02-35：使用range()循环输出列表中的元素 ··· 47
实例02-36：循环输出10～20的各数是否是质数 ·· 47
实例02-37：获取两个整数之间的所有素数 ··· 48
实例02-38：输出九九乘法表 ·· 48
实例02-39：使用while循环输出整数0～4 ··· 49
实例02-40：while循环的死循环问题 ·· 49
实例02-41：输出10以内的所有素数 ··· 50
实例02-42：在for循环和while循环中使用break语句 ····································· 50
实例02-43：在for循环和while循环中使用continue语句 ······························· 51
实例02-44：输出指定英文单词中的每个英文字母 ··· 52

2.6　函数 ·· 52
实例02-45：使用函数计算序列内元素的和 ··· 52
实例02-46：创建一个可以接受任意数量参数的函数 ····································· 53
实例02-47：减少函数的参数 ·· 54
实例02-48：家政公司的打扫服务 ·· 54

2.7　类和对象 ·· 55
实例02-49：创建并使用类和对象 ·· 55
实例02-50：定义并使用类方法 ·· 56
实例02-51：我的宠物狗 ·· 56
实例02-52：使用属性函数创建可以扩展功能的属性 ····································· 57

实例 02-53：使用 Python 描述符实现延迟初始化·················59
实例 02-54：在宝马汽车中使用继承·····························60
实例 02-55：在子类中扩展属性·································61
实例 02-56：模仿 Linux 操作系统中的文件读写接口···············62
2.8 迭代器和生成器··63
实例 02-57：创建并使用迭代器·································63
实例 02-58：创建并使用生成器·································64
实例 02-59：实现委托迭代处理·································64
实例 02-60：使用生成器创建新迭代模式·························65
实例 02-61：使用生成器函数实现一个迭代协议···················66
实例 02-62：使用函数 reversed()实现反转迭代···················67
实例 02-63：实现有额外状态的生成器函数·······················68
实例 02-64：实现迭代器切片处理·······························69
实例 02-65：迭代出所有可能的组合·····························69
实例 02-66：使用函数 enumerate()创建索引迭代序列··············70
实例 02-67：使用函数 zip()或 zip_longest()迭代多个序列·········71

第 3 章 标准库函数实战···72
3.1 字符串处理函数··73
实例 03-01：使用函数 split()分割指定的字符串··················73
实例 03-02：使用函数 re.split()分割指定字符串··················73
实例 03-03：字符串开头和结尾处理·····························73
实例 03-04：实现字符串匹配处理·······························74
实例 03-05：文本模式匹配和查找·······························75
实例 03-06：文本查找和替换···································76
实例 03-07：实现最短文本匹配·································78
实例 03-08：处理 Unicode 字符串······························78
实例 03-09：删除字符串中的字符·······························79
实例 03-10：字符过滤和清理···································80
实例 03-11：字符串对齐处理···································81
实例 03-12：字符串连接·······································81
实例 03-13：重新格式化字符串·································83
实例 03-14：在字符串中处理 HTML 和 XML 标记················84
实例 03-15：在字节串中实现基本文本处理·······················84
3.2 文件和 I/O 处理函数··85
实例 03-16：刷新缓冲区·······································85
实例 03-17：检测文件是否连接到一个终端设备···················86
实例 03-18：返回文件各行内容·································86
实例 03-19：返回文件 3 个字节的内容···························87
实例 03-20：返回文件中所有行·································87
实例 03-21：重复读取文件中的第 1 行内容······················88
实例 03-22：获取当前文件位置·································88
实例 03-23：截取文件中前 3 个字符·····························88

目　录

实例 03-24：向文件中写入多行字符串 89
实例 03-25：获取文件操作权限 89
实例 03-26：修改当前工作目录到指定路径 90
实例 03-27：修改文件或目录权限 90
实例 03-28：遍历显示某个目录中所有文件夹和文件列表 90
实例 03-29：修改一个目录名字 91
实例 03-30：读取两个文本文件内容 91
实例 03-31：字符串 I/O 操作 92
实例 03-32：读写压缩文件 92
实例 03-33：对二进制文件做内存映射 93
实例 03-34：检测某个文件或目录是否存在 94
实例 03-35：获取某个目录中的文件列表信息 94
实例 03-36：获取目录的详细信息 95
实例 03-37：绕过文件名编码设置编码格式 95
实例 03-38：创建并读取临时文件信息 96
实例 03-39：实现数据序列化 97

3.3　数字处理函数 98

实例 03-40：使用函数 abs()返回绝对值 98
实例 03-41：返回最小整数 98
实例 03-42：返回参数指数值 99
实例 03-43：返回参数的绝对值 99
实例 03-44：返回指定数字的下舍整数 99
实例 03-45：计算指定数字自然对数 99
实例 03-46：计算指定数字以 10 为基数的对数 100
实例 03-47：获取参数最大值 100
实例 03-48：获取参数最小值 100
实例 03-49：获取参数的整数部分和小数部分 101
实例 03-50：计算 x 的 y 次方的结果 101
实例 03-51：计算指定数字的四舍五入值 101
实例 03-52：使用格式化方式设置数字精度 102
实例 03-53：计算指定数字的平方根 102
实例 03-54：分别实现无穷大数和 NaN 验证处理 102
实例 03-55：实现误差运算和精确运算 103
实例 03-56：将整数转换为二进制、八进制或十六进制数据 104
实例 03-57：实现复数运算 104
实例 03-58：使用 fractions 模块处理分数 105
实例 03-59：使用 NumPy 模块分别创建一维数组和二维数组 106
实例 03-60：使用函数 choice()创建随机数 106

3.4　日期和时间函数 106

实例 03-61：使用函数 time.clock()处理时间 106
实例 03-62：使用函数 time.tzset()操作时间 107
实例 03-63：使用 calendar 模块函数操作日期 107

实例 03-64：使用类 date 的实例方法和属性实现日期操作……109
实例 03-65：使用类 time 实现日期操作……110
实例 03-66：使用类 datetime 实现日期操作……110
实例 03-67：使用类 datetime 格式化日期……110
实例 03-68：使用类 datetime 实现时间换算……111
实例 03-69：获取某一周中某一天的日期……112
实例 03-70：输出当月每一天的日期……113
实例 03-71：循环输出当月每一天的日期……113
实例 03-72：将字符串转换为日期……114

第 4 章 进程通信和联网操作实战……115

4.1 使用 Socket 网络接口库……116
实例 04-01：分别创建简单 Socket 服务器和客户端……116
实例 04-02：使用 Socket 建立 TCP "客户端/服务器" 连接……116
实例 04-03：TCP "客户端/服务器" 模式的机器人聊天程序……117
实例 04-04：实现一个文件上传系统……118
实例 04-05：使用 Socket 建立 UDP "客户端/服务器" 连接……119

4.2 实现安全 Socket 编程……120
实例 04-06：创建 SSL Socket 连接……120
实例 04-07：实现客户端和服务器 SSL 安全交互……121

4.3 实现 I/O 多路复用……123
实例 04-08：使用 select 同时监听多个端口……123
实例 04-09：模拟多线程并实现读写分离……125
实例 04-10：使用 select 实现一个可并发的服务器……126
实例 04-11：实现一个可并发的服务器……126
实例 04-12：实现高级 I/O 多路复用……127

4.4 实现异步 I/O 处理……128
实例 04-13：使用 asyncio 实现 Hello world 代码……128
实例 04-14：使用 asyncio 获取网站首页信息……129
实例 04-15：以动画的方式显示文本式旋转指针……130

4.5 实现异步 Socket 处理……131
实例 04-16：使用模块 asyncore 实现一个基本的 HTTP 客户端……131
实例 04-17：使用模块 asyncore 响应客户端发送数据……132

4.6 实现内存映射……133
实例 04-18：读取文件 test.txt 的内容……133
实例 04-19：读取整个文件 test.txt 的内容……134
实例 04-20：逐步读取文件 test.txt 中的指定字节数内容……134

4.7 socketserver 编程……135
实例 04-21：使用 socketserver 创建 TCP "客户端/服务器" 程序……135
实例 04-22：使用 ThreadingTCPServer 创建 "客户端/服务器" 通信程序……135

第 5 章 结构化标记处理实战……137

5.1 使用内置模块 html……138
实例 05-01：使用 html.parser 创建 HTML 解析器……138

目　录

实例 05-02：使用 html.entities 解析 HTML ……………………………………………… 139
5.2　使用内置模块解析 XML …………………………………………………………… 140
实例 05-03：使用模块 xml.etree.ElementTree 读取 XML 文件 ………………………… 140
实例 05-04：使用 SAX 方法解析 XML 文件 …………………………………………… 141
实例 05-05：使用 DOM 解析 XML 文件 ………………………………………………… 143
实例 05-06：使用 DOM 获取 XML 文件中指定元素 …………………………………… 144
实例 05-07：使用模块 xml.sax.saxutils 创建一个指定元素的 XML 文件 ……………… 146
实例 05-08：使用模块 xml.parsers.expat 解析 XML 文件 ……………………………… 148
5.3　使用第三方库解析 HTML 和 XML ………………………………………………… 149
实例 05-09：使用库 Beautiful Soup 解析 HTML 代码 …………………………………… 149
实例 05-10：使用库 Beautiful Soup 解析指定 HTML 标签 ……………………………… 150
实例 05-11：将 p 标签下的所有子标签存入一个列表中 ………………………………… 150
实例 05-12：获取 p 标签下的所有子节点内容 …………………………………………… 151
实例 05-13：处理标签中的兄弟节点和父节点 …………………………………………… 151
实例 05-14：根据标签名查找文件 ………………………………………………………… 152
实例 05-15：使用函数 find_all() 根据属性查找文件 …………………………………… 153
实例 05-16：用函数 find_all() 根据 text 查找文件 ……………………………………… 153
实例 05-17：使用其他标准选择器 ………………………………………………………… 153
实例 05-18：使用 select() 直接传入 CSS 选择器 ………………………………………… 154
实例 05-19：使用库 bleach 过滤 HTML 代码 …………………………………………… 155
实例 05-20：使用方法 bleach.clean() 不同参数实现过滤处理 …………………………… 155
实例 05-21：使用方法 bleach.linkify() 添加指定属性 …………………………………… 156
实例 05-22：使用 callback 参数删除指定属性 …………………………………………… 157
实例 05-23：使用 bleach.linkifier.Linker 处理链接 ……………………………………… 157
实例 05-24：使用 bleach.linkifier.LinkifyFilter 处理链接 ………………………………… 158
实例 05-25：使用库 cssutils 处理 CSS 标记 ……………………………………………… 158
实例 05-26：使用 html5lib 解析 HTML 代码 ……………………………………………… 159
实例 05-27：使用 html5lib 解析 HTML 中的指定标签 …………………………………… 159
实例 05-28：使用库 MarkupSafe 构建安全 HTML ……………………………………… 160
实例 05-29：使用库 MarkupSafe 实现格式化 …………………………………………… 161
实例 05-30：使用库 pyquery 实现字符串初始化 ………………………………………… 161
实例 05-31：使用 pyquery 解析 HTML 内容 ……………………………………………… 162
实例 05-32：使用库 pyquery 解析本地 HTML 文件和网络页面 ………………………… 163
实例 05-33：使用库 pyquery 实现基于 CSS 选择器查找 ………………………………… 163
实例 05-34：使用库 pyquery 查找子节点 ………………………………………………… 164
实例 05-35：使用库 pyquery 查找父节点 ………………………………………………… 165
实例 05-36：使用库 pyquery 获取兄弟节点信息 ………………………………………… 165
第 6 章　应用程序开发实战 ………………………………………………………………… 167
6.1　使用 webbrowser 实现浏览器操作 ………………………………………………… 168
实例 06-01：分别调用 IE 和谷歌浏览器打开百度网主页 ……………………………… 168
实例 06-02：调用默认浏览器每隔 5s 打开一次指定网页 ……………………………… 168
6.2　使用 urllib 包 ………………………………………………………………………… 168

实例 06-03：在百度搜索关键词中得到第一页链接 ··· 168
实例 06-04：使用 urllib 实现 HTTP 身份验证 ·· 169
6.3 使用内置模块 http ··· 170
实例 06-05：访问指定的网站 ··· 170
实例 06-06：使用 http.client 模块中 GET 方式获取数据 ································· 171
实例 06-07：综合使用模块 http 和 urllib ··· 171
实例 06-08：发送 HTTP GET 请求到远端服务器 ·· 172
实例 06-09：使用 POST 方法在请求主体中发送查询参数 ································ 172
实例 06-10：在发出的请求中提供自定义的 HTTP 头 ······································ 173
6.4 FTP 传输、SMTP 服务器和 XML-RPC 服务器 ································· 173
实例 06-11：创建一个 FTP 文件传输客户端 ··· 174
实例 06-12：使用模块 smtpd 创建一个 SMTP 服务器 ···································· 175
实例 06-13：使用模块 xmlrpc.server 实现 XML-RPC 客户端和服务器相互通信 ······ 175
6.5 开发电子邮件系统 ··· 176
实例 06-14：获取指定电子邮箱中最新两封电子邮件的主题和发件人 ················ 176
实例 06-15：向指定电子邮箱发送电子邮件 ·· 177
实例 06-16：发送带附件功能的电子邮件 ··· 178
实例 06-17：使用库 envelopes 向指定电子邮箱发送电子邮件 ·························· 179
实例 06-18：使用库 envelopes 构建 Flask Web 电子邮件发送程序 ··················· 179
实例 06-19：创建一个带有 HTTP REST 接口的 SMTP 服务器 ························ 180
6.6 解析 JSON 数据 ··· 180
实例 06-20：将 Python 字典转换为 JSON 对象 ··· 180
实例 06-21：将 JSON 编码的字符串转换为 Python 数据结构 ···························· 181
实例 06-22：编写自定义类解析 JSON 数据 ·· 181
6.7 实现数据编码和解码 ··· 185
实例 06-23：实现数据"编码/解码"操作 ·· 185
实例 06-24：实现 bytes 类型和 base64 类型的相互转换 ··································· 186
实例 06-25：生成由某地址可表示的全部 IP 地址的范围 ································· 187
6.8 实现身份验证 ··· 188
实例 06-26：获取指定字符串的数据指纹 ··· 188
实例 06-27：利用 hmac 模块实现简单且高效的身份验证 ································· 188
实例 06-28：Socket 服务器和客户端的加密认证 ··· 189
6.9 使用第三方库处理 HTTP ··· 190
实例 06-29：使用库 aiohttp 实现异步处理 ·· 190
实例 06-30：使用库 aiohttp 爬取指定 CSDN 博客中技术文章地址 ···················· 191
实例 06-31：使用库 requests 返回指定 URL 请求 ·· 192
实例 06-32：提交的数据是向指定地址传送的 data 里面的数据 ························ 192
实例 06-33：使用 GET 和 POST 方式处理 JSON 数据 ···································· 193
实例 06-34：添加 headers 获取知乎页面信息 ·· 193
实例 06-35：使用自定义的编码格式进行解码 ·· 194
实例 06-36：访问远程页面信息 ·· 194
实例 06-37：使用库 grequests 同时处理一组请求 ·· 195

目 录

　　实例06-38：使用库grequests提升访问请求性能 195
　　实例06-39：使用库httplib2获取网页数据 196
　　实例06-40：使用库httplib2处理网页缓存数据 196
　　实例06-41：使用POST发送构造数据 197
　　实例06-42：使用库urllib3中的request()方法创建请求 198
　　实例06-43：在request()方法中添加head头创建请求 198
　　实例06-44：使用库urllib3中的post()方法创建请求 198
　　实例06-45：使用库urllib3发送JSON数据 199
　6.10 使用第三方库处理URL 200
　　实例06-46：使用库furl优雅地处理URL分页 200
　　实例06-47：使用库furl处理URL参数 200
　　实例06-48：使用内联方法处理URL参数 201
　　实例06-49：使用库purl处理3种构造类型URL 201
　　实例06-50：使用库purl返回各个URL对象 201
　　实例06-51：使用库purl修改URL参数值 202
　　实例06-52：在当前路径末尾添加字段 202
　　实例06-53：使用库webargs处理URL参数 203
　　实例06-54：在aiohttp程序中使用库webargs 203
　　实例06-55：在Tornado程序中使用库webargs 204

第7章 数据持久化操作实战 205
　7.1 操作SQLite3数据库 206
　　实例07-01：使用方法cursor.execute()执行指定SQL语句 206
　　实例07-02：使用方法cursor.executemany()执行指定的SQL命令 206
　　实例07-03：同时执行多个SQL语句 207
　　实例07-04：使用方法create_function()执行指定函数 207
　　实例07-05：创建用户定义的聚合函数 207
　　实例07-06：用自定义排序规则以"错误方式"进行排序 208
　　实例07-07：生成一个 SQLite Shell 209
　　实例07-08：返回数据库中的列名称列表 209
　　实例07-09：操作SQLite3数据库 210
　　实例07-10：将自定义类Point适配SQLite3数据库 211
　　实例07-11：使用函数register_adapter()注册适配器函数 212
　　实例07-12：将datetime.datetime对象保存为UNIX时间戳 212
　　实例07-13：将自定义Python类型转换成SQLite类型 212
　　实例07-14：使用默认适配器和转换器 213
　　实例07-15：使用isolation_level开启智能commit 214
　　实例07-16：手动开始commit（提交执行）操作 215
　　实例07-17：使用模块apsw创建并操作SQLite数据库数据 215
　　实例07-18：同时批处理上千条数据 215
　7.2 操作MySQL数据库 217
　　实例07-19：显示PyMySQL数据库版本号 217
　　实例07-20：创建新表 218

实例 07-21：向数据库中插入数据 ·· 218
实例 07-22：查询数据库中的数据 ·· 219
实例 07-23：更新数据库中的数据 ·· 220
实例 07-24：删除数据库中的数据 ·· 220
实例 07-25：通过执行事务删除表中的数据 ·· 221
实例 07-26：足球俱乐部球员管理系统 ··· 221

7.3 使用 MariaDB 数据库 ·· 224
实例 07-27：搭建 MariaDB 数据库环境 ·· 224
实例 07-28：在 Python 程序中使用 MariaDB 数据库 ····························· 226
实例 07-29：使用 MariaDB 创建 MySQL 数据库 ································· 228

7.4 使用 MongoDB 数据库 ··· 229
实例 07-30：搭建 MongoDB 环境 ··· 229
实例 07-31：使用 PyMongo 操作 MongoDB 数据库 ······························ 230
实例 07-32：使用 mongoengine 操作 MongoDB 数据库 ·························· 232

7.5 使用 ORM 操作数据库 ··· 234
实例 07-33：使用 SQLAlchemy 操作两种数据库 ·································· 234
实例 07-34：使用 Peewee 操作 SQLite 数据库 ···································· 237
实例 07-35：更新和删除指定数据库中数据 ·· 238
实例 07-36：查询数据库中指定范围内的数据 ····································· 238
实例 07-37：使用 Peewee 在 MySQL 数据库中创建两个表 ····················· 240
实例 07-38：使用 Pony 创建一个 SQLite 数据库 ································· 241
实例 07-39：使用 Pony 向数据库的指定表中添加新数据 ······················· 242
实例 07-40：使用 Pony 查询并修改数据库中指定数据 ·························· 244
实例 07-41：使用 Pony 删除数据库的表中的某条数据 ·························· 245
实例 07-42：在指定 MySQL 数据库中创建指定的表 ···························· 245
实例 07-43：在 MySQL 数据库中实现一对多和继承操作 ······················· 246
实例 07-44：下载并安装 PostgreSQL ··· 248
实例 07-45：连接指定 PostgreSQL 数据库 ·· 248
实例 07-46：在 PostgreSQL 数据库中创建指定表 ································ 249
实例 07-47：创建 PostgreSQL 表并插入新数据 ··································· 249
实例 07-48：查询显示指定表中的数据 ··· 250
实例 07-49：向 PostgreSQL 数据库中插入新数据并更新数据 ·················· 251
实例 07-50：删除 PostgreSQL 数据库中的指定数据 ····························· 252
实例 07-51：创建 PostgreSQL 表并实现插入、查询、更新和删除数据 ······· 253
实例 07-52：使用模块 queries 查询 PostgreSQL 数据库中的数据 ············· 254
实例 07-53：查询并显示 PostgreSQL 数据库中的数据 ·························· 254

7.6 连接 SQL Server 数据库 ·· 255
实例 07-54：连接并操作 SQL Server 数据库 ······································ 255
实例 07-55：创建 SQL Server 表并查询其数据 ··································· 256

7.7 使用 Redis 存储 ·· 257
实例 07-56：使用 Redis 连接服务器 ·· 257
实例 07-57：使用 ConnectionPool 创建连接池 ···································· 257

目　录

实例 07-58：实现"发布-订阅"模式 258
实例 07-59：在 Redis 中使用 delete 命令和 exists 命令 258
实例 07-60：使用 expire 命令和 expireat 命令 259
实例 07-61：使用 persist 命令、keys 命令和 move 命令 259

第 8 章　特殊文本格式处理实战 261

8.1　Tablib 模块实战演练 262

实例 08-01：操作数据集中的指定行和列 262
实例 08-02：删除指定数据并导出不同文本格式的数据 262
实例 08-03：将 Tablib 数据集导出到新建 Excel 文件 263
实例 08-04：将多个 Tablib 数据集导出到 Excel 文件 264
实例 08-05：使用标签过滤 Tablib 数据集 266
实例 08-06：将两组数据分离导入 Excel 文件 266

8.2　Office 处理实战 267

实例 08-07：使用 openpyxl 读取 Excel 文件 267
实例 08-08：将 4 组数据导入 Excel 文件中 267
实例 08-09：在 Excel 文件中检索某关键字 268
实例 08-10：将数据导入 Excel 文件并生成图表 269
实例 08-11：使用 pyexcel 读取并写入 CSV 文件 270
实例 08-12：使用 pyexcel 读取 Excel 文件中的每个单元格内容 271
实例 08-13：按列读取并显示 Excel 文件中的每个单元格内容 271
实例 08-14：读取并显示 Excel 文件中的所有数据 272
实例 08-15：将 3 组数据导入新建的 Excel 文件中 272
实例 08-16：使用 pyexcel 以多种方式获取 Excel 数据 273
实例 08-17：将数据分别导入 Excel 文件和 SQLite 数据库 274
实例 08-18：使用 python-docx 创建 Word 文档 274
实例 08-19：在 Word 文档中插入 20 个实心图形 275
实例 08-20：向 Word 文档中添加指定段落样式的内容 276
实例 08-21：得到英文的样式名称 277
实例 08-22：获取 Word 文档中的文本样式名称 278
实例 08-23：获取 Word 文档中的文本内容 278
实例 08-24：在 Word 文档中创建表格 278
实例 08-25：创建表格并合并其中的单元格 279
实例 08-26：调整 Word 表格宽度 280
实例 08-27：获取 python-docx 内部的表格样式名称 280
实例 08-28：使用指定样式修饰表格 281
实例 08-29：创建样式和设置字体 281
实例 08-30：使用 Run.font 设置字体样式 282
实例 08-31：设置段落递进的左对齐样式 282
实例 08-32：自定义创建 Word 样式 283
实例 08-33：使用库 xlrd 读取 Excel 文件的内容 284
实例 08-34：将指定内容写入 Excel 文件并创建 Excel 文件 285
实例 08-35：使用库 xlsxwriter 创建一个指定内容的 Excel 文件 285

实例 08-36：向 Excel 文件中批量写入内容………………………………………………286
实例 08-37：设置表格样式………………………………………………………………286
实例 08-38：向 Excel 文件中插入图像…………………………………………………288
实例 08-39：向 Excel 文件中插入数据并绘制柱状图…………………………………289
实例 08-40：向 Excel 文件中插入数据并绘制散点图…………………………………290
实例 08-41：向 Excel 文件中插入数据并绘制柱状图和饼图…………………………291

8.3 PDF 处理实战………………………………………………………………………293
实例 08-42：将 PDF 文件中的内容转换为 TXT 文本…………………………………293
实例 08-43：解析某个在线 PDF 文件的内容…………………………………………295
实例 08-44：使用 PyPDF2 读取 PDF 文件……………………………………………297
实例 08-45：使用 PyPDF2 将 PDF 文件中写入另一个 PDF 文件内容………………298
实例 08-46：将两个 PDF 文件合并为一个 PDF 文件…………………………………298
实例 08-47：分割某个指定 PDF 文件…………………………………………………299
实例 08-48：合并 3 个 PDF 文件………………………………………………………299
实例 08-49：向指定 PDF 文件中写入文本……………………………………………300
实例 08-50：向 PDF 文件中写入指定样式的文本……………………………………300
实例 08-51：在 PDF 文件中绘制矢量图形……………………………………………301
实例 08-52：在 PDF 文件中绘制图像…………………………………………………301
实例 08-53：分别在 PDF 文件和 PNG 文件中绘制饼图………………………………302
实例 08-54：在 PDF 文件中分别生成条形图和二维码…………………………………303

第 9 章 图形化界面开发实战……………………………………………………………305
9.1 使用内置库 tkinter……………………………………………………………………306
实例 09-01：创建第一个 GUI 程序……………………………………………………306
实例 09-02：向窗口中添加组件…………………………………………………………306
实例 09-03：使用 Frame()布局窗体界面………………………………………………307
实例 09-04：向窗口中添加按钮控件……………………………………………………307
实例 09-05：使用文本框控件……………………………………………………………308
实例 09-06：实现会员注册界面效果……………………………………………………309
实例 09-07：使用菜单控件………………………………………………………………310
实例 09-08：在窗口中创建标签…………………………………………………………310
实例 09-09：在 tkinter 窗口中创建单选按钮和复选框…………………………………311
实例 09-10：在窗口中绘制图形…………………………………………………………312
实例 09-11：使用事件机制创建一个"英尺/米"转换器………………………………313
实例 09-12：实现一个动态绘图程序……………………………………………………314
实例 09-13：实现一个简单计算器程序…………………………………………………316
实例 09-14：创建消息对话框……………………………………………………………317
实例 09-15：创建输入对话框……………………………………………………………318
实例 09-16：创建打开/保存文件对话框…………………………………………………319
实例 09-17：创建选择颜色对话框………………………………………………………319
实例 09-18：创建自定义对话框…………………………………………………………320
实例 09-19：开发一个记事本程序………………………………………………………321
实例 09-20：使用偏函数模拟实现交通标志……………………………………………323

目 录

实例 09-21：创建桌面天气预报程序 ··· 325
实例 09-22：创建精简版资源管理器 ··· 326

9.2 使用 tkinter 的扩展小部件 tkinter.tix ··· 328

实例 09-23：使用 Balloon 组件 ··· 328
实例 09-24：使用 DirList 组件 ··· 329
实例 09-25：使用分组列表组件 ··· 330
实例 09-26：使用管理组件 ··· 332
实例 09-27：实现一个日历程序 ··· 333

9.3 Pmw 库开发实战 ··· 335

实例 09-28：下载并安装 Pmw ··· 335
实例 09-29：使用 ButtonBox 组件 ··· 336
实例 09-30：使用 ComboBox 组件 ··· 337
实例 09-31：使用 Counter 组件 ··· 339
实例 09-32：使用 Group 组件 ··· 340
实例 09-33：使用 LabeledWidget 组件 ··· 341
实例 09-34：使用 MainMenuBar 组件 ··· 342
实例 09-35：使用 MessageBar 组件 ··· 344
实例 09-36：使用 OptionMenu 组件 ··· 345
实例 09-37：使用 RadioSelect 组件 ··· 345
实例 09-38：使用 ScrolledCanvas 组件 ··· 347
实例 09-39：使用 AboutDialog 组件 ··· 348
实例 09-40：使用 Balloon 组件 ··· 349
实例 09-41：使用 PyQt 创建第一个 GUI 程序 ··· 351
实例 09-42：在 PyQt 窗体中创建一个图标 ··· 352
实例 09-43：在 PyQt 窗体中实现一个提示信息 ··· 353
实例 09-44：在 PyQt 窗体中创建状态栏信息 ··· 354
实例 09-45：在 PyQt 窗体中同时创建菜单栏和状态栏信息 ··· 354
实例 09-46：在 PyQt 窗体中创建工具栏 ··· 356
实例 09-47：在 PyQt 窗体中使用绝对定位方式 ··· 356
实例 09-48：使用箱布局 ··· 357
实例 09-49：使用网格布局模拟实现一个计算器界面 ··· 358
实例 09-50：使用表单布局实现一个留言板界面 ··· 359
实例 09-51：使用单击按钮事件处理程序 ··· 359
实例 09-52：在 PyQt5 中使用信号和槽 ··· 360
实例 09-53：重新实现按键盘按键后的操作功能 ··· 361
实例 09-54：重新实现按住方向键后的操作功能 ··· 361
实例 09-55：实现人机对战"石头、剪刀、布"小游戏 ··· 362
实例 09-56：发送自定义信号 ··· 363
实例 09-57：使用对话框获取用户名信息 ··· 364
实例 09-58：使用颜色选择对话框设置背景颜色 ··· 365
实例 09-59：使用字体选择对话框设置字体 ··· 366
实例 09-60：使用文件选择对话框选择一个文件 ··· 367

实例 09-61：使用 QCheckBox 实现复选框功能 ··· 368
实例 09-62：使用 QRadioButton 实现单选按钮功能ㆍ·· 369
实例 09-63：使用 QPushButton 实现切换按钮功能 ·· 370
实例 09-64：使用 QSlider 实现一个音量控制器 ·· 371
实例 09-65：使用 QProgressBar 实现一个进度条效果 ··· 372
实例 09-66：使用 QCalendarWidget 实现一个日历 ·· 373
实例 09-67：在窗口中显示一个图片 ··· 374
实例 09-68：创建一个单行文本编辑框 ·· 375
实例 09-69：创建两个分割框组件 ·· 376
实例 09-70：使用 Eric6 提高开发效率 ··· 377

9.4 使用 pyglet 库 ··· 378

实例 09-71：创建第一个 pyglet 程序 ··· 378
实例 09-72：在窗体中显示指定图片 ··· 378
实例 09-73：使用库 pyglet 处理键盘事件程序 ·· 379
实例 09-74：在屏幕上绘制一个三角形 ·· 379
实例 09-75：使用顶点数组绘制三角形 ·· 380
实例 09-76：开发一个 Minecraft 游戏 ·· 381

9.5 使用 Toga 库 ·· 385

实例 09-77：使用 Toga 创建第一个 GUI 程序 ·· 385
实例 09-78：创建一个温度转换器 ·· 386
实例 09-79：使用组件 ScrollContainer 实现滚动功能 ··· 387
实例 09-80：使用绘图组件 ·· 388

9.6 wxPython 实战 ·· 389

实例 09-81：开发第一个 wxPython 程序 ·· 389
实例 09-82：使用 StaticText 组件在窗体中显示文本 ·· 389
实例 09-83：创建 4 种不同样式的文本框 ·· 391
实例 09-84：使用 RadioButton 组件 ··· 392

9.7 GUI 高级实战 ··· 393

实例 09-85：实现 tkinter+ SQLite3 图书馆系统 ·· 393
实例 09-86：实现 tkinter + SQLite3 多线程计时器系统 ··· 395

第 1 章

搭建开发环境实战

古人云:"工欲善其事,必先利其器。"我们在使用 Python 进行项目开发时,需要先搭建开发环境,并准备开发工具。本章将详细讲解搭建 Python 开发环境的知识,为读者学习本书后面的知识打下基础。

1.1 安装 Python 环境

实例 01-01：在 Windows 操作系统中下载并安装 Python

因为 Python 可以在 Windows、Linux 和 macOS 这当今三大主流的计算机操作系统中执行，所以本书将详细讲解在这 3 种操作系统中安装 Python 的方法，接下来首先讲解在 Windows 操作系统中下载并安装 Python 的过程。

（1）登录 Python 官方网站，单击顶部导航中的 Downloads 链接，进入图 1-1 所示的下载页面。

（2）因为当前计算机的操作系统是 Windows 操作系统，所以单击 Looking for Python with a different OS? Python for 后面的 Windows 链接，出现图 1-2 所示的 Windows 版下载页面。

图 1-1　下载页面

图 1-2　Windows 版下载页面

图 1-2 所示的是 Windows 操作系统的 Python 安装包，其中 x86 对应的安装包适合 32 位操作系统，x86-64 对应的安装包适合 64 位操作系统。可以通过如下 3 种途径获取 Python。

❑ web-based installer：通过联网完成安装。
❑ executable installer：通过可执行文件（*.exe）安装。
❑ embeddable zip file：嵌入式版本，可以集成到其他应用程序中。

（3）因为作者的计算机操作系统是 64 位操作系统，所以需要选择一个 x86-64 对应的安装包——当前（作者写稿时）最新版本 Windows x86-64 executable installer。弹出图 1-3 所示的下载对话框，单击"下载"按钮后开始下载。

（4）下载成功后得到一个扩展名为.exe 的可执行文件，双击此文件开始安装。在第一个安装界面中勾选下面两个复选框，然后单击 Install Now 链接，如图 1-4 所示。

> 注意：勾选 Add Python x.x to PATH 复选框的目的是把 Python 的安装路径添加到系统路径。勾选这个复选框后，在 CMD 控制台输入"python"后就可调用 python.exe。如果不勾选这个复选框，在 CMD 控制台输入"python"，操作系统会报错。

（5）弹出图 1-5 所示的安装进度界面，进行安装。

（6）安装完成的界面如图 1-6 所示，单击 Close 按钮完成安装。

1.1 安装 Python 环境

图 1-3　下载对话框

图 1-4　第一个安装界面

图 1-5　安装进度界面

图 1-6　安装完成的界面

（7）选择"开始"→"运行"，先在弹出的"运行"对话框中输入"cmd"并按 Enter 键进入 CMD 控制台，然后输入"python"并按 Enter 键验证 Python 是否安装成功。控制台输出图 1-7 所示的界面，表示安装成功。

图 1-7　安装成功

实例 01-02：在 macOS 操作系统中下载并安装 Python

macOS 操作系统已经默认安装了 Python，开发者只需要安装一个文本编辑器来编写 Python 程序即可，并且需要确保其配置信息正确无误。要想检查当前使用的 macOS 操作系统是否安装了 Python，需要完成如下工作。

（1）打开终端窗口（和 Windows 操作系统中的 CMD 控制台类似）。

打开 Applications/Utilities 文件夹，选择 Terminal 并打开，这样可以打开一个终端窗口。另外，也可以按 Command + Space 组合键，再输入"terminal"并按 Enter 键打开终端窗口。

（2）输入"python"命令。

为了确定是否安装了 Python，接下来需要执行"python"命令（注意，其中的 p 是小写的）。如果输出类似于下面的内容(指出了安装的 Python 版本)，则表示 Python 安装成功；最后的">>>"是一个提示符，用于进一步输入 Python 命令。

```
$ python
Python 3.10.4 (default, Mar 9 2022, 22:15:05)
[GCC 4.2.1 Compatible Apple LLVM 5.0 (clang-500.0.68)] on darwin
```

```
Type "help", "copyright", "credits", or "license" for more information.
>>>
```

上述输出表明，当前计算机默认使用的 Python 为 Python 3.10.4。看到上述输出后，如果要退出 Python 并返回终端窗口，可按 Ctrl + D 组合键或执行 exit()命令。

实例 01-03：在 Linux 操作系统中下载并安装 Python

在众多开发者的眼中，Linux 操作系统是专门为开发者所设计的。大多数的 Linux 操作系统已经默认安装了 Python。要在 Linux 操作系统中编写 Python 程序，开发者几乎不用安装什么软件，也几乎不用修改配置。要想检查当前使用的 Linux 操作系统是否安装了 Python，需要完成如下工作。

（1）在操作系统中执行应用程序 Terminal（如果使用的是 Ubuntu，可以按 Ctrl + Alt + T 组合键），打开一个终端窗口。

（2）为了确定是否安装了 Python，需要执行"python"命令（请注意，其中的 p 是小写的）。如果输出类似于下面的内容（指出了 Python 版本），则表示已经安装了 Python；最后的">>>"是一个提示符，用于继续输入 Python 命令。

```
$ python
Python 2.7.6 (default, Mar 22 2014, 22:59:38)
[GCC 4.8.2] on linux2
Type "help", "copyright", "credits" or "license" for more information.
>>>
```

上述输出表明，当前计算机默认使用的 Python 为 Python 2.7.6。看到上述输出后，如果要退出 Python 并返回终端窗口，可按 Ctrl + D 组合键或执行 exit()命令。要想检查操作系统是否安装了 Python 3，可能需要指定相应的版本，例如，尝试执行命令"python3"。

```
$ python3
Python 3.10.4 (default, Sep 17 2022, 13:05:18)
[GCC 4.8.4] on linux
Type "help", "copyright", "credits" or "license" for more information.
>>>
```

上述输出表明，当前 Linux 操作系统也安装了 Python 3，所以开发者可以使用这两个版本中的任何一个。在这种情况下，需要将本书中的命令"python"都替换为"python3"。在大多数情况下，Linux 操作系统默认安装了 Python。

1.2 安装开发工具

实例 01-04：使用 Python 自带工具 IDLE

IDLE 是 Python 自带开发工具，它是应用 Python 第三方库的图形接口库 tkinter 开发的一个包含图形界面的开发工具，其主要特点如下。

- 跨平台，包括 Windows、Linux、UNIX 和 macOS。
- 智能缩进。
- 代码着色。
- 自动提示。
- 可以实现断点设置、单步执行等调试功能。
- 具有智能化菜单。

当在 Windows 操作系统下安装 Python 时，IDLE 会被自动安装，在"开始"菜单的 Python 3.x 子菜单中就可以找到它，如图 1-8 所示。

在 Linux 操作系统下需要使用 yum 或 apt-get 命令进行单独安装。在 Windows 操作系统下，IDLE 的界面效果如图 1-9 所示，其标题栏与普通的 Windows 应用程序相同，而其中所写的代码是被自动着色的。

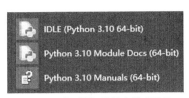

图 1-8 "开始"菜单中的 IDLE

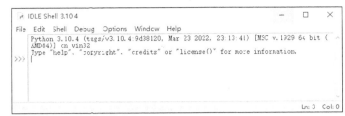

图 1-9 IDLE 的界面

IDLE 常用快捷键如表 1-1 所示。

表 1-1　　　　　　　　　　　　IDLE 常用快捷键

快　捷　键	功　　能
Ctrl+]	缩进代码
Ctrl+[	取消缩进
Alt+3	注释代码
Alt+4	去除注释
F5	执行代码
Ctrl+Z	撤销一步

实例 01-05：安装 PyCharm

PyCharm 是一款著名的 Python IDE，拥有一整套可以帮助用户在使用 Python 开发时提高其效率的工具，具备基本的调试、语法高亮、项目管理、代码跳转、智能提示、自动完成、单元测试、版本控制等功能。下载、安装并设置 PyCharm 的流程如下。

（1）登录 PyCharm 官方页面，单击 DOWNLOAD NOW 按钮，如图 1-10 所示。

图 1-10　PyCharm 官方页面

（2）打开的新页面中显示了可以下载 PyCharm 的如下两个版本，如图 1-11 所示。
- Professional：专业版，可以使用 PyCharm 的全部功能，但是收费。
- Community：社区版，可以满足 Python 开发的大多数功能，完全免费。

图 1-11 专业版和社区版

另外,在该页面还可以选择操作系统,PyCharm 分别提供了 Windows、macOS 和 Linux 三大主流操作系统相应的下载版本,并且它们都分为专业版和社区版两种。

(3)作者使用的操作系统是 Windows 专业版,所以这里依次单击 Windows→Professional→DOWNLOAD 项,在弹出的下载对话框中单击"下载"按钮开始下载 PyCharm。

(4)下载成功后将会得到一个类似 pycharm-professional-201x.x.x.exe 的可执行文件,双击打开这个可执行文件,弹出图 1-12 所示的欢迎安装界面。

图 1-12 欢迎安装界面

(5)单击 Next 按钮后弹出安装目录界面,在此设置 PyCharm 的安装目录。

(6)单击 Next 按钮后弹出安装选项界面,在此根据自己计算机的配置勾选对应的复选框,因为作者使用的操作系统是 64 位操作系统,所以此处勾选 64-bit launcher 复选框。然后勾选 Create associations 选项组中的.py 复选框,如图 1-13 所示。

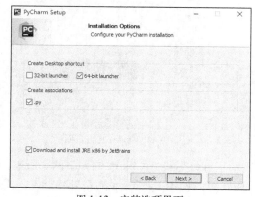

图 1-13 安装选项界面

（7）单击 Next 按钮后弹出创建启动菜单界面，如图 1-14 所示。

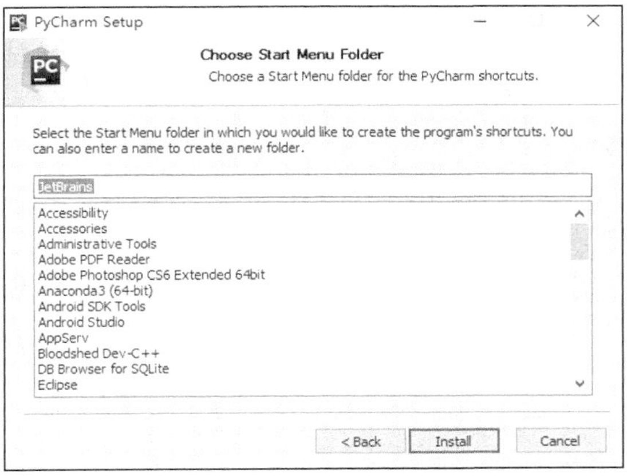

图 1-14　创建启动菜单界面

（8）单击 Install 按钮后弹出安装进度界面，这一步需要读者耐心等待。
（9）安装完成后弹出完成安装界面，单击 Finish 按钮完成 PyCharm 的全部安装工作。
（10）单击桌面快捷方式图标或选择"开始"菜单中的对应选项启动 PyCharm，因为是第一次打开 PyCharm，系统会询问是否要导入先前的设置（默认不导入）。因为这次安装是第一次安装，所以这里直接单击 OK 按钮即可。接着 PyCharm 会让我们设置主题和代码编辑器的样式，读者可以根据自己的喜好进行设置，例如，有 Vsual Studio.NET 开发经验的读者可以选择 Vsual Studio 风格。完全启动 PyCharm 后的界面效果如图 1-15 所示。

图 1-15　完全启动 PyCharm 后的界面效果

- 左侧区域面板：列表显示过去创建或使用过的项目，因为这是第一次安装，所以这里暂时显示为空白。
- 中间的 Create New Project 按钮：单击此按钮后将弹出新建项目对话框，开始新建项目。
- 中间的 Open 按钮：单击此按钮后将弹出打开对话框，用于打开已经创建的项目。
- 中间的 Check out from Version Control 下拉按钮：单击此下拉按钮后弹出项目的地址来源列表，里面有 CVS、GitHub、Git 等常见的版本控制分支渠道。

- 右下角的 Configure 下拉按钮：单击此下拉按钮后弹出和设置相关的列表，可以实现基本的设置功能。
- 右下角的 Get Help 下拉按钮：单击此下拉按钮后弹出和使用帮助相关的列表，可以帮助使用者快速入门。

实例 01-06：安装 Eclipse

有 Java、Android 或 C/C++基础的读者对 Eclipse 应该十分熟悉了，这是一个开放源代码的软件开发项目，是一个基于 Java 的可扩展开发平台。就其本身而言，Eclipse 只是一个框架和一组服务，能够通过插件和组件来构建开发环境。下载并安装 Eclipse 的流程如下。

（1）在浏览器中打开 Eclipse 官网首页，然后单击右上角的 DOWNLOAD 按钮，如图 1-16 所示。

图 1-16　Eclipse 官网首页

（2）官网会自动检测当前计算机的操作系统，并提供对应版本的下载链接。例如，作者的计算机操作系统是 64 位 Windows 操作系统，所以会自动显示 64 位 Eclipse 的下载按钮，如图 1-17 所示。

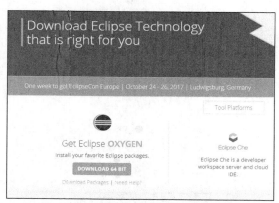

图 1-17　自动显示 64 位 Eclipse 的下载按钮

（3）单击 DOWNLOAD 64 BIT 按钮弹出一个新的下载页面，在下载页面上单击 Select Another Mirror 链接后会在下方弹出很多镜像下载地址。

（4）读者可以选择一个镜像下载地址，也可以直接单击上方的 DOWNLOAD 按钮进行下载。下载完成后会得到一个扩展名为.exe 的可执行文件，双击这个文件开始安装 Eclipse。

（5）首先会弹出启动界面，如图 1-18 所示。然后会显示一个选择列表对话框，列表中显示了不同版本的 Eclipse，在此读者需要选择要下载的版本，如图 1-19 所示。

1.2 安装开发工具

图 1-18 启动界面

图 1-19 不同版本的 Eclipse

（6）我们只需选择安装第一项即可，选择 Eclipse IDE for Java Developers 选项后弹出安装目录对话框，在此可以设置 Eclipse 的安装目录，如图 1-20 所示。

（7）单击 INSTALL 按钮后开始安装，首先会弹出 Eclipse 安装协议界面，单击 Accept Now 按钮，如图 1-21 所示。

图 1-20 设置 Eclipse 的安装目录

图 1-21 单击 Accept Now 按钮

（8）此时会弹出一个安装进度条界面，这说明开始正式安装 Eclipse。这个安装过程通常会比较慢，需要读者耐心等待。

（9）安装完成后会在界面下方显示一个 LAUNCH 按钮。单击 LAUNCH 按钮后会执行已经安装成功的 Eclipse，首次执行时会弹出一个设置 workspace 的对话框，在此可以设置一个本地目录作为 workspace。

注意："workspace"通常被翻译为工作空间，这个目录用于保存 Java 程序文件。workspace 是 Eclipse 的硬性规定，每次启动 Eclipse 时，都要将 workspace 中的所有 Java 项目加载到 Eclipse

中去。如果没有设置 workspace，Eclipse 会弹出图 1-19 所示的选择列表对话框，只有设置一个目录后才能启动 Eclipse。设置一个本地目录为 workspace 后，会在这个目录中自动创建子目录.metadata，里面生成了一些文件夹和文件。

（10）设置完 workspace，单击 OK 按钮后会弹出图 1-18 所示的启动界面。启动完成后会弹出欢迎使用界面。

在安装 PyDev 插件之前需要先确保已经安装了 Python，然后按照如下步骤安装 PyDev 插件。

（1）打开 Eclipse，依次选择菜单 Help→Install New Software，弹出的 Install 对话框如图 1-22 所示。

（2）单击 Add 按钮，在弹出的 Add Repository 对话框中设置安装插件的标签名（Name）和 URL（Location），如图 1-23 所示。其中，Name 可以随意设置，例如，可以将 Name 设置为 PyDev，但是 Location 必须设置为 http://pydev.org/updates。

图 1-22　Install 对话框　　　　　　　　图 1-23　Add Repository 对话框

（3）单击 OK 按钮后开始获取远程 URL 的安装信息，并在 Install 对话框中间位置显示获取的插件列表信息。建议勾选所有插件列表信息对应的复选框，如图 1-24 所示。

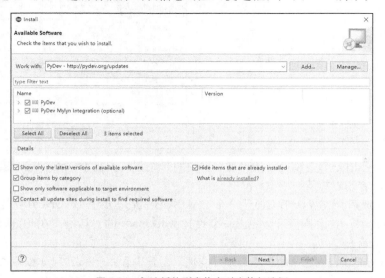

图 1-24　勾选插件列表信息对应的复选框

1.2 安装开发工具

（4）单击 Next 按钮后弹出 Install Details 界面。

（5）单击 Next 按钮后弹出 Review Licenses 界面，如图 1-25 所示。选中 I accept the terms of the license agreements 单选按钮，单击 Finish 按钮开始下载并安装 PyDev 插件，下载过程可能会有一点儿慢，并且可能需要重启 Eclipse，需要读者耐心等待。

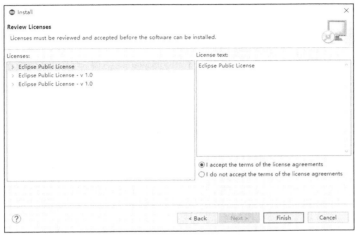

图 1-25　Review Licenses 界面

（6）选择 Eclipse 顶部菜单栏中的 Window→Preferences 打开 Preferences 窗口，如图 1-26 所示。先在左侧列表中依次选择 PyDev→Interpreters→Python Interpreter，然后单击右上角的 Quick Auto-Config 按钮，这样 Eclipse 自动将当前计算机中安装的 Python 和前面刚安装的 Eclipse+ PyDev 绑定。如果当前计算机中安装了多个版本的 Python，则可以多次单击 Quick Auto-Config 按钮实现多次绑定，单击右下角的 Apply and Close 按钮完成绑定功能。这个过程可能有点儿慢，需要读者耐心等待。

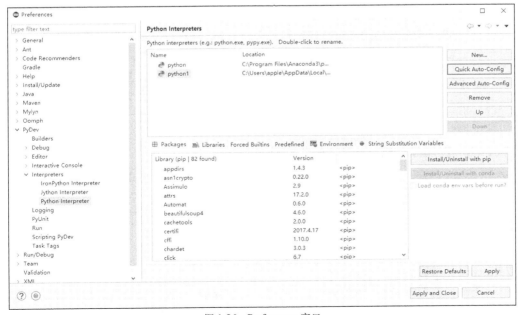

图 1-26　Preferences 窗口

> **注意**：如果不进行上面的步骤（6）所示的操作，在使用 Eclipse+PyDev 新建 Python 项目时，Eclipse 将会提示 Project interpreter not specified 错误。

1.3 编写并运行 Python 程序

实例 01-07：使用 IDLE 编写并运行 Python 程序

使用 IDLE 编写并运行 Python 程序的过程如下。

（1）打开 IDLE，依次选择 File→New File，在弹出的新建文件对话框中输入如下代码：

```
print('同学们好,我的名字是——Python!')
print('这就是我的代码，简单吗？')
```

在 Python 中，print() 是一个输出函数，用于在命令行界面中输出指定的内容，和 C 中的 printf() 函数、Java 中的 println() 函数类似。本实例在 IDLE 编辑器中输入代码，如图 1-27 所示。

（2）依次选择 File→Save，将其保存为 first.py 文件，如图 1-28 所示。

图 1-27　输入代码　　　　　　　　图 1-28　保存为 first.py 文件

（3）按 F5 键，或依次选择 Run→Run Module 选项运行当前代码，如图 1-29 所示。

（4）本实例运行后会使用 print() 函数输出两行文本，运行效果如图 1-30 所示。

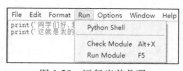
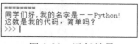

图 1-29　运行当前代码　　　　　　　　图 1-30　运行效果

实例 01-08：使用命令行运行 Python 程序

在 Windows 操作系统下还可以直接使用鼠标双击的方式来运行 Python 程序。双击运行上面编写的程序文件 first.py，一个 CMD 控制台开启后又关闭，这个过程很快，我们看不到输出的内容，因为程序运行结束后立即退出了。为了能看到程序的输出内容，可以按以下步骤进行操作。

（1）单击"开始"菜单，在"搜索程序和文件"文本框中输入"cmd"，并按 Enter 键，打开 Windows 的 CMD 控制台。

（2）先输入文件 first.py 的绝对路径及文件名，再按 Enter 键运行程序。也可以使用 cd 命令进入文件 first.py 所在的目录，如 D:\lx，然后在命令行提示符下输入"first.py"或者"python first.py"，按 Enter 键即可运行。

> **注意**：在 Linux 操作系统中，在终端窗口的命令提示符下可以使用 python first.py 命令来运行 Python 程序。

实例 01-09：交互式运行 Python 程序

Python 程序的交互式运行方式是指一边输入程序，一边运行程序。具体操作步骤如下。

（1）打开 IDLE，在命令行中输入如下代码。

```
print('同学们好,我的名字是——Python!')
```

按 Enter 键后即可立即运行上述代码，运行效果如图 1-31 所示。

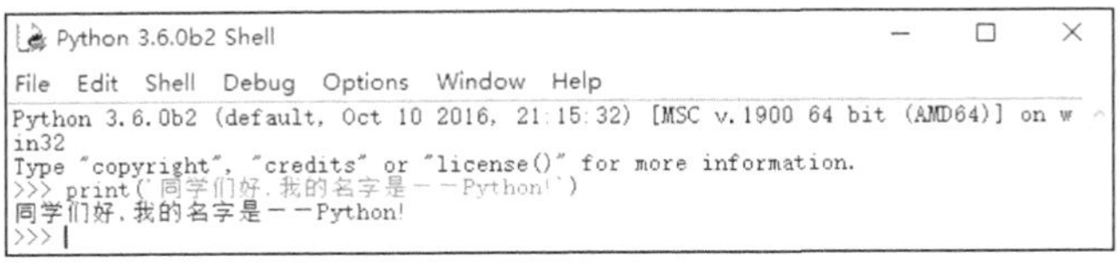

图 1-31　运行效果（一）

（2）继续输入如下代码。

```
print('这就是我的代码，简单吗？')
```

按 Enter 键后即可立即运行上述代码，运行效果如图 1-32 所示。

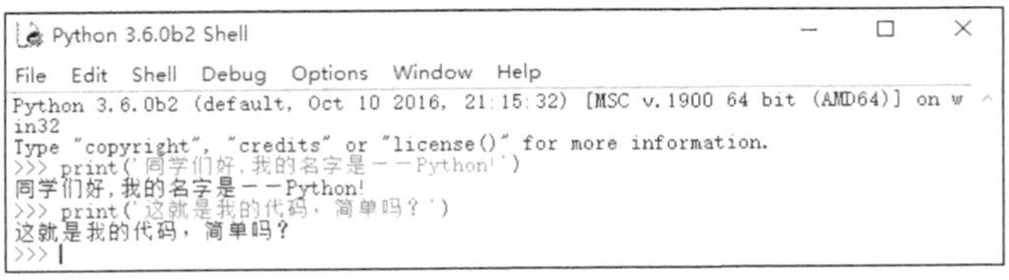

图 1-32　运行效果（二）

> **注意**：在 Linux 中也可以通过在终端窗口的命令提示符下运行命令"python"来启动 Python 的交互式运行环境，实现边输入程序边运行程序。

实例 01-10：使用 PyCharm 编写并运行 Python 程序

（1）打开 PyCharm，单击图 1-15 所示的 Create New Project 按钮，弹出 New Project 界面，选择左侧列表中的 Pure Python 选项，如图 1-33 所示。

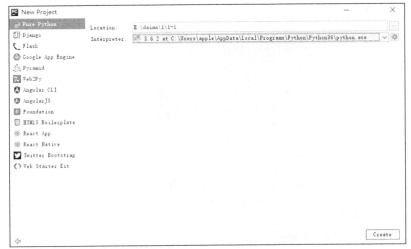

图 1-33　New Project 界面

❑ Location:Python 项目的保存路径。
❑ Interpreter:选择 Python 的版本,很多开发者在计算机中安装了多个版本的 Python,如 Python 2.7、Python3.5 或 Python 3.6 等。这一功能十分人性化,因为在不同版本之间切换十分方便。

(2)单击 Create 按钮后将创建一个 Python 项目,如图 1-34 所示。在图 1-34 所示的 PyCharm 项目界面中,依次单击菜单栏中的 File→New Project 选项,也可以实现创建 Python 项目的功能。

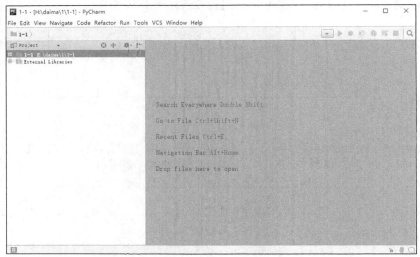

图 1-34 创建 Python 项目

(3)右击左侧项目名,在弹出的快捷菜单中依次选择 New→Python File 选项,如图 1-35 所示。

(4)弹出 New Python file 对话框,在 Name 文本框中输入将要创建的 Python 文件的名称,如"first",如图 1-36 所示。

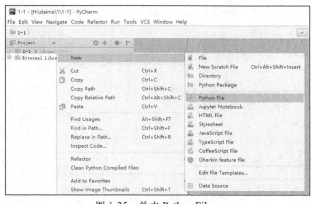
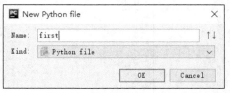

图 1-35 单击 Python File　　　　　　　　图 1-36 输入 Python 文件的名称

(5)单击 OK 按钮后将创建一个名为"first.py"的 Python 文件,选择左侧列表中的 first.py,如图 1-37 所示,在 PyCharm 右侧代码编辑界面中编写 Python 代码,例如,编写如下代码。

```
# if True 是一个固定语句,后面的代码总是运行
if True:
        print("Hello 这是第一个Python程序!")        #缩进4个空白的占位
else:
        print("Hello Python!")                      #缩进4个空白的占位
```
                                                    #与if对齐

1.3 编写并运行 Python 程序

图 1-37　Python 文件 first.py

（6）运行文件 first.py，在运行该文件之前会发现 PyCharm 顶部菜单中的功能按钮是灰色的，处于不可用状态。这时，需要我们对控制台进行配置，方法是单击 ▼ 按钮，然后在打开的下拉列表框中选择 Edit Configurations 选项（或者依次选择 PyCharm 顶部菜单栏中的 Run→Edit Configurations 选项）进入 Run/Debug Configurations 界面，如图 1-38 所示。

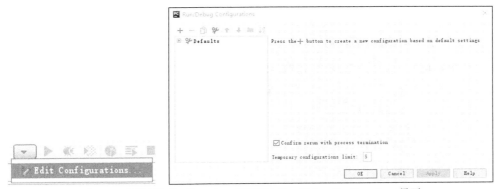

图 1-38　选择 Edit Configurations 选项进入 Run/Debug Configurations 界面

（7）单击左上角的绿色加号，在弹出的列表中选择 Python 选项，设置右侧窗格中的 Script 文本框内容为我们前面刚刚编写的文件 first.py 的路径，如图 1-39 所示。

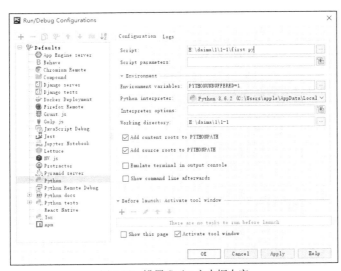

图 1-39　设置 Script 文本框内容

（8）单击 OK 按钮返回 PyCharm 代码编辑界面，此时会发现功能按钮全部处于可用状态，单击后即可运行文件 first.py。也可以右击左侧列表中的文件名"first.py",在弹出的快捷菜单中选择 Run 'first'命令来运行文件 first.py,如图 1-40 所示。

在 PyCharm 底部的调试面板中将会显示文件 first.py 的运行效果,如图 1-41 所示。

图 1-40 选择 Run 'first'命令运行文件 first.py 　　图 1-41 文件 first.py 的运行效果

实例 01-11:使用 Eclipse 编写并运行 Python 程序

(1)打开 Eclipse,在顶部菜单栏中依次选择 File→New→Project 选项创建一个项目,如图 1-42 所示。

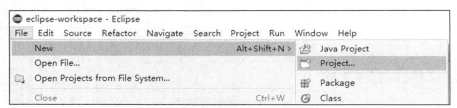

图 1-42 选择命令

(2)在打开的 New Project 窗口中选择 PyDev Project 选项,然后单击 Next 按钮,如图 1-43 所示。

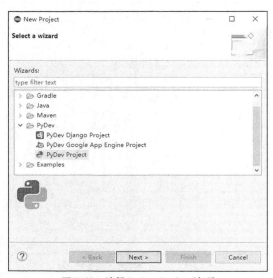

图 1-43 选择 PyDev Project 选项

（3）弹出 PyDev Project 窗口，在 Project name 文本框中输入项目名称，如"second"，在 Directory 文本框中设置保存项目的路径，在 Grammar Version 下拉列表框中设置 Python 的版本为 3.6，在 Interpreter 下拉列表框中设置本地绑定的 Python 环境，其他设置保持默认，如图 1-44 所示。

（4）单击 Finish 按钮后成功创建了一个名为"second"的 Python 项目，在 Eclipse 左侧的 PyDev Package Explorer 面板中，右击项目名称"second"，在弹出的快捷菜单中依次选择 New→PyDev Module 命令，如图 1-45 所示。

图 1-44 创建项目

图 1-45 依次选择 New→PyDev Module 命令

（5）弹出 Create a new Python module 窗口，在 Name 文本框中输入将要创建的 Python 文件名，如图 1-46 所示。

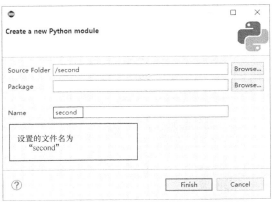

图 1-46 输入 Python 文件名

（6）单击 Finish 按钮后弹出 Template 窗口，在此可以选择快速创建 Python 程序的模板。因为我们的例子比较简单，所以这里选择<Empty>模板，如图 1-47 所示。

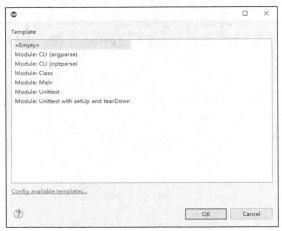

图 1-47　选择<Empty>模板

（7）单击 OK 按钮后返回 Eclipse 代码编辑界面，Eclipse 会自动打开刚刚创建的文件 second.py，如图 1-48 所示。读者此时会发现 Eclipse 会自动创建一些 Java 代码，提高了开发效率。

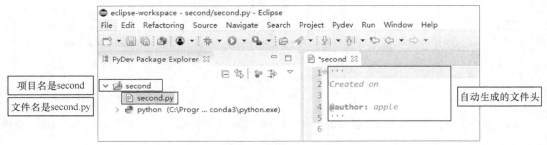

图 1-48　自动打开 second.py

（8）选择左侧列表中的文件 second.py，在右侧代码编辑界面编写 Python 代码，如图 1-49 所示。

图 1-49　编写 Python 代码

（9）运行 Python 文件 second.py。单击 Eclipse 顶部的 ▶ 按钮即可运行当前的整个 Python 项目 second。如果一个项目中有多个.py 文件，而我们只想编译调试其中的某一个文件，这时应该怎样实现呢？可以先右击要运行的 Python 文件，如前面编写的 second.py，然后在弹出的快捷菜单中依次选择 Run As→1 Python Run 命令，如图 1-50 所示。此时便会运行文件 second.py，运行效果如图 1-51 所示。

图 1-50　依次选择 Run As→1 Python Run 命令　　　　图 1-51　运行效果

第 2 章

基础语法实战

语法是任何一门编程语言的核心，Python 正是通过本身的语法知识实现具体功能的。本章将详细讲解 Python 的基本语法知识，包括列表、元组、字典、集合、循环和面向对象等，为读者学习本书后面的知识打下基础。

2.1 使用列表

Python 内置了多种数据结构，如列表、元组、字典和集合等。可通过数据结构来保存项目需要的数据信息。

实例 02-01：创建数字列表

在 Python 程序中，可以使用方法 range() 创建数字列表。下面的实例文件 num.py 使用方法 range() 创建了一个包含 3 个数字的列表。

源码路径：daima\2\02-01\num.py

```
numbers = list(range(1,4))    #使用方法range()创建列表
print(numbers)
```

在上述代码中，一定要注意方法 range() 的结尾参数是 4，这样才能创建 3 个列表元素。执行效果如图 2-1 所示。

```
[1, 2, 3]
```

图 2-1　执行效果

实例 02-02：访问列表中的值

在 Python 程序中，因为列表是一个有序集合，所以要想访问列表中的任何元素，只需将该元素的位置或索引告诉 Python 即可。要想访问列表元素，可以先指出列表的名称，再指出元素的索引，并将其放在方括号内。例如，下面的代码可以从列表 car 中提取第一款汽车。

```
car = ['audi', 'bmw', 'benchi', 'lingzhi']
print(car[0])
```

上述代码演示了访问列表元素的语法。当发出获取列表中某个元素的请求时，Python 只会返回该元素，而不包括方括号和引号，上述代码执行后只会输出：

```
audi
```

开发者还可以通过方法 title() 获取任何列表元素，例如，获取元素 "audi" 的代码如下。

```
car = ['audi', 'bmw', 'benchi', 'lingzhi']
print(car[0].title())
```

上述代码执行后的输出结果与前面的代码相同，只是首字母 a 变为大写的，上述代码执行后只会输出：

```
Audi
```

在 Python 程序中，字符串还可以通过序号（序号从 0 开始）来取出其中的某个字符，例如，abcde.[1] 取得的值是 'b'。

再看下面的实例文件 fang.py，功能是访问并显示列表中元素的值。

源码路径：daima\2\02-02\fang.py

```
list1 = ['Google', 'baidu', 1997, 2000];    #定义第1个列表list1
list2 = [1, 2, 3, 4, 5, 6, 7 ];             #定义第2个列表list2
print ("list1[0]: ", list1[0])              #输出列表list1中的第1个元素
print ("list2[1:5]: ", list2[1:5])          #输出列表list2中的第2个到第5个元素
```

上述代码分别定义了两个列表 list1 和 list2，执行效果如图 2-2 所示。

```
list1[0]:  Google
list2[1:5]:  [2, 3, 4, 5]
```

图 2-2　执行效果

在 Python 程序中，第一个列表元素的索引为 0，而不是 1。在大多数编程语言中，数组也

是如此，这与列表操作的底层实现相关。自然而然地，第二个列表元素的索引为 1。根据这种简单的计数方式，要访问列表的任何元素，都可将其位置减 1，并将结果作为索引。例如，要访问列表中的第 4 个元素，可使用索引 3 实现。例如，下面的代码演示了显示列表中第 2 个和第 4 个元素的方法。

```
car = ['audi', 'bmw', 'benchi', 'lingzhi']    #定义一个拥有4个元素的列表
print(car[1])                                  #输出列表中的第2个元素
print(car[3])                                  #输出列表中的第4个元素
```

执行后会输出：

```
bmw
lingzhi
```

实例 02-03：删除列表中的重复元素并保持顺序不变

在 Python 程序中，我们可以删除列表中重复出现的元素，并且保持剩下元素的显示顺序不变。如果序列中保存的元素是可哈希（hashable）的，那么上述功能可以使用集合和生成器实现。下面的实例文件 delshun.py 演示了在可哈希情况下删除重复元素的过程。

源码路径： daima\2\02-03\delshun.py

```python
def dedupe(items):
    seen = set()
    for item in items:
        if item not in seen:
            yield item
            seen.add(item)

if __name__ == '__main__':
    a = [5, 5, 2, 1, 9, 1, 5, 10]
    print(a)
    print(list(dedupe(a)))
```

如果一个对象是可哈希的，那么该对象在其生存期内必须是不可变的，这需要有一个 __hash__()方法。在 Python 程序中，整数、浮点数、字符串和元组都是不可变的。在上述代码中，函数 dedupe()实现了可哈希情况下的删除重复元素功能，并且保持剩下元素的显示顺序不变。执行效果如图 2-3 所示。

```
[5, 5, 2, 1, 9, 1, 5, 10]
[5, 2, 1, 9, 10]
```

图 2-3　执行效果

上述实例文件 delshun.py 有一个缺陷，只有当序列中的元素是可哈希的时才能这么做。如果想在不可哈希的序列中去除重复项，并保持顺序不变应该如何实现呢？下面的实例文件 buhaxi.py 演示了上述功能的实现过程。

源码路径： daima\2\02-03\buhaxi.py

```python
def buha(items, key=None):
    seen = set()
    for item in items:
        val = item if key is None else key(item)
        if val not in seen:
            yield item
            seen.add(val)

if __name__ == '__main__':
    a = [
        {'x': 2, 'y': 3},
        {'x': 1, 'y': 4},
        {'x': 2, 'y': 3},
        {'x': 2, 'y': 3},
        {'x': 10, 'y': 15}
```

```
    ]
    print(a)
    print(list(buha(a, key=lambda a: (a['x'],a['y']))))
```

在上述代码中，函数 buha() 中的参数 key 的功能是设置一个函数，将序列中的元素转换为可哈希的类型，这样做的目的是检测重复选项。执行效果如图 2-4 所示。

```
[{'x': 2, 'y': 3}, {'x': 1, 'y': 4}, {'x': 2, 'y': 3}, {'x': 2, 'y': 3}, {'x': 10, 'y': 15}]
[{'x': 2, 'y': 3}, {'x': 1, 'y': 4}, {'x': 10, 'y': 15}]
```

图 2-4　执行效果

实例 02-04：找出列表中出现次数最多的元素

在 Python 程序中，如果想找出列表中出现次数最多的元素，则可以考虑使用 collections 模块中的 Counter 类，通过调用 Counter 类中的函数 most_common() 来实现上述功能。下面的实例文件 most.py 演示了使用函数 most_common() 找出列表中出现次数最多的元素的过程。

源码路径：daima\2\02-04\most.py

```python
words = [
'look', 'into', 'my', 'AAA', 'look', 'into', 'my', 'AAA',
'the', 'AAA', 'the', 'AAA', 'the', 'eyes', 'not', 'BBB', 'the',
'AAA', "don't", 'BBB', 'around', 'the', 'AAA', 'look', 'into',
'BBB', 'AAA', "BBB", 'under'
]
from collections import Counter
word_counts = Counter(words)
top_three = word_counts.most_common(3)
print(top_three)
```

上述代码预先定义了一个列表 words，其中保存了一系列的英文单词，使用函数 most_common() 找出哪些单词出现的次数最多。执行效果如图 2-5 所示。

```
[('AAA', 7), ('the', 5), ('BBB', 4)]
```

图 2-5　执行效果

实例 02-05：对类的实例进行排序

在 Python 程序中，我们可以对一个类定义的多个实例进行排序。使用内置函数 sorted() 可以接受一个用来传递可调用（callable）对象的参数 key，而这个可调用对象会返回待排序对象中的某些值，sorted() 函数则利用这些值来比较对象。假设程序中存在多个 User 类的实例，如果想通过属性 user_id 来对这些实例进行排序，则可以提供一个可调用对象将 User 类的实例作为输入，然后返回 user_id。下面的实例文件 leishili.py 演示了排序上述 User 类的实例的过程。

源码路径：daima\2\02-05\leishili.py

```python
class User:
    def __init__(self, user_id):
        self.user_id = user_id
    def __repr__(self):
        return 'User({})'.format(self.user_id)

# 原来的顺序
users = [User(19), User(17), User(18)]
print(users)

# 根据user_id排序
①print(sorted(users, key=lambda u: u.user_id))
from operator import attrgetter
②print(sorted(users, key=attrgetter('user_id')))
```

在上述代码中，在①处使用 lambda 表达式进行了处理，在②处使用内置函数 operator.attrgetter() 进行了处理。执行效果如图 2-6 所示。

```
[User(19), User(17), User(18)]
[User(17), User(18), User(19)]
[User(17), User(18), User(19)]
```

图 2-6 执行效果

实例 02-06：使用列表推导式

在 Python 程序中，列表推导式（list comprehension）是一种简化代码的优美方法。Python 官方文档描述：列表推导式提供了一种创建列表的简洁方法。列表推导式能够非常简洁地构造一个新列表，只需用一条简洁的表达式即可对得到的元素进行转换变形。使用 Python 列表推导式的语法格式如下。

```
variable = [out_exp_res for out_exp in input_list if out_exp == 2]
```

- out_exp_res：列表生成元素表达式，可以是有返回值的函数。
- for out_exp in input_list：迭代 input_list，将 out_exp 传入 out_exp_res 表达式中。
- if out_exp == 2：根据条件可以过滤值。

例如，想创建一个包含 1~10 的二次方的列表，下面的实例文件 chuantong.py 分别演示了传统方法和列表推导式方法的实现过程。

源码路径：daima\2\02-06\chuantong.py

```
①squares = []
  for x in range(10):
      squares.append(x**2)
②print(squares)

③squares1 = [x**2 for x in range(10)]
  print(squares1)
```

在上述代码中，①~②是通过传统方式实现的，③处的代码是通过列表推导式方法实现的。执行后将会输出：

```
[0, 1, 4, 9, 16, 25, 36, 49, 64, 81]
[0, 1, 4, 9, 16, 25, 36, 49, 64, 81]
```

假如想输出 30 以内能够整除 3 的整数，使用传统方法的实现代码如下。

```
numbers = []
for x in range(100):
    if x % 3 == 0:
        numbers.append(x)
```

而通过列表推导式方法的实现代码如下。

```
multiples = [i for i in range(30) if i % 3 is 0]
print(multiples)
```

上述两种方式执行后都会输出：

```
[0, 3, 6, 9, 12, 15, 18, 21, 24, 27]
```

再看下面的代码，首先获取 30 以内能够整除 3 的整数，然后依次输出获取的整数的二次方。

```
def squared(x):
    return x*x
multiples = [squared(i) for i in range(30) if i % 3 is 0]
print (multiples)
```

执行后会输出：

```
[0, 9, 36, 81, 144, 225, 324, 441, 576, 729]
```

下面的实例文件 shaixuan.py 使用列表推导式筛选了列表中的数据。

源码路径：daima\2\02-06\shaixuan.py

```
mylist = [1, 4, -5, 10, -7, 2, 3, -1]

#大于0的元素
zheng = [n for n in mylist if n > 0]
print(zheng)
```

```
#小于0的元素
fu = [n for n in mylist if n < 0]
print(fu)
```
通过上述代码,分别筛选出列表 mylist 中大于 0 和小于 0 的元素,执行效果如图 2-7 所示。

```
[1, 4, 10, 2, 3]
[-5, -7, -1]
```
图 2-7 执行效果

在 Python 程序中,有时候筛选的标准无法简单地表示在列表推导式或生成器表达式中。例如,假设筛选过程涉及异常处理或其他复杂的细节,此时可以考虑先将处理筛选功能的代码放到单独的功能函数中,然后使用内建的 filter()函数进行处理。下面的实例文件 dandu.py 演示了这一功能。

源码路径:daima\2\02-06\dandu.py

```
values = ['1', '2', '-3', '-', '4', 'N/A', '5']
def is_int(val):
    try:
        x = int(val)
        return True
    except ValueError:
        return False
ivals = list(filter(is_int, values))
print(ivals)
```
在上述代码中,因为使用函数 filter()创建了一个迭代器,所以如果想要得到一个列表形式的结果,需确保在 filter()前面加上 list()函数。执行后会输出:

```
['1', '2', '-3', '4', '5']
```

实例 02-07:命名切片操作

在 Python 程序中,有时会发现编写的代码变得杂乱无章而无法阅读(到处都是硬编码的切片索引),此时需要将它们清理干净。代码中存在过多硬编码的切片索引,还会减弱代码的可维护性。在 Python 程序中,使用函数 slice()可以实现切片对象,它能够在切片操作函数中实现参数传递功能,该函数可以被用在任何允许进行切片操作的地方。使用函数 slice()的语法格式如下。

```
class slice(stop)
class slice(start, stop[, step])
```
- start:起始位置。
- stop:结束位置。
- step:间距。

下面的实例文件 qie.py 演示了使用函数 slice()实现切片操作的过程。

源码路径:daima\2\02-07\qie.py

```
items = [0, 1, 2, 3, 4, 5, 6]
a = slice(2, 4)
print(items[2:4])
print(items[a])
items[a] = [10, 11]
print(items)
print(a.start)
print(a.stop)
print(a.step)
s = 'HelloWorld'
①print(a.indices(len(s)))
```

```
for i in range(*a.indices(len(s))):
    print(s[i])
```

在上述代码中，slice 对象实例 a 可以分别通过属性 a.start、a.stop 和 a.step 来获取该对象的信息。在①处使用 indices(size)函数可将切片映射到特定大小的序列上，这将会返回一个(start, stop, step)元组，所有的值已经正好被限制在边界以内，这样当进行索引操作时可以避免出现 IndexError 异常。执行效果如图 2-8 所示。

```
[2, 3]
[2, 3]
[0, 1, 10, 11, 4, 5, 6]
2
4
None
(2, 4, 1)
1
1
```

图 2-8 执行效果

2.2 使用元组

在 Python 程序中，可以将元组看作一种特殊的列表。唯一与列表不同的是，元组内的数据元素不能发生改变，不但不能改变其中的数据元素，而且也不能添加和删除数据元素。当开发者需要创建一组不可改变的数据时，通常会把这些数据放到一个元组中。

实例 02-08：创建并访问元组

在 Python 程序中，创建元组的基本形式是以圆括号"()"将数据元素括起来，各个元素之间用逗号","隔开。元组的索引从 0 开始，并且可以进行截取和组合等操作。下面的实例文件 zu.py 演示了创建并访问元组的过程。

源码路径：daima\2\02-08\zu.py

```
tup1 = ('Google', 'toppr', 1997, 2000)    #创建元组tup1
tup2 = (1, 2, 3, 4, 5, 6, 7)              #创建元组tup2
#显示元组tup1中索引为0的元素的值
print ("tup1[0]: ", tup1[0])
#显示元组tup2中索引为1～4的元素的值
print ("tup2[1:5]: ", tup2[1:5])
```

上述代码定义了两个元组，即 tup1 和 tup2，然后在第 4 行代码中读取了元组 tup1 中索引为 0 的元素的值，然后在第 6 行代码中读取了元组 tup2 中索引为 1～4 的元素的值。执行效果如图 2-9 所示。

```
tup1[0]:  Google
tup2[1:5]:  (2, 3, 4, 5)
```

图 2-9 执行效果

实例 02-09：修改元组的值

在 Python 程序中，元组一旦被创建后就是不可被修改的。但是在现实应用中，开发者可以对元组进行连接组合。下面的实例文件 lian.py 演示了连接组合两个元组的过程。

源码路径：daima\2\02-09\lian.py

```python
tup1 = (12, 34.56);          #定义元组tup1
tup2 = ('abc', 'xyz')        #定义元组tup2
#下面一行代码修改元组元素的操作是非法的
# tup1[0] = 100
tup3 = tup1 + tup2;          #创建一个新的元组tup3
print (tup3)                 #输出元组tup3中的值
```

上述代码定义了两个元组，即 tup1 和 tup2，然后将这两个元组进行连接组合，并将连接组合后的值赋给新元组 tup3，执行后会输出新元组 tup3 中的值，执行效果如图 2-10 所示。

```
(12, 34.56, 'abc', 'xyz')
```

图 2-10　执行效果

实例 02-10：删除元组

在 Python 程序中，虽然不允许删除一个元组中的元素，但是可以使用 del 语句来删除整个元组。下面的实例文件 shan.py 演示了使用 del 语句来删除整个元组的过程。

源码路径：daima\2\02-10\shan.py

```python
#定义元组tup
tup = ('Google', 'Toppr', 1997, 2000)
print (tup)                         #输出元组tup中的元素
del tup;                            #删除元组tup
#因为元组tup已经被删除，所以不能显示里面的元素
print ("元组tup被删除后，系统会出错！")
print (tup)                         #这行代码会出错
```

上述代码定义了一个元组 tup，然后使用 del 语句来删除整个元组的过程。删除元组 tup 后，最后一行代码使用 print (tup) 输出元组 tup 的值时会出现系统错误。执行效果如图 2-11 所示。

```
Traceback (most recent call last):
('Google', 'Toppr', 1997, 2000)
  File "H:/daima/2/2-2/shan.py", line 7, in <module>
元组tup被删除后，系统会出错！
    print (tup)           #这行代码会出错
NameError: name 'tup' is not defined
```

图 2-11　执行效果

实例 02-11：使用内置方法操作元组

在 Python 程序中，可以使用内置方法来操作元组，其中较为常用的方法如下。
- len(tuple)：计算元组的元素个数。
- max(tuple)：返回元组中元素的最大值。
- min(tuple)：返回元组中元素的最小值。
- tuple(seq)：将列表转换为元组。

下面的实例文件 neizhi.py 演示了使用内置方法操作元组的过程。

源码路径：daima\2\02-11\neizhi.py

```python
car = ['奥迪', '宝马', '奔驰', '雷克萨斯']      #创建列表car
print(len(car))                                #输出列表car的长度
tuple2 = ('5', '4', '8')                       #创建元组tuple2
print(max(tuple2))                             #显示元组tuple2中元素的最大值
tuple3 = ('5', '4', '8')                       #创建元组tuple3
print(min(tuple3))                             #显示元组tuple3中元素的最小值
list1= ['Google', 'Taobao', 'Toppr', 'Baidu']  #创建列表list1
tuple1=tuple(list1)                            #将列表list1的值赋予元组tuple1
print(tuple1)                                  #再次输出元组tuple1中的元素
```

执行效果如图2-12所示。

```
4
8
4
('Google', 'Taobao', 'Toppr', 'Baidu')
```

图2-12 执行效果

实例02-12：将序列分解为单独的变量

在Python程序中，可以将一个包含多个元素的元组或其他序列分解为多个单独的变量。这是因为Python语法允许任何序列（或可迭代的对象）通过一个简单的赋值操作来分解为单独的变量，唯一的要求是变量的总数和结构要与序列相吻合。下面的实例文件fenjie.py演示了将序列分解为单独的变量的过程。

源码路径：daima\2\02-12\fenjie.py

```python
p = (4, 5)
x, y = p
print(x)
print(y)
data = [ 'ACME', 50, 91.1, (2012, 12, 21) ]
name, shares, price, date = data
print(name)
print(date)
```

执行效果如图2-13所示。

对于分解未知或任意长度的可迭代对象，上述分解操作是为其量身定做的工具。通常在这类可迭代对象中会有一些已知的组件或模式（例如，元素1之后的所有内容都是电话号码），利用"*"星号表达式分解可迭代对象后，使得开发者能够轻松利用这些模式，而无须在可迭代对象中做复杂的操作才能得到相关的元素。

图2-13 执行效果

在Python程序中，星号表达式在迭代一个元素长度有变化的序列时十分有用。下面的实例文件xinghao.py演示了分解一个带标记序列的过程。

源码路径：daima\2\02-12\xinghao.py

```python
records = [
    ('AAA', 1, 2),
    ('BBB', 'hello'),
    ('CCC', 5, 3)
]

def do_foo(x, y):
    print('AAA', x, y)

def do_bar(s):
    print('BBB', s)

for tag, *args in records:
    if tag == 'AAA':
        do_foo(*args)
    elif tag == 'BBB':
        do_bar(*args)

line = 'guan:ijing234://wef:678d:guan'
uname, *fields, homedir, sh = line.split(':')
print(uname)
print(homedir)
```

执行效果如图 2-14 所示。

```
AAA 1 2
BBB hello
guan
678d
```

图 2-14 执行效果

实例 02-13：将序列的最后几项作为历史记录

在 Python 程序中迭代处理列表或元组等序列时，有时需要统计最后几项记录以实现历史记录统计的功能。下面的实例文件 lishi.py 演示了将序列的最后几项作为历史记录的过程。

源码路径：daima\2\02-13\lishi.py

```python
from _collections import deque
def search(lines, pattern, history=5):
    previous_lines = deque(maxlen=history)
    for line in lines:
        if pattern in line:
            yield line, previous_lines
        previous_lines.append(line)
if __name__ == '__main__':
    with open('123.txt') as f:
        for line, prevlines in search(f, 'python', 5):
            for pline in prevlines:
                print(pline)
            print(line)
            print('-' * 20)
q = deque(maxlen=3)
q.append(1)
q.append(2)
q.append(3)
print(q)
q.append(4)
print(q)
```

在上述代码中，对一系列文本行实现了简单的文本匹配操作，当发现有合适的匹配行时就输出当前的匹配行及最后检查过的多行文本。deque(maxlen=N)创建了一个固定长度的队列。当新记录加入而使得队列变成已满状态时，最旧的那条记录会被自动移除。当编写搜索某项记录的代码时，通常会用到含有 yield 关键字的生成器函数，能够将处理搜索过程的代码和使用搜索结果的代码成功解耦。执行效果如图 2-15 所示。

```
pythonpythonpythonpython
deque([1, 2, 3], maxlen=3)
deque([2, 3, 4], maxlen=3)
```

图 2-15 执行效果

实例 02-14：实现优先级队列

在 Python 程序中，使用内置模块 heapq 可以实现一个简单的优先级队列。下面的实例文件 youxianpy.py 演示了实现一个简单的优先级队列的过程。

源码路径：daima\2\02-14\youxianpy.py

```python
import heapq
class PriorityQueue:
```

```python
    def __init__(self):
        self._queue = []
        self._index = 0

    def push(self, item, priority):
        heapq.heappush(self._queue, (-priority, self._index, item))
        self._index += 1

    def pop(self):
        return heapq.heappop(self._queue)[-1]
class Item:
    def __init__(self, name):
        self.name = name

    def __repr__(self):
        return 'Item({!r})'.format(self.name)
q = PriorityQueue()
q.push(Item('AAA'), 1)
q.push(Item('BBB'), 4)
q.push(Item('CCC'), 5)
q.push(Item('DDD'), 1)
print(q.pop())
print(q.pop())
print(q.pop())
```

上述代码利用 heapq 模块实现了一个简单的优先级队列，第一次执行 pop() 操作时返回的元素具有最高的优先级。拥有相同优先级的两个元素返回的顺序，同插入队列时的顺序相同。函数 heapq.heappush() 和函数 heapq.heappop() 分别用于对列表 self._queue 中的元素实现插入和移除操作，并且保证列表中第一个元素的优先级最低。函数 heappop() 总是返回"最小"的元素，并且因为 push() 操作和 pop() 操作的复杂度都是 O(logN)，其中 N 代表堆中元素的数量，因此即使 N 的值很大，这些操作的效率也非常高。上述代码中的队列以元组(-priority, self._index, item)的形式组成，priority 取负值是为了让队列能够按元素的优先级从高到低的顺序排列。这和正常的堆排列顺序相反，在一般情况下，堆是按从小到大的顺序排列的。变量 self._index 的作用是将具有相同优先级的元素以适当的顺序排列。通过维护一个不断递增的索引，元素将以它们入队列时的顺序来排列。index 在对具有相同优先级的元素进行比较操作时，同样扮演了一个重要的角色。执行效果如图 2-16 所示。

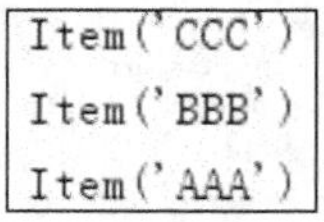

```
Item('CCC')
Item('BBB')
Item('AAA')
```

图 2-16 执行效果

在 Python 程序中，如果以元组（priority, item）的形式来存储元素，那么只要它们的优先级不同，就可以对它们进行比较。但是如果两个元组的优先级相同，对其进行比较操作会失败。这时可以考虑引入一个额外的索引值，以（prioroty, index, item）的方式建立元组，因为没有哪两个元组会有相同的 index 值，所以这样就可以完全避免上述问题。一旦比较操作的结果可以确定，Python 就不会再去比较剩下的元组元素了。下面的实例文件 suoyin.py 演示了实现一个简单的优先级队列的过程。

源码路径：daima\2\02-14\suoyin.py

```python
import heapq
class PriorityQueue:
    def __init__(self):
        self._queue = []
        self._index = 0

    def push(self, item, priority):
        heapq.heappush(self._queue, (-priority, self._index, item))
        self._index += 1

    def pop(self):
        return heapq.heappop(self._queue)[-1]
```

```
class Item:
    def __init__(self, name):
        self.name = name

    def __repr__(self):
        return 'Item({!r})'.format(self.name)

a = Item('AAA')①
b = Item('BBB')
#a < b为False③
a = (1, Item('AAA'))
b = (5, Item('BBB'))
print(a < b)
c = (1, Item('CCC'))
#a < c为False
a = (1, 0, Item('AAA'))
b = (5, 1, Item('BBB'))
c = (1, 2, Item('CCC'))
print(a < b)
print(a < c)
```

在上述代码中，因为代码中未注明①～③行代码中没有添加索引，所以两个元组的优先级值相同时会出错。而代码中未注明①～③行代码中添加了索引，这样就不会出错了。执行效果如图 2-17 所示。

图 2-17 执行效果

2.3 使用字典

在 Python 程序中，字典是一种比较特别的数据结构，字典中的每个元素以"键:值"(key:value)的形式成对存在，即键值对。字典是以花括号"{}"包围，并且以键值对的方式声明和存在的数据集合。字典与列表相比，最大的不同在于字典是无序的，其元素位置只是象征性的，在字典中通过键来访问元素，而不能通过位置来访问元素。

实例 02-15：创建并访问字典

在 Python 程序中，字典可以存储任意类型对象。字典的每个键值对的键和值之间必须用冒号":"分隔，每个键值对之间用逗号","分隔，整个字典被包括在花括号"{}"中。要想获取某个键的值，可以通过访问键的方式实现。下面的实例文件 fang.py 演示了获取字典中 3 个键的值的过程。

源码路径：daima\2\02-15\fang.py

```
dict = {'数学': '99', '语文': '99', '英语': '99' }   #创建字典dict
print ("语文成绩是： ",dict['语文'])                  #输出语文成绩
print ("数学成绩是： ",dict['数学'])                  #输出数学成绩
print ("英语成绩是： ",dict['英语'])                  #输出英语成绩
```

执行效果如图 2-18 所示。

```
语文成绩是：    99
数学成绩是：    99
英语成绩是：    99
```

图 2-18 执行效果

如果调用的字典中没有这个键，则代码执行后会输出执行错误的提示信息。例如，在下面的代码中，字典 dict 中并没有键为 Alice 的值。

```
dict = {'Name': 'Toppr', 'Age': 7, 'Class': 'First'};   #创建字典dict
print ("dict['Alice']: ", dict['Alice'])                 #输出字典dict中键为Alice的值
```

所以，上述代码执行后会输出如下错误提示信息：

```
Traceback (most recent call last):
    File "test.py", line 5, in <module>
        print ("dict['Alice']: ", dict['Alice'])
KeyError: 'Alice'
```

实例 02-16：向字典中添加数据

在 Python 程序中，字典是一种动态结构，可以随时在其中添加键值对。在添加键值对时，需要首先指定字典名，然后用方括号将键括起来，最后写明这个键的值。下面的实例文件 add.py 定义了字典 dict，先在字典中设置 3 科的成绩，然后又通过上面介绍的方法添加了两个键值对。

源码路径：daima\2\02-16\add.py

```
dict = {'数学': '99', '语文': '99', '英语': '99' }    #创建字典dict
dict['物理'] =100                                    #添加键值对
dict['化学'] =98                                     #添加键值对
print (dict)                                        #输出字典dict中的值
print ("物理成绩是：",dict['物理'])                  #显示物理成绩
print ("化学成绩是：",dict['化学'])                  #显示化学成绩
```

通过上述代码，向字典中添加两个键值对，分别表示物理成绩和化学成绩。其中第 2 行代码设置在字典 dict 中新增一个键值对，其中的键为"物理"，而值为 100。而在第 3 行代码中重复了上述操作，设置新添加的键为"化学"，而对应的值为 98。执行效果如图 2-19 所示。

```
{'数学': '99', '语文': '99', '英语': '99', '物理': 100, '化学': 98}
物理成绩是： 100
化学成绩是： 98
```

图 2-19　执行效果

注意：键值对的排列顺序与添加顺序不同。Python 不关心键值对的添加顺序，而只关心键和值之间的关联关系而已。

实例 02-17：修改字典中的数据

在 Python 程序中，要想修改字典中的值，需要首先指定字典名，然后使用方括号把将要修改的键和新值对应起来。下面的实例文件 xiu.py 演示了在字典中实现修改和添加功能的过程。

源码路径：daima\2\02-17\xiu.py

```
#创建字典 "dict"
dict = {'Name': 'Toppr', 'Age': 7, 'Class': 'First'}
dict['Age'] = 8;                              #更新Age的值
dict['School'] = "Python教程"                 #添加新的键值对
print ("dict['Age']: ", dict['Age'])          #输出键Age的值
print ("dict['School']: ", dict['School'])    #输出键School的值
print (dict)                                  #显示字典dict中的元素
```

上述代码先更新了字典中键 Age 的值为 8，然后添加了新键 School。执行效果如图 2-20 所示。

```
dict['Age']:  8
dict['School']:  Python教程
```

图 2-20　执行效果

实例 02-18：删除字典中的元素

在 Python 程序中，对于字典不再需要的信息，可以使用 del 语句将相应的键值对彻底删除。在使用 del 语句时，必须指定字典名和要删除的键。下面的实例文件 del.py 演示了删除

字典中某个元素的过程。

源码路径：daima\2\02-18\del.py

```
#创建字典"dict"
dict = {'Name': 'Toppr', 'Age': 7, 'Class': 'First'}
del dict['Name']              #删除键为Name的元素
print (dict)                  #显示字典dict中的元素
```

在上述代码中，使用 del 语句删除了字典中键为 Name 的元素。执行效果如图 2-21 所示。

```
{'Age': 7, 'Class': 'First'}
```
图 2-21 执行效果

实例 02-19：创建一键多值字典

在 Python 程序中，可以创建将某个键映射到多个值的字典，即一键多值字典（multidict）。为了能方便地创建映射多个值的字典，可以使用内置模块 collections 中的 defaultdict 类来实现。defaultdict 类的一个主要特点是会自动初始化第一个值，这样只需关注添加元素即可。下面的实例文件 yingshe.py 演示了创建一键多值字典的过程。

源码路径：daima\2\02-19\yingshe.py

```
  d = {
      'a': [1, 2, 3],
      'b': [4, 5]
① }
  e = {
      'a': {1, 2, 3},
      'b': {4, 5}
② }
  from collections import defaultdict
③ d = defaultdict(list)
  d['a'].append(1)
  d['a'].append(2)
④ d['a'].append(3)
  print(d)

⑤ d = defaultdict(set)
  d['a'].add(1)
  d['a'].add(2)
  d['a'].add(3)
⑥ print(d)

⑦ d = {}
  d.setdefault('a', []).append(1)
  d.setdefault('a', []).append(2)
  d.setdefault('b', []).append(3)
⑧ print(d)

  d = {}
⑨ for key, value in d:   # pairs:
      if key not in d:
          d[key] = []
      d[key].append(value)
  d = defaultdict(list)
⑩ print(d)

⑪ for key, value in d:   # pairs:
      d[key].append(value)
⑫ print(d)
```

上述代码用到了内置函数 setdefault()，如果键不存在于字典中，则会添加键并将值设为默认值。首先在①～②部分创建了一个字典，③～④和⑤～⑥部分分别利用两种方式为字典中的键创建了相同的多个值。因为函数 defaultdict()会自动创建字典元素以待稍后的访问，若不想要这个功能，可以在普通的字典上调用函数 setdefault()来取代 defaultdict()，如⑦～⑧所示。⑨～⑩和⑪～⑫分别演示了两种对一键多值字典中第一个值进行初始化的过程。可以看出，⑨～⑩使用 defaultdict()函数实现的方式比较清晰明了。执行效果如图 2-22 所示。

```
defaultdict(<class 'list'>, {'a': [1, 2, 3]})
defaultdict(<class 'set'>, {'a': {1, 2, 3}})
{'a': [1, 2], 'b': [3]}
defaultdict(<class 'list'>, {})
defaultdict(<class 'list'>, {})
```

图 2-22　执行效果

实例 02-20：使用 OrderedDict 类创建和修改有序字典

在 Python 程序中创建一个字典后，不但可以对字典进行迭代或序列化操作，而且能控制其中元素的排列顺序。下面的实例文件 youxu.py 演示了创建有序字典的过程。

源码路径：daima\2\02-20\youxu.py

```python
import collections

dic = collections.OrderedDict()
dic['k1'] = 'v1'
dic['k2'] = 'v2'
dic['k3'] = 'v3'
print(dic)
```

执行后会输出：

```
OrderedDict([('k1', 'v1'), ('k2', 'v2'), ('k3', 'v3')])
```

下面的实例文件 qingkong.py 演示了清空有序字典的过程。

源码路径：daima\2\02-20\qingkong.py

```python
import collections

dic = collections.OrderedDict()
dic['k1'] = 'v1'
dic['k2'] = 'v2'
dic.clear()
print(dic)
```

执行后会输出：

```
OrderedDict()
```

再看下面的实例文件 xianjin.py，其功能是使用函数 popitem()按照后进先出原则，删除最后加入的元素并返回键值对。

源码路径：daima\2\02-20\xianjin.py

```python
import collections

dic = collections.OrderedDict()
dic['k1'] = 'v1'
dic['k2'] = 'v2'
dic['k3'] = 'v3'
print(dic.popitem(),dic)
print(dic.popitem(),dic)
```

执行后会输出：

```
('k3', 'v3') OrderedDict([('k1', 'v1'), ('k2', 'v2')])
('k2', 'v2') OrderedDict([('k1', 'v1')])
```

❀ **注意**：Python 的 OrderedDict 内部维护了一个双向链表，它会根据元素加入的顺序来排列键的位置。第 1 个新加入的元素被放置在链表的末尾，接下来会对已存在的键做重新赋值而不会改变键的顺序。开发者需要注意的是，OrderedDict 的大小是普通字典的 2 倍多，这是由它额外创建的链表所致。因此，如果想构建一个涉及大量 OrderedDict 实例的数据结构（例如，从 CSV 文件中读取 100 000 行内容到 OrderedDict 列表中），那么需要认真对应用做需求分析，从而推断出使用 OrderedDict 所带来的好处是否能超越因额外的内存开销所带来的坏处。

实例 02-21：获取字典中的最大值和最小值

在 Python 程序中，函数 zip()可以将可迭代的对象作为参数，将对象中对应的元素打包成元组，然后返回由这些元组组成的列表。如果各个迭代器的元素个数不一致，则返回列表长度与最短的对象的长度相同。利用星号"*"操作符，可以将元组解压为列表。使用函数 zip()的语法格式如下。

```
zip([iterable, ...])
```

其中，参数 iterabl 表示一个或多个迭代器。下面的实例文件 jisuan.py 演示了分别获取字典中最大值和最小值的过程。

源码路径：daima\2\02-21\jisuan.py

```python
price = {
    '小米': 899,
    '华为': 1999,
    '三星': 3999,
    '谷歌': 4999,
    '酷派': 599,
    '苹果': 5000,
}

min_price = min(zip(price.values(), price.keys()))
print(min_price)

max_price = max(zip(price.values(), price.keys()))
print(max_price)

price_sorted = sorted(zip(price.values(), price.keys()))
print(price_sorted)

price_and_names = zip(price.values(), price.keys())
print((min(price_and_names)))

# print (max(price_and_names))    error   zip()创建了迭代器，内容只能被消费一次
print(min(price))
print(max(price))
print(min(price.values()))
print(max(price.values()))
print(min(price, key=lambda k: price[k]))
print(max(price, key=lambda k: price[k]))
```

执行效果如图 2-23 所示。

```
(599, '酷派')
(5000, '苹果')
[(599, '酷派'), (899, '小米'), (1999, '华为'), (3999, '三星'), (4999, '谷歌'), (5000, '苹果')]
(599, '酷派')
iPhone
苹果
599
5000
酷派
苹果
```

图 2-23 执行效果

实例 02-22：获取两个字典中的相同键值对

在 Python 程序中，我们可以寻找并获取两个字典中相同的键值对，此功能只需通过 keys() 或 items()这两个函数执行基本的集合操作即可实现。下面的实例文件 same.py 演示了获取两个

字典中的相同键值对的过程。

源码路径：daima\2\02-22\same.py

```
a = {
    'x': 1,
    'y': 2,
    'z': 3
}

b = {
    'x': 11,
    'y': 2,
    'w': 10
}
①print(a.keys() & b.keys())    # {'x','y'}
  print(a.keys() - b.keys())    # {'z'}
②print(a.items() & b.items())  # {('y', 2)}

③c = {key: a[key] for key in a.keys() - {'z', 'w'}}
④print(c)   # {'x':1, 'y':2}
```

在上述代码中，①～②通过 keys()和 items()执行集合操作实现获取两个字典的相同键值对。③～④是使用字典推导式实现的，能够修改或过滤字典中的内容。如果想创建一个新的字典，则可能会去掉其中某些键。执行效果如图 2-24 所示。

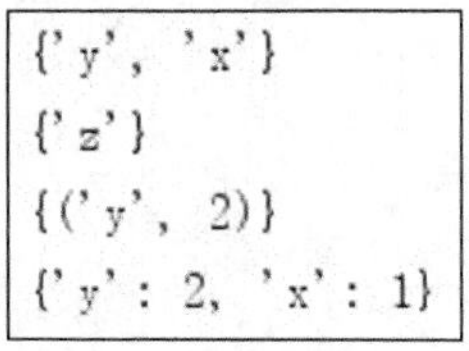

```
{'y', 'x'}
{'z'}
{('y', 2)}
{'y': 2, 'x': 1}
```

图 2-24　执行效果

实例 02-23：使用函数 itemgetter()对字典进行排序

在 Python 程序中，如果存在一个字典列表，该如何根据一个或多个字典中的值来对列表进行排序呢？建议使用 operator 模块中的内置函数 itemgetter()。函数 itemgetter()的功能是获取对象的数据，参数为一些序号（即需要获取的数据在对象中的序号）。下面的实例文件 wei.py 的功能是获取对象中指定域的值。

源码路径：daima\2\02-23\wei.py

```
from operator import itemgetter
a = [1,2,3]
b=itemgetter(1)              #定义函数b，获取对象的第1个域的值
print(b(a))

b=itemgetter(1,0)            #定义函数b，获取对象的第1个域和第0个域的值
print(b(a))
```

函数 itemgetter()获取的不是值，而是定义了一个函数，通过该函数作用到对象上才能获取值。执行后会输出：

```
2
(2, 1)
```

再看下面的实例文件 pai.py，功能是使用函数 itemgetter()排序字典中的值。

源码路径：daima\2\02-23\pai.py

```
from operator import itemgetter
①rows = [
    {'fname': 'AAA', 'lname': 'ZHANG', 'uid': 1001},
    {'fname': 'BBB', 'lname': 'ZHOU', 'uid': 1002},
    {'fname': 'CCC', 'lname': 'WU', 'uid': 1004},
    {'fname': 'DDD', 'lname': 'LI', 'uid': 1003}
]
```

```
②rows_by_fname = sorted(rows, key=itemgetter('fname'))
    rows_by_uid = sorted(rows, key=itemgetter('uid'))
    print(rows_by_fname)
③print(rows_by_uid)

④rows_by_lfname = sorted(rows, key=itemgetter('lname', 'fname'))
    print(rows_by_lfname)

⑤rows_by_fname = sorted(rows, key=lambda r: r['fname'])
⑥rows_by_lfname = sorted(rows, key=lambda r: (r['fname'], r['lname']))
    print(rows_by_fname)
    print(rows_by_lfname)
⑦print(min(rows, key=itemgetter('uid')))
⑧print(max(rows, key=itemgetter('uid')))
```

- 在①中，定义了一个保存用户信息的字典 rows。
- 在②~③中，根据字典中的共有的字段来对 rows 中的记录进行排序。
- 在④中，itemgetter()函数接受了多个键。
- 在⑤~⑥中，使用 lambda 表达式来代替 itemgetter()函数。在此提醒读者，少用 lambda 表达式方式，使用 itemgetter()函数方式会执行得更快一些。如果需要考虑程序的性能问题，则建议使用 itemgetter()函数。
- 在⑦~⑧中，函数 itemgetter()同样可以操作 min()函数和 max()函数。

执行后会输出：

```
[{'fname': 'AAA', 'lname': 'ZHANG', 'uid': 1001}, {'fname': 'BBB', 'lname': 'ZHOU', 'uid': 1002}, {'fname': 'CCC', 'lname': 'WU', 'uid': 1004}, {'fname': 'DDD', 'lname': 'LI', 'uid': 1003}]
[{'fname': 'AAA', 'lname': 'ZHANG', 'uid': 1001}, {'fname': 'BBB', 'lname': 'ZHOU', 'uid': 1002}, {'fname': 'DDD', 'lname': 'LI', 'uid': 1003}, {'fname': 'CCC', 'lname': 'WU', 'uid': 1004}]
[{'fname': 'DDD', 'lname': 'LI', 'uid': 1003}, {'fname': 'CCC', 'lname': 'WU', 'uid': 1004}, {'fname': 'AAA', 'lname': 'ZHANG', 'uid': 1001}, {'fname': 'BBB', 'lname': 'ZHOU', 'uid': 1002}]
[{'fname': 'AAA', 'lname': 'ZHANG', 'uid': 1001}, {'fname': 'BBB', 'lname': 'ZHOU', 'uid': 1002}, {'fname': 'CCC', 'lname': 'WU', 'uid': 1004}, {'fname': 'DDD', 'lname': 'LI', 'uid': 1003}]
[{'fname': 'AAA', 'lname': 'ZHANG', 'uid': 1001}, {'fname': 'BBB', 'lname': 'ZHOU', 'uid': 1002}, {'fname': 'CCC', 'lname': 'WU', 'uid': 1004}, {'fname': 'DDD', 'lname': 'LI', 'uid': 1003}]
{'fname': 'AAA', 'lname': 'ZHANG', 'uid': 1001}
{'fname': 'CCC', 'lname': 'WU', 'uid': 1004}
```

实例 02-24：使用字典推导式

在 Python 程序中，字典推导式和本章讲解的列表推导式的用法类似，只是将列表中的方括号修改为字典中的花括号而已。在下面的实例文件 zitui.py 中，演示了使用字典推导式实现大写键、小写键合并的过程。

源码路径：daima\2\02-24\zitui.py

```
mcase = {'a': 10, 'b': 34, 'A': 7, 'Z': 3}
mcase_frequency = {
    k.lower(): mcase.get(k.lower(), 0) + mcase.get(k.upper(), 0)
    for k in mcase.keys()
    if k.lower() in ['a','b']
}
print (mcase_frequency)
```

执行后会输出：

```
{'a': 17, 'b': 34}
```

再看下面的实例文件 tiqu.py，其功能是使用字典推导式从字典中提取子集。

源码路径：daima\2\02-24\tiqu.py

```
  prices = {'ASP.NET': 49.9, 'Python': 69.9, 'Java': 59.9, 'C语言': 45.9, 'PHP': 79.9}
①p1 = {key: value for key, value in prices.items() if value > 50}
  print(p1)
  tech_names = {'Python', 'Java', 'C语言'}

②p2 = {key: value for key, value in prices.items() if key in tech_names}
  print(p2)
```

```
p3 = dict((key, value) for key, value in prices.items() if value > 50)    #慢
print(p3)

tech_names = {'Python', 'Java', 'C语言'}
p4 = {key: prices[key] for key in prices.keys() if key in tech_names}    #慢
print(p4)
```

在 Python 程序中，虽然大部分可以用字典推导式解决的问题也可以通过先创建元组，然后将它们传给 dict()函数来完成，如上述代码中①的做法。但是使用字典推导式的方案更加清晰，而且程序实际执行起来也要快很多，以上述代码②中的字典 prices 来测试，效率要提高 2 倍左右。执行后会输出：

```
{'Python': 69.9, 'Java': 59.9, 'PHP': 79.9}
{'Python': 69.9, 'Java': 59.9, 'C语言': 45.9}
{'Python': 69.9, 'Java': 59.9, 'PHP': 79.9}
{'Python': 69.9, 'Java': 59.9, 'C语言': 45.9}
```

实例 02-25：根据字段进行分组

在 Python 程序中，可以将字典或对象实例中的信息根据某个特定的字段（如日期）来分组。Python 的 itertools 模块提供了内置函数 groupby()，能够方便地对数据进行分组处理。使用函数 groupby()的语法格式如下。

```
groupby(iterable [,key]):
```

下面的实例文件 fen.py 演示了使用函数 groupby()分组数据的过程。

源码路径：daima\2\02-25\fen.py

```
from itertools import groupby
from operator import itemgetter
things = [('2018-05-21', 11), ('2018-05-21', 3), ('2018-05-22', 10),
          ('2018-05-22', 4), ('2018-05-22', 22),('2018-05-23', 33)]
for key, items in groupby(things, itemgetter(0)):
    print(key)

    for subitem in items:
        print(subitem)
print('-' * 20)
```

执行后会输出：

```
2018-05-21
2018-05-22
2018-05-23
('2018-05-23', 33)
```

再看下面的实例文件 fenzu.py，其演示了使用函数 groupby()分组复杂数据的过程。

源码路径：daima\2\02-25\fenzu.py

```
①rows = [
    {'address': '5412 N CLARK', 'data': '07/01/2018'},
    {'address': '5232 N CLARK', 'data': '07/04/2018'},
    {'address': '5542 E 58ARK', 'data': '07/02/2018'},
    {'address': '5152 N CLARK', 'data': '07/03/2018'},
    {'address': '7412 N CLARK', 'data': '07/02/2018'},
    {'address': '6789 w CLARK', 'data': '07/03/2018'},
    {'address': '9008 N CLARK', 'data': '07/01/2018'},
    {'address': '2227 W CLARK', 'data': '07/04/2018'}
]
②from operator import itemgetter
  from itertools import groupby

  rows.sort(key=itemgetter('data'))
  for data, items in groupby(rows, key=itemgetter('data')):
      print(data)
      for i in items:
③        print(' ', i)

④from collections import defaultdict
  rows_by_date = defaultdict(list)
```

```
    for row in rows:
⑤       rows_by_date[row['data']].append(row)

⑥for r in rows_by_date['07/04/2018']:
    print(r)
```

在①中，创建了包含时间和地址的一系列字典数据。

在②~③中，根据日期以分组的方式迭代数据，首先以目标字段 data 对数据进行排序，然后使用函数 groupby()进行分组。这里的重点是首先要根据感兴趣的字段对数据进行排序，因为函数 groupby()只能检查连续的项，如果不首先排序的话，则会无法按照所想的方式对数据进行分组。

如果只想简单地根据日期将数据分组，并将其放进一个大的数据结构中以允许进行随机访问，那么建议像④~⑤那样使用函数 defaultdict()构建一个一键多值字典。

⑥访问每一个日期对应的数据。

执行后会输出：

```
07/01/2018
    {'address': '5412 N CLARK', 'data': '07/01/2018'}
    {'address': '9008 N CLARK', 'data': '07/01/2018'}
07/02/2018
    {'address': '5542 E 58ARK', 'data': '07/02/2018'}
    {'address': '7412 N CLARK', 'data': '07/02/2018'}
07/03/2018
    {'address': '5152 N CLARK', 'data': '07/03/2018'}
    {'address': '6789 w CLARK', 'data': '07/03/2018'}
07/04/2018
    {'address': '5232 N CLARK', 'data': '07/04/2018'}
    {'address': '2227 W CLARK', 'data': '07/04/2018'}
{'address': '5232 N CLARK', 'data': '07/04/2018'}
{'address': '2227 W CLARK', 'data': '07/04/2018'}
```

实例 02-26：转换并换算数据

在 Python 程序中，我们可以对字典或列表中的数据同时进行转换和换算操作。此时需要先对数据进行转换或筛选操作，然后调用换算（reduction）函数（例如 sum()、min()、max()）进行处理。下面的实例文件 zuixiaoda.py 演示了使用函数 min()和 max()获取最小值和最大值的过程。

源码路径：daima\2\02-26\zuixiaoda.py

```
print ("min(80, 100, 1000) : ", min(80, 100, 1000))
print ("min(-20, 100, 400) : ", min(-20, 100, 400))
print ("min(-80, -20, -10) : ", min(-80, -20, -10))
print ("min(0, 100, -400) : ", min(0, 100, -400))

print ("max(80, 100, 1000) : ", max(80, 100, 1000))
print ("max(-20, 100, 400) : ", max(-20, 100, 400))
print ("max(-80, -20, -10) : ", max(-80, -20, -10))
print ("max(0, 100, -400) : ", max(0, 100, -400))
```

执行后会输出：

```
min(80, 100, 1000) :   80
min(-20, 100, 400) :   -20
min(-80, -20, -10) :   -80
min(0, 100, -400) :   -400
max(80, 100, 1000) :   1000
max(-20, 100, 400) :   400
max(-80, -20, -10) :   -10
max(0, 100, -400) :   100
```

再看下面的实例文件 huansuan.py，其演示了同时对数据做转换和换算的过程。

源码路径：daima\2\02-26\huansuan.py

```
nums = [1, 2, 3, 4, 5]
s = sum( x*x for x in nums )
print(s)
```

```
import os
files = os.listdir('.idea')
if any(name.endswith('.py') for name in files):
    print('这是一个Python文件!')
else:
    print('这里没有Python文件!')
s = ('RMB', 50, 128.88)
print(','.join(str(x) for x in s))

portfolio = [
    {'name': 'AAA', 'shares': 50},
    {'name': 'BBB', 'shares': 65},
    {'name': 'CCC', 'shares': 40},
    {'name': 'DDD', 'shares': 35}
]

min_shares = min(s['shares'] for s in portfolio)
```

上述代码以一种非常优雅的方式将数据转换和换算结合在一起，具体方法是在函数参数中使用生成器表达式。执行效果如图 2-25 所示。

```
55
这里没有Python文件!
RMB,50,128.88
35
```

图 2-25　执行效果

实例 02-27：将多个映射合并为单个映射

如果 Python 程序中有多个字典或映射，要想在逻辑上将它们合并为一个单独的映射结构，并且以此来执行某些特定的操作，如查找某个值或检查某个键是否存在，则需要考虑将多个映射合并为单个映射。下面的实例文件 hebing.py 演示了将多个映射合并为单个映射的过程。

源码路径：daima\2\02-27\hebing.py

```
①a = {'x': 1, 'z': 3 }
  b = {'y': 2, 'z': 4 }

  from collections import ChainMap
  c = ChainMap(a,b)

  print(c['x'])
  print(c['y'])
②print(c['z'])

③print(len(c))
  print(list(c.keys()))
④print(list(c.values()))

⑤c['z'] = 10
  c['w'] = 40
  del c['x']
⑥print(a)
```

①～②在执行查找操作之前必须先检查这两个字典（例如，先在 a 中查找，如果没找到再去 b 中查找）。上述代码演示了一种非常简单的方法，就是利用 collections 模块中的 ChainMap 类来解决这个问题。

ChainMap 可以接受多个映射，这样在逻辑上使它们表现为一个单独的映射结构。但是这些映射在字面上并不会合并在一起。相反，ChainMap 只是简单地维护一个记录底层映射关系的列表，然后重定义常见的字典操作来扫描这个列表。③～④演示了这一特性。

⑤～⑥中，如果有重复的键，那么将会采用第一个映射中所对应的值。所以上述代码中的 c['z'] 总是引用字典 a 中的值，而不是字典 b 中的值。实现修改映射的操作总会作用在列出的第

一个映射上。

执行效果如图 2-26 所示。

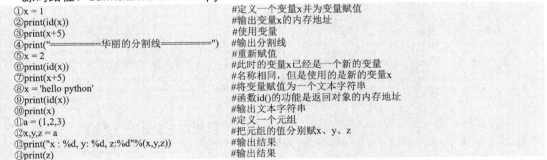

图 2-26 执行效果

2.4 变量

变量是指其值可以随着时间的变化、程序的执行而发生变化的量。变量和常量是相对的，在计算机编程语言中，其值永远不会发生变化的量称为常量。因为种种原因，Python 并没有提供常量，这也是 Python 跟 C、C++、Java 等开发语言的一大区别。

实例 02-28：实现变量赋值

在给 Python 变量赋值时，变量名由字母、数字、下划线组成，不能以数字开头，并且不能和内置关键字重名。下面的实例代码演示了赋值变量并输出变量值的过程。实例文件 bian.py 的具体实现代码如下。

源码路径：daima\2\02-28\bian.py

```
①x = 1                              #定义一个变量x并为变量赋值
②print(id(x))                       #输出变量x的内存地址
③print(x+5)                         #使用变量
④print("=========华丽的分割线=========")  #输出分割线
                                    #重新赋值
⑤x = 2                              #此时的变量x已经是一个新的变量
⑥print(id(x))                       #名称相同，但是使用的是新的变量x
⑦print(x+5)
⑧x = 'hello python'                 #将变量赋值为一个文本字符串
⑨print(id(x))                       #函数id()的功能是返回对象的内存地址
⑩print(x)                           #输出文本字符串
⑪a = (1,2,3)                        #定义一个元组
⑫x,y,z = a                          #把元组的值分别赋x、y、z
⑬print("x : %d, y: %d, z:%d"%(x,y,z))  #输出结果
⑭print(z)                           #输出结果
```

①定义了一个变量 x，并赋值为 1。由此可见，Python 变量不用在变量名 x 前声明一个数据类型，这是和 C、C++、Java、C#等语言的一个重要区别，也体现了 Python 的灵活性。

②通过内置函数 print()输出 id(x)的值，id()是 Python 的一个内置函数，id(x)的功能是返回变量 x 的内存地址。

③通过内置函数 print()输出 x+5 的值，x+5=1+5=6。

④输出 "=========华丽的分割线========="。

⑤给变量 x 重新赋值为 2，这就是变量的作用，随着程序的执行，x 由原来的 1 变为 2。

⑥通过内置函数 print()输出 id(x)的值，此时的 x 值是 2。

⑦通过内置函数 print()输出 x+5 的值，x+5=2+5=7。

⑧给变量 x 重新赋值为一个文本字符串 "hello python"。

⑨通过内置函数 print()输出 id(x)的值，此时的 x 值是 "hello python"。

⑩通过内置函数 print()输出变量 x 的值，此时的 x 值是“hello python”。

⑪定义一个元组 a，里面包含 3 个整数，即 1、2、3。

⑫同时给变量 x、y、z 赋值，把元组 a 中的值 1、2、3 分别赋给变量 x、y、z，也就是 x 赋值为 1，y 赋值为 2，z 赋值为 3。

⑬分别输出变量 x、y 和 z 的值。

⑭单独输出变量 z 的值。

执行效果如图 2-27 所示。

```
1774175920
6
══════华丽的分割线══════
1774175952
7
2191753813488
hello python
x : 1, y: 2, z:3
3
```

图 2-27　执行效果

实例 02-29：使用局部变量

文件 file01.py 中定义了函数 fun()，函数 fun()定义了一个局部变量 local_var，并赋值为 100。局部变量 local_var 只在函数 fun()内有效，只能被函数 fun()访问，即使文件 file01.py 中定义的函数 fun2()也不能使用 local_var。

源码路径：daima\2\02-29\file01.py

```
#fileName:file1
def fun():
    local_var = 100                     #定义一个局部变量
    print(local_var)                    #这行代码可以成功执行，输出变量值100
def fun2():
    zero = local_var - 100              #在函数fun2()中使用局部变量local_var是非法的
    print("get zero : %d"%zero)
fun()
#fun2()
print("local_var -1 = %d"%(local_var - 1)) #file01.py中使用局部变量(不可以)
```

执行上述代码后会出错，执行效果如图 2-28 所示。

```
C:\Users\apple\AppData\Local\Programs\Python\Python36\python.exe H:/daima/2/2-2/file01.py
100
Traceback (most recent call last):
  File "H:/daima/2/2-2/file01.py", line 10, in <module>
    print("local_var -1 = %d"%(local_var - 1)) #file01.py 中使用局部变量(不可以)
NameError: name 'local_var' is not defined
```

图 2-28　执行效果

而在另外一个实例文件 file02.py 中，即使使用 import 语句导入了文件 file01.py 中的功能，但是因为变量 local_var 是一个局部变量，所以其仍不能在文件 file02.py 中生效。实例文件 file02.py 的具体实现代码如下。

源码路径：daima\2\02-29\file02.py

```
import file01
file01.fun()
print(local_var)
```

实例 02-30：使用全局变量

在 Python 程序中，在函数之外定义的变量叫作全局变量。全局变量能够被不同的函数、类或文件所共享，可以被文件内的任何函数和外部文件所访问。下面的实例文件 quan.py 演示了使用全局变量的过程。

源码路径：daima\2\02-30\quan.py

```
g_num1 = 1                  #定义全局变量
g_num2 = 2                  #定义全局变量
```

```
def add_num():
    global g_num1              #引用全局变量
    g_num1 = 3                 #修改全局变量的值
    result = g_num1 + 1
    print("result : %d"%result)
def sub_num():
    global g_num2              #使用global关键字
    g_num2 = 5
    result = g_num2 - 3
    print("result : %d"%result)
①add_num()
②sub_num()
③print("g_num1:%d "%g_num1)
④print("g_num2:%d "%g_num2)
```

在上述代码中,在函数外部分别定义了两个全局变量 g_num1 和 g_num2,并分别设置初始值为 1 和 2。在函数 add_num()内部使用了全局变量 g_num1,在使用时用到了关键字 global。

在①中,在调用函数 add_num()时,result 为局部变量,执行后会输出"4"。

在②中,在调用函数 sub_num()时,result 为局部变量,执行后会输出"2"。

在③中,在执行 add_num()函数时,使用关键字 global 改变了全局变量 g_num1 的值,执行后会输出"3"。

在④中,在执行 sub_num()函数时,使用关键字 global 改变了全局变量 g_num2 的值,执行后会输出"5"。

实例文件 quan.py 的执行效果如图 2-29 所示。

```
result : 4
result : 2
g_num1:3
g_num2:5
```

图 2-29 执行效果

再看实例文件 wai.py,使用 import 调用了上面的文件 quan.py,具体实现代码如下。

源码路径: daima\2\02-30\wai.py

```
import quan                    #调用文件quan.py中的功能
quan.add_num()                 #调用文件quan.py中的函数add_num(),此时g_num1被改变
test = quan.g_num1 + 1         #调用文件quan.py中的全局变量g_num1
print("test :%d"%test)         #输出test的值
```

在上述代码中,第 2 行调用了文件 quan.py 中的函数 dd_num(),在此函数中全局变量 g_num1 的值被修改为 3,所以第 3 行变量 test 的值为 3+1=4。实例文件 wai.py 的执行效果如图 2-30 所示。

```
result : 4
result : 2
g_num1:3
g_num2:5
result : 4
test :4
```

图 2-30 执行效果

实例 02-31:体验关键字 global 的作用

在 Python 中,关键字 global 用于引用(使用)一个全局变量。如果不使用 global 关键字,

则在函数中再为 g_num1 赋值时，其将会被解释为定义了一个局部变量。实例文件 quan1.py 的具体实现代码如下。

源码路径：daima\2\02-31\quan1.py

```
g_num1 = 1                              #定义全局变量
g_num2 = 2                              #定义全局变量
def add_num():
    g_num1 = 3                          #修改全局变量的值，此时已经变成了局部变量
    result = g_num1 + 1
    print("result : %d"%result)
def sub_num():
    g_num2 = 5                          #修改全局变量的值，此时已经变成了局部变量
    result = g_num2 - 3
    print("result : %d"%result)
①add_num()
②sub_num()
③print("g_num1:%d "%g_num1)
④print("g_num2:%d "%g_num2)
```

上述代码没有用到关键字 global，当在函数中使用前面定义的全局变量 g_num1 和 g_num2 时，Python 编译器将函数中的 g_num1 和 g_num2 认为是同名局部变量。

在①中，在函数 add_num() 内部，g_num1 被认为是局部变量，值被修改为 3，所以执行后会输出"4"。

在②中，在函数 sub_num() 内部，g_num2 被认为是局部变量，值被修改为 5，所以执行后会输出"2"。

在③中，在执行 add_num() 函数时，因为没有使用关键字 global，所以全局变量 g_num1 的值不变，还是原来的"1"。

在④中，在执行 sub_num() 函数时，因为没有使用关键字 global，所以全局变量 g_num2 的值不变，还是原来的"2"。

实例文件 quan1.py 的执行效果如图 2-31 所示。

```
result : 4
result : 2
g_num1:1
g_num2:2
```

图 2-31　执行效果

> **注意**：在 Python 程序中，应该尽量避免使用全局变量。不同的模块可以自由地访问全局变量，可能会产生全局变量的不可预知性。全局变量会减弱函数或者模块之间的通用性，不同的函数或模块都要依赖于全局变量。同样，全局变量减弱了代码的可读性，阅读者可能不知道调用的某个变量是全局变量。

实例 02-32：定义常量功能类_const

常量是指一旦初始化之后就不能被修改的固定值，常量在整个文件、整个项目中都不能被修改。尽管 Python 没有提供定义常量的关键字，但是可以通过定义常量类的方式实现常量功能。例如，下面的实例文件 chang.py 定义常量功能类_const，设置 name 是一个常量，其值固定不变。如果修改 name 的值，则输出"不能重新定义常量（name）"的提示。文件 chang.py 的具体实现代码如下。

源码路径：daima\2\02-32\chang.py

```
import sys
class _const:
```

```
class ConstError(TypeError):pass
def __setattr__(self,name,value):
    if name in self.__dict__:
        raise self.ConstError("不能重新定义常量(%s)"%name)
    self.__dict__[name] = value
sys.modules[__name__] = _const()
```

在实例文件 chang2.py 中测试上面的常量功能类，具体实现代码如下。

源码路径：daima\2\02-32\chang2.py

```
import chang
chang.name='张三'
chang.name='lisi'
```

在上述代码中，首先设置常量 name 的值为"张三"，然后重新设置常量 name 的值为"李四"，程序执行后会输出"不能重新定义常量（name）"的提示，如图 2-32 所示。

```
Traceback (most recent call last):
  File "H:/daima/2/2-5/chang2.py", line 3, in <module>
    chang.name='李四'
  File "H:\daima\2\2-5\chang.py", line 6, in __setattr__
    raise self.ConstError("不能重新定义常量(%s)"%name)
chang.ConstError: 不能重新定义常量(name)
```

图 2-32　输出错误提示

2.5　条件语句和循环流程控制

条件语句和循环流程控制是 Python 语法的重要组成部分，在其他开发语言中也占据了重要的地位。本节将详细讲解 Python 条件语句和循环流程控制的知识。

实例 02-33：使用条件语句判断年龄

在 Python 中，条件语句通过一条或多条语句的输出结果（True 或者 False）来决定执行的代码块。下面的实例文件 if2.py 演示了简单使用 if...elif 语句判断年龄的过程。

源码路径：daima\2\02-33\if2.py

```
age = int(input("请输入你家狗狗的年龄: "))
print("")
if age < 0:
    print("你是在逗我吧!")
elif age == 1:
    print("相当于 14 岁的人。")
elif age == 2:
    print("相当于 22 岁的人。")
elif age > 2:
    human = 22 + (age - 2) * 5
    print("对应人类年龄: ", human)
### 退出提示
input("单击Enter键退出")
```

在上述代码中，通过 if...elif 语句划分了 4 个年龄区间，能够根据用户在命令行中输入的整数来输出小狗相当于人类的年龄。例如，输入整数 3 后的执行效果如图 2-33 所示。

```
请输入你家狗狗的年龄：3

对应人类年龄： 27
单击 Enter 键退出
```

图 2-33　执行效果

和其他主流编程语言一样，Python 中的条件语句也可以嵌套使用。例如，在下面的代码中，外侧的条件语句包含 2 个分支；第 1 个分支包含一行简短的语句；第 2 个分支则包含了另一个条件语句，它自己的代码下也有 2 个分支。这两个分支都是非常容易理解的简单语句。

```
if x == y:
    print ("x 等于 y")
else:
    if < y:
        print ("x 小于 y")
    else:
        print ("x 大于 y")
```

虽然 Python 程序严格的代码缩进格式让上述嵌套代码的结构变得非常清晰，但嵌套条件语句会随着它嵌套数量的增加而变得非常难理解和阅读。在 Python 程序中，通过使用逻辑操作符可以简化 Python 嵌套条件语句的使用。请看下面的演示代码。

```
if 0 < x:
    if x < 10:
        print ("x是一个正整数")
```

我们可以将上述嵌套代码简写为如下格式。

```
if 0 < x and x < 10:
    print ("x是一个正整数")
```

实例文件 if3.py 演示了嵌套条件语句的执行流程，具体实现代码如下。

源码路径：daima\2\02-33\if3.py

```
num=int(input("输入一个数字："))
①if num%2==0:
    if num%3==0:
        print ("你输入的数字可以整除 2 和 3")
    else:
        print ("你输入的数字可以整除 2，但不能整除 3")
②else:
    if num%3==0:
        print ("你输入的数字可以整除 3，但不能整除 2")
    else:
        print ("你输入的数字不能整除 2 和 3")
```

在上述代码中，①和②是同级并列的 if...else 语句。在①中，if 语句的后面嵌套了一个 if...else 子语句；在②中，else 语句的后面嵌套了一个 if...else 子语句。幸运的是，Python 有着严格的代码缩进规定，所以以上嵌套代码的结构一目了然。执行后可以根据用户输入的整数输出对应的判断结果。例如，输入整数 6 后的执行效果如图 2-34 所示。

```
输入一个数字：6
你输入的数字可以整除 2 和 3
```

图 2-34 执行效果

实例 02-34：使用 for 循环遍历单词 "Python" 中的各个字母

与 Java、C、C++等其他语言相比，Python 中的 for 循环有很大的不同，其他高级语言的 for 循环需要用循环控制变量来控制循环；而 Python 中的 for 循环是通过循环遍历某一序列（如元组、列表、字典等）来构建循环的，循环结束的条件就是对象被遍历完成。下面的实例文件 for01.py 演示了使用 for 循环的基本过程。

源码路径：daima\2\02-34\for01.py

```
for letter in 'Python':              #第一个实例，定义一个字符
    print ('当前字母 :', letter)      #循环输出单词 "Python" 中的各个字母
fruits = ['banana', 'apple',  'mango']  #定义一个列表
for fruit in fruits:
    print ('当前单词 :', fruit)       #循环输出列表fruits中的3个值
print ("Good bye!")
```

在上述代码中，使用第 1 个 for 循环遍历了单词 "Python" 中的各个字母，并输出了各个字母；使用第 2 个 for 循环遍历了列表 fruits 中的 3 个元素，并输出了列表 fruits 中的 3 个值。执行效果如图 2-35 所示。

```
当前字母 : P
当前字母 : y
当前字母 : t
当前字母 : h
当前字母 : o
当前字母 : n
当前单词 : banana
当前单词 : apple
当前单词 : mango
Good bye!
```

图 2-35 执行效果

实例 02-35：使用 range()循环输出列表中的元素

在 Python 中，还可以通过序列索引迭代的方式实现循环功能，在具体实现时，可以借助于内置函数 range()实现。因为在 Python 的 for 循环中，对象可以是列表、字典及元组等，所以可以通过函数 range()产生一个整数列表，这样可以完成计数循环功能。下面的实例文件 for02.py 演示了通过序列索引迭代的方式循环输出列表中的元素的过程。

源码路径：daima\2\02-35\for02.py

```python
fruits = ['banana', 'apple',    'mango']         #定义一个数组
for index in range(len(fruits)):                 #使用函数range()遍历数组
    print ('当前水果 :', fruits[index])          #输出遍历数组后的结果
print ("Good bye!")
```

执行效果如图 2-36 所示。

```
当前水果 : banana
当前水果 : apple
当前水果 : mango
Good bye!
```

图 2-36 执行效果

实例 02-36：循环输出 10~20 的各数是否是质数

在 Python 程序中，for...else 表示的意思是，for 后的语句和普通的没有区别，else 后的语句会在循环正常执行完（即 for 循环不是通过 break 跳出而中断的）的情况下执行。下面的实例文件 for03.py 演示了使用 for...else 循环语句的过程。

源码路径：daima\2\02-36\for03.py

```python
for num in range(10,20):                    #循环迭代10~20中的数字
    for i in range(2,num):                  #根据因子迭代
        if num%i == 0:                      #确定第1个因子
            j=num/i                         #计算第2个因子
            print ('%d 等于 %d * %d' % (num,i,j))
            break                           #跳出当前循环
    else:                                   #如果上面的条件不成立，则执行循环中的else部分
        print (num, '是一个质数')           #输出这是一个质数
```

通过上述代码，循环输出 10~20 的各数是否是质数。执行效果如图 2-37 所示。

```
10 等于 2 * 5
11 是一个质数
12 等于 2 * 6
13 是一个质数
14 等于 2 * 7
15 等于 3 * 5
16 等于 2 * 8
17 是一个质数
18 等于 2 * 9
19 是一个质数
```

图 2-37 执行效果

实例 02-37：获取两个整数之间的所有素数

当在 Python 程序中使用 for 循环时，for 循环可以嵌套。也就是说，可以在一个 for 循环中使用另一个 for 循环。下面实例文件 for04.py 的功能是，使用嵌套 for 循环获取两个整数之间的所有素数。

源码路径：daima\2\02-37\for04.py

```python
#提示我们输入一个整数
x = (int(input("请输入一个整数作为开始：")),int(input("请输入一个整数作为结尾：")))
x1 = min(x)                     #获取输入的第1个整数
x2 = max(x)                     #获取输入的第2个整数
for n in range(x1,x2+1):        #使用外循环语句生成要判断是否为素数的序列
    for i in range(2,n-1):      #使用内循环语句生成测试的因子
        if n % i == 0:          #生成测试的因子能够整除，则不是素数
            break
    else:                       #上述条件不成立，则说明是素数
        print("你输入的",n,"是素数。")
```

在上述代码中，首先使用输入函数获取用户指定的序列开始和结束，然后使用 for 循环构建了两层嵌套的循环语句用来获取素数并输出结果。使用外循环语句生成要判断是否为素数的序列，使用内循环语句生成测试的因子。else 子句通过缩进来表示它属于内嵌的 for 循环：如果其多缩进一个单位，则表示其属于 if 语句；如果其少缩进一个单位，则表示其属于外层的 for 循环。因此，Python 中的缩进是整个程序的重要构成部分。程序执行后将提示用户输入两个整数作为范围，例如，分别输入 "10" 和 "20" 后的执行效果如图 2-38 所示。

```
请输入一个整数值作为开始：10
请输入一个整数值作为结尾：20
你输入的 11 是素数。
你输入的 13 是素数。
你输入的 17 是素数。
你输入的 19 是素数。
```

图 2-38 执行效果

实例 02-38：输出九九乘法表

下面的实例文件 for05.py 的功能是使用嵌套 while 循环输出九九乘法表。

源码路径：daima\2\02-38\for05.py

```python
for j in range(1, 10):
    for i in range(1, j+1):
        print('%d*%d=%d' % (i, j, i*j), end='\t')
```

```
        i += 1
    print()
    j += 1
```
执行效果如图 2-39 所示。

```
1*1=1
1*2=2   2*2=4
1*3=3   2*3=6    3*3=9
1*4=4   2*4=8    3*4=12   4*4=16
1*5=5   2*5=10   3*5=15   4*5=20   5*5=25
1*6=6   2*6=12   3*6=18   4*6=24   5*6=30   6*6=36
1*7=7   2*7=14   3*7=21   4*7=28   5*7=35   6*7=42   7*7=49
1*8=8   2*8=16   3*8=24   4*8=32   5*8=40   6*8=48   7*8=56   8*8=64
1*9=9   2*9=18   3*9=27   4*9=36   5*9=45   6*9=54   7*9=63   8*9=72   9*9=81
```

图 2-39 执行效果

实例 02-39：使用 while 循环输出整数 0～4

在 Python 程序中，while 循环用于循环执行某段程序，以处理需要重复处理的相同任务。在 Python 中，虽然绝大多数的循环结构是用 for 循环来完成的，但是 while 循环也可以完成 for 循环的功能，只不过不如 for 循环简单明了。在 Python 程序中，while 循环主要用于构建比较特别的循环。while 循环的最大的特点就是不知道循环多少次，当不知道语句块或语句需要重复多少次时，使用 while 循环是最好的选择。当 while 循环的表达式是真时，while 循环重复执行一条语句或语句块。

实例文件 while01.py 的功能是循环输出整数 0～4，具体实现代码如下。

源码路径：daima\2\02-39\while01.py

```
count = 0                                #设置count的初始值为0
while (count < 5):                       #如果count小于5，则执行下面的while循环
    print ('The count is:', count)
    count = count + 1                    #每次循环时，count值递增1
print ("Good bye!")
```

执行效果如图 2-40 所示。

```
The count is: 0
The count is: 1
The count is: 2
The count is: 3
The count is: 4
Good bye!
```

图 2-40 执行效果

实例 02-40：while 循环的死循环问题

死循环也称为无限循环，是指这个循环将一直执行下去。在 Python 程序中，while 循环不像 for 循环那样可以遍历某一个对象。在使用 while 循环构造循环时，最容易出现的问题就是条件永远为真，导致死循环。因此，在使用 while 循环时应仔细检查 while 的循环条件，避免出现死循环。下面的实例文件 while03.py 演示了使用 while 循环的死循环问题。

源码路径：daima\2\02-40\while03.py

```
var = 1              #设置var的初始值为1
#当var为1时执行循环，实际上var的值确实为1
```

```
while var == 1 : #所以该条件永远为true，循环将一直执行下去
    num = input("亲，请输入一个整数，谢谢！")#提示输入一个整数
    print ("亲，您输入的是：", num)   #显示一个整数
print ("再见，Good bye!")
```

在上述代码中，因为循环条件变量 var 的值永远为 1，所以该条件永远为 true，所以循环将无限执行下去，这就形成了死循环。所以程序执行后将一直提示用户输入一个整数，在用户输入一个整数后还继续无限次地提示用户输入一个整数，如图 2-41 所示。

亲，请输入一个整数，谢谢！1
亲，您输入的是：1
亲，请输入一个整数，谢谢！2
亲，您输入的是：2
亲，请输入一个整数，谢谢！3
亲，您输入的是：3
亲，请输入一个整数，谢谢！

图 2-41　无限次地提示用户输入一个整数

实例 02-41：输出 10 以内的所有素数

和使用 for 循环嵌套语句一样，在 Python 程序中也可以嵌套使用 while 循环，并且还可以在循环体内嵌入其他的循环体。例如，在 while 循环中嵌入 for 循环；反之，也可以在 for 循环中嵌入 while 循环。下面的实例文件 while04.py 演示了使用 while 循环嵌套的过程。

源码路径：daima\2\02-41\while04.py

```
i = 2                     #设置i的初始值为2
while(i < 10):            #如果i的值小于10则进行循环
    j = 2                 #设置j的初始值为2
    while(j <= (i/j)):    #如果j的值小于等于i/j，则进行循环
        if not(i%j): break #如果能整除，则用break语句停止执行
        j = j + 1          #将j的值加1
    if (j > i/j) : print (i, " 是素数")  #如果j > i/j，则输出i的值
    i = i + 1              #循环输出素数i的值
print ("谢谢使用，Good bye!")
```

通过上述代码，输出了 10 以内的所有素数，执行效果如图 2-42 所示。

2 是素数
3 是素数
5 是素数
7 是素数
谢谢使用，Good bye!

图 2-42　执行效果

实例 02-42：在 for 循环和 while 循环中使用 break 语句

在 Python 程序中，break 语句的功能是终止循环语句，当循环条件不为 False 或序列还没有被完全遍历时，循环语句也会停止执行。break 语句通常用在 while 循环和 for 循环中。下面的实例文件 br1.py 分别演示了在 for 循环和 while 循环中使用 break 语句的过程。

源码路径：daima\2\02-42\br1.py

```
for letter in 'Python':              #第1个例子，用在for循环中，设置字符串"Python"
    if letter == 'h':                #如果找到字母"h"
        break                        #则停止遍历
    print ('Current Letter :', letter) #显示遍历的字母
var = 10                             #第2个例子，用在while循环中，设置var的初始值是10
```

```
    while var > 0:
        print ('Current variable value :', var)      #如果var大于0,则下一行代码输出当前var的值
        var = var -1                                  #然后循环设置var的值减1
        if var == 5:                                  #如果var的值递减到5,则使用break语句停止循环
            break
    print ("执行完毕，Good bye!")
```

执行效果如图 2-43 所示。

```
Current Letter : P
Current Letter : y
Current Letter : t
Current variable value : 10
Current variable value : 9
Current variable value : 8
Current variable value : 7
Current variable value : 6
执行完毕, Good bye!
```

图 2-43　执行效果

实例 02-43：在 for 循环和 while 循环中使用 continue 语句

在 Python 程序中，continue 语句通常用在 while 循环和 for 循环中，功能是跳出本次循环。这和 break 语句是有区别的，break 语句的功能是跳出整个循环。通过使用 continue 语句，可以告诉 Python 跳过当前循环的剩余语句，然后继续进行下一次循环。下面的实例文件 con1.py 演示了在 for 循环和 while 循环中使用 continue 语句的过程。

源码路径：daima\2\02-43\con1.py

```
for letter in 'Python':                              #第1个例子,用在for循环中,设置字符串"Python"
    if letter == 'h':                                #如果找到字母"h"
        continue                                     #则使用continue语句跳出当前循环,然后进行后面的循环
    print ('当前字母 :', letter)                      #循环显示字母
var = 10                                             #第2个例子,用在while循环中,设置var的初始值是10
while var > 0:                                       #如果var的值大于0
    var = var -1                                     #则循环设置var的值减1
    if var == 5:                                     #如果var的值递减到5
        continue                                     #则使用continue语句跳出当前循环,然后进行后面的循环
    print ('当前变量值 :', var)                       #循环显示数字
print ("执行完毕，游戏结束，Good bye!")
```

执行效果如图 2-44 所示。

```
当前字母 : P
当前字母 : y
当前字母 : t
当前字母 : o
当前字母 : n
当前变量值 : 9
当前变量值 : 8
当前变量值 : 7
当前变量值 : 6
当前变量值 : 4
当前变量值 : 3
当前变量值 : 2
当前变量值 : 1
当前变量值 : 0
执行完毕, 游戏结束, Good bye!
```

图 2-44　执行效果

实例02-44：输出指定英文单词中的每个英文字母

在Python程序中，pass是一个空语句，是为了保持程序结构的完整性而推出的语句。在程序中，pass语句不做任何事情，一般只用作占位语句。下面的实例文件kong.py演示了在程序中使用pass语句的过程，实例文件的功能是输出指定英文单词中的每个英文字母。

源码路径：daima\2\02-44\kong.py

```python
for letter in 'Python':          #从字符串"Python"中遍历每一个字母
    if letter == 'h':            #如果遍历到字母"h"，则使用pass语句
        pass
        print ('这是pass语句，是一个空语句，什么都不执行！')
    print ('当前字母 :', letter)  #输出"Python"的每个字母
print ("程序执行完毕，Good bye!")
```

执行效果如图2-45所示。

```
当前字母 : P
当前字母 : y
当前字母 : t
这是pass语句，是一个空语句，什么都不执行！
当前字母 : h
当前字母 : o
当前字母 : n
程序执行完毕, Good bye!
```

图2-45 执行效果

2.6 函数

函数是Python程序的基本构成模块，通过对函数的调用能够实现软件项目需要的特定功能。在一个Python软件项目中，几乎所有的基本功能都是通过函数实现的。

实例02-45：使用函数计算序列内元素的和

在编写Python程序的过程中，可以将完成某个指定功能的语句提取出来，将其编写为函数。这样，在程序中可以方便地调用函数来完成这个功能，并且可以多次调用完成这个功能，而不必重复地复制、粘贴代码。使用函数，也可以使程序结构更加清晰，使程序更容易维护。由此可见，函数的最基本用法包括函数定义和函数调用。下面的实例文件he.py演示了定义并调用自定义函数的过程。

源码路径：daima\2\02-45\he.py

```python
def tpl_sum( T ):                         #定义函数tpl_sum()
    result = 0                            #定义result的初始值为0
    for i in T:                           #遍历T中的每一个元素
        result += i                       #计算各个元素i的和
    return result                         #函数tpl_sum()最终返回计算的和
print("(1,2,3,4)元组中元素的和为：",tpl_sum((1,2,3,4)))      #使用函数tpl_sum()计算元组内元素的和
print("[3,4,5,6]列表中元素的和为：",tpl_sum([3,4,5,6]))      #使用函数tpl_sum()计算列表内元素的和
print("[2.7,2,5.8]列表中元素的和为：",tpl_sum([2.7,2,5.8]))  #使用函数tpl_sum()计算列表内元素的和
print("[1,2,2.4]列表中元素的和为：",tpl_sum([1,2,2.4]))      #使用函数tpl_sum()计算列表内元素的和
```

在上述代码中定义了函数tpl_sum()，函数的功能是先计算序列内元素的和；然后在最后的4行代码中分别调用了4次函数，并且这4次调用的参数不一样。执行效果如图2-46所示。

```
(1,2,3,4)元组中元素的和为： 10
[3,4,5,6]列表中元素的和为： 18
[2.7,2,5.8]列表中元素的和为： 10.5
[1,2,2.4]列表中元素的和为： 5.4
```

图 2-46 执行效果

实例 02-46：创建一个可以接受任意数量参数的函数

在 Python 程序中，参数是函数的重要组成元素。Python 中的函数参数有多种形式，例如，在调用某个函数时，既可以向其传递参数，也可以不传递参数，但是这都不影响函数的正常调用。下面的实例文件 any.py 演示了创建一个可以接受任意数量参数的函数的过程。

源码路径：daima\2\02-46\any.py

```
① def avg(first, *rest):
②     return (first + sum(rest)) / (1 + len(rest))
   print(avg(1, 2))
   print(avg(1,2,3,4))

   import html

③ def make_element(name,value,**attrs):
④     keyvals = [' %s="%s"' % item for item in attrs.items()]
       attr_str = ''.join(keyvals)
⑤     element = '<{name} {attrs}>{value}</{name}>'.format(
                   name=name,
                   attrs=attr_str,
⑥                 value=html.escape(value))
       return element
   print(make_element('商品', '小鹰登山包', size='大号', quantity=6))
   print(make_element('p','<spam>'))
```

在①中，要想在 Python 程序中实现一个可接受任意数量参数的函数，可以使用以星号"*"开头的参数。其中，用星号标识的参数 rest 是一个元组，加了星号"*"的变量名会存放所有未命名的变量参数。如果在函数调用时没有指定参数，则它就是一个空元组，开发者也可以不向函数传递未命名的变量参数。由此可见，在自定义函数时，如果参数名前加上一个星号"*"，则表示该参数就是一个可变长参数。在调用该函数时，当依次对所有的其他变量都赋值之后，剩下的参数将会被收集在一个元组中，元组的名称就是前面带星号的参数名。

在②中可变参数 rest 会被作为一个序列来处理。

在③中定义函数 make_element()，使用双星号"**"设置可以接受任意数量的关键字参数 attrs。

在④中将参数 attrs 设置为一个字典，它包含了所有传递过来的关键字参数（如果有的话）；生成标签属性列表的 keyvals 字典变量。使用函数 items()以列表的形式返回可遍历的(键,值)元组数组。%s 用于格化式一个对象为字符，在④中设置两个%s 对应一个 item。

在⑤～⑥中设置生成 HTML 标签的格式。

执行效果如图 2-47 所示。

```
1.5
2.5
<商品 size="大号" quantity="6">小鹰登山包</商品>
<p>&lt;spam&gt;</p>
```

图 2-47 执行效果

实例 02-47：减少函数的参数

在现实应用中，可调用的对象可能会以回调函数的形式同其他的 Python 程序进行交互。但是当这个可调用对象需要过多的参数时，如果直接进行调用就会发生异常。这时候需要考虑将函数参数进行减少处理，并且函数的参数越少，函数越容易被理解。在 Python 程序中，使用 functools 模块的内置函数 partial()减少函数的参数。通过函数 partial()可以给一个或多个参数设定一个固定的值，以此减少需要提供给函数调用的参数。下面的实例文件 shao.py 演示了使用函数 partial()减少函数的参数的过程。

源码路径：daima\2\02-47\shao.py

```
①def spam(a, b, c, d):
    print(a, b, c, d)
    from functools import partial
②s1 = partial(spam, 1)                    #默认给a赋值为1
    s1(2, 3, 4)
③s2 = partial(spam, d=58)                 #给d赋值为58
    s2(1, 2, 3)
    s2(4, 5, 5)
④s3 = partial(spam, 1, 2, d=58)           #分别赋值：a = 1，b = 2，d = 58
    s3(3)
    s3(4)
    s3(5)
```

在①中，定义函数 spam()时设置了 4 个参数 a、b、c 和 d。

在②中，使用函数 partial()将 spam()函数中的第 1 个参数 a 设置为 1。

在③中，使用函数 partial()将 spam()函数中的第 4 个参数 d 设置为 58。

在④中，使用函数 partial()给 spam()函数中的③个参数分别赋值：a = 1，b = 2，d = 58。

执行效果如图 2-48 所示。

```
1 2 3 4
1 2 3 58
4 5 5 58
1 2 3 58
1 2 4 58
1 2 5 58
```

图 2-48　执行效果

实例 02-48：家政公司的打扫服务

假设存在一个场景：家政公司提供了一个服务 API，功能是打扫房间，而且还能提供各种打扫功能，如扫地、擦家具、清洗油烟机等。如果我们把打扫房间看做一个库函数，那么打扫房间的方式是由我们自己决定的，具体怎么打扫就要预约并执行相应回调函数，预约并执行服务的行为叫作登记回调函数。这时我们编写实例文件 huidiao.py 演示上述场景，具体实现代码如下。

源码路径：daima\2\02-48\huidiao.py

```
def clean1(times):
    """
    参数times: 清洗次数
    返回值None
    """
    print('已完成扫地次数:', str(times))
def clean2(times):
    """
    :param times: 清洗次数
    :return: None
```

```
    """
    print('已洗抽油烟机次数', str(times))
def call_clean(times, function_name):
    """
    这个函数很重要,它就是家政公司的业务系统,要什么业务都得在此处说明
    它是实现回调函数的核心
    :param times:清洗次数
    :param function_name:回调函数名
    :return:调用的函数结果
    """
    return function_name(times)
if __name__ == '__main__':
    call_clean(10, clean2)    #在函数call_clean()中将函数clean2()作为参数,功能是清洗10次油烟机
```

在上述代码的最后一行中,函数 call_clean()将函数 clean2()作为参数来实现具体功能,函数 clean2()就是一个回调函数。由此可见,回调函数就是一个通过函数指针调用的函数。如果把函数的指针(地址)作为参数传递给另一个函数,当这个指针被用来调用其所指向的函数时,我们就说该函数是回调函数。执行效果如图 2-49 所示。

已洗抽油烟机次数 10

图 2-49 执行效果

2.7 类和对象

在面向对象编程语言中,具有相同属性或能力的模型是使用类进行定义和表示的。在程序中需要编写能够表示现实世界中的事物和情景的类,并基于这些类来创建对象。在使用创建的类时,必须先进行类的实例化操作,根据类来创建对象的过程被称为实例化。在实例化后,我们才能够使用类的实例。在 Python 程序中,类被实例化后就生成了一个对象,通过对象实现程序的具体功能。

实例 02-49:创建并使用类和对象

在 Python 3 中,所有类都默认继承自 object 类,只要是继承了 object 类的子类,和该子类的子类,都被称为新式类(在 Python 3 中所有的类都是新式类)。没有继承 object 类的子类称为经典类(在 Python 2 中,没有继承 object 的类及它的子类都是经典类)。在 Python 程序中,把具有相同属性和方法的对象归为一个类,例如,可以将人类、动物和植物看作不同的"类"。

在 Python 程序中,类实例化后就生成了一个对象。类对象支持两种操作,分别是属性引用和实例化。属性引用的使用方法和 Python 中所有的属性引用的方法一样,都使用 "obj.name" 格式。在类对象被创建后,类命名空间中所有的命名都是有效属性名。下面的实例文件 dui.py 演示了使用类和对象的基本过程。

源码路径: daima\2\02-49\dui.py
```
class MyClass:                  #定义类MyClass
    """一个简单的类"""
    i = 12345                   #设置变量i的初始值
    def f(self):                #定义类方法f()
        return 'hello world'    #输出文本
x = MyClass()                   #实例化类,x就是一个对象
#下面两行代码分别访问类的属性和方法
print("类MyClass中的属性i为: ", x.i)
print("类MyClass中的方法f()输出为: ", x.f())
```

上述代码创建了一个新的类并将该类的对象赋给局部变量 x。x 的初始值是一个空的 MyClass 对象,通过最后两行代码分别对 x 对象成员进行了赋值。执行效果如图 2-50 所示。

```
类MyClass中的属性i为： 12345
类MyClass中的方法f()输出为： hello world
```

图 2-50　执行效果

实例 02-50：定义并使用类方法

在 Python 程序中，可以使用关键字 def 在类的内部定义一个方法。在定义类的方法后，可以让类具有一定的功能，在类外部调用该类的方法时就可以完成相应的功能，或者改变类的状态，或达到其他目的。下面的实例文件 fang.py 演示了定义并使用类方法的过程。

源码路径：daima\2\02-50\fang.py

```python
class SmplClass:                        #定义类SmplClass
    def info(self):                     #定义类方法info()
        print('我定义的类！')            #输出文本
    def mycacl(self,x,y):               #定义类方法mycacl()
        return x + y                    #返回参数x和y的和
sc = SmplClass()                        #实例化类SmplClass
print('调用info()方法的结果：')
sc.info()                               #调用对象sc中的方法info()
print('调用mycacl()方法的结果：')
print(sc.mycacl(3,4))                   #调用对象sc中的方法mycacl()
```

在上述代码中，首先定义了一个具有两个方法 info() 和 mycacl() 的类，然后实例化该类，并调用这两个方法。其中，第 1 个方法的功能是直接输出信息，第 2 个方法的功能是计算参数 x 和 y 的和。执行效果如图 2-51 所示。

```
调用info()方法的结果：
我定义的类！
调用mycacl()方法的结果：
7
```

图 2-51　执行效果

实例 02-51：我的宠物狗

在 Python 程序中，可以将类看作创建对象的一个模板。只有在类中创建对象，这个类才变得有意义。下面的实例文件 duo.py 演示了在类中创建多个对象的过程。

源码路径：daima\2\02-51\duo.py

```python
class Dog():
    """小狗"""
    def __init__(self, name, age):
        """初始化属性name和age"""
        self.name = name
        self.age = age
    def wang(self):
        """模拟小狗汪汪叫"""
        print(self.name.title() + " 汪汪")
    def shen(self):
        """模拟小狗伸舌头"""
        print(self.name.title() + "伸舌头")
my_dog = Dog('大狗', 6)
your_dog = Dog('小狗', 3)
print("我爱犬的名字是" + my_dog.name.title() + ".")
print("我的爱犬已经" + str(my_dog.age) + "岁了！")
my_dog.wang()
print("\n你爱犬的名字是 " + your_dog.name.title() + ".")
print("你的爱犬已经" + str(your_dog.age) + "岁了！")
your_dog.wang()
```

在上述实例代码中，使用了 Dog 类，在第 13 行代码中创建了一个 name 为"大狗"、age 为"6"的小狗，当执行这行代码时，Python 会使用实参"大狗"和"6"调用 Dog 类中的方法 __init__()。方法 __init__()会创建一个表示特定小狗的对象，并使用我们提供的值来设置属性 name 和 age。另外，虽然方法 __init__()中并没有显式地包含 return 语句，但是 Python 会自动返回一个表示这条小狗的对象。在上述代码中，这个对象存储在变量 my_dog 中。

而在第 14 行代码中创建了一个新的对象，其中 name 为"小狗"，age 为"3"。第 13 行中的小狗对象和第 14 行中的小狗对象各自独立，各自有自己的属性，并且能够执行相同的操作。例如，在第 15、第 16、第 17 行代码中，独立输出了对象"my_dog"的信息。而第 18、第 19、第 20 行代码独立输出了对象"your_dog"的信息。执行效果如图 2-52 所示。

```
我爱犬的名字是大狗.
我的爱犬已经6岁了！
大狗 汪汪

你爱犬的名字是 小狗.
你的爱犬已经3岁了！
小狗 汪汪
```

图 2-52　执行效果

实例 02-52：使用属性函数创建可以扩展功能的属性

在 Python 程序中获取和设定对象属性时，有时可能需要增加一些额外的功能，如类型检查或者验证，这时候就需要创建一个可管理的属性。创建可管理属性的最佳方法是在类中定义一个函数，然后将这个函数作为属性来使用。上述函数就是属性（property）函数，它可以扩展实现一些有用的功能。

在 Python 程序中，使用标记@property 来创建属性函数。使用属性函数的最简单的方法之一是将其作为一个方法的装饰器来使用，这可以让开发者将一个类方法转换成一个类属性。属性和方法是 Python 对象中十分重要的两个元素，在调用两者的时候，两者是有区别的。下面的实例文件 qubie.py 演示了调用属性和方法的区别。

源码路径：daima\2\02-52\qubie.py

```python
class People:
    def __init__(self,first_name,last_name):
        self.first_name = first_name
        self.last_name  = last_name

    def get_first_name(self):
        return self.first_name

a = People('laoguan','xijing')
print(a.get_first_name())
print(a.first_name)
```

执行效果如图 2-53 所示。

```
laoguan
laoguan
```

图 2-53　执行效果

第 2 章 基础语法实战

由此可见,调用属性和方法的执行效果是一样的,但是具体的调用过程不一样(虽然只是多一个括号而已)。那么有没有一种办法,使在调用属性时就会自动调用相应的方法呢?也就是增加一些额外的处理过程(如类型检查或验证),这时候属性函数(由@property 标记)就能给我们提供很好的解决方案。请看下面简单的实例文件 zidong.py,功能是自动调用 getter()、setter()和 deleter()函数对属性进行处理。

源码路径:daima\2\02-52\zidong.py

```
class People:
    def __init__(self,name):
        self.name = name
    #getter()函数
    @property                #属性函数
    def name(self):
        return self._name

    #setter()函数
    @name.setter
    def name(self,name):
        self._name = name

    #deleter()函数
    @name.deleter
    def name(self):
        raise AttributeError('不能删除name')

a = People('laoguan')
print(a.name)            #自动调用getter()函数处理属性
a.name = 'laoguan'       #自动调用setter()函数处理属性
print(a.name)
del a.name
```

执行效果如图 2-54 所示。

```
laoguan
Traceback (most recent call last):
laoguan
  File "H:/daima/3/3-19/zidong.py", line 23, in <module>
    del a.name
  File "H:/daima/3/3-19/zidong.py", line 17, in name
    raise AttributeError('不能删除name')
AttributeError: 不能删除name
```

图 2-54 执行效果

正如上述代码演示的这样,要定义对属性的访问,一种最简单的方法就是将其定义为属性函数。下面的实例文件 zeng.py 增加了对属性的类型检查功能,演示了创建可以扩展功能的属性的过程。

源码路径:daima\2\02-52\zeng.py

```
class Person:
    def __init__(self, first_name):
        self.first_name = first_name

    # getter()函数
    @property
    def first_name(self):
①       return self._first_name

    # setter()函数
    @first_name.setter
```

```
            def first_name(self, value):
                if not isinstance(value, str):
                    raise TypeError('出错了！')
②               self._first_name = value

            # deleter()函数(可选的)
            @first_name.deleter
            def first_name(self):
                raise AttributeError("不能删除属性！")
        if __name__ == '__main__':
            a = Person('Li')
            print(a.first_name)
            a.first_name = 'Lei'
            print(a.first_name)
            try:
                a.first_name = 42
            except TypeError as e:
                print(e)
            del a.first_name
```

上述代码创建了一个功能扩展属性，扩展功能用于实现对属性的类型检查。上述代码一共创建了 3 个功能互相关联的方法 first_name()，这 3 个方法必须有相同的名称 first_name。其中第 1 个方法 first_name(self)是一个 getter()函数，并且将方法名 first_name 定义为属性。其他两个方法分别将可选的 setter()和 deleter()函数附加到了 first_name 属性上。在此需要注意的是，除非方法 first_name()已经通过@property 方式定义为了属性，否则不能定义@first_name.setter 和 @first_name.deleter 装饰器。当在 Python 程序中实现一个属性时，底层的数据（如果有的话）仍然需要被保存到某个地方。所以在 getter()函数和 setter()函数的①、②中是直接对_first_name 进行操作的，这就是实际保存数据的地方。有些读者可能会问为什么在方法__init__()中设置的是 self.first_name，而不是 self._first_name 呢？在上述代码中，属性的核心意义是在设置属性时可以执行类型检查功能。因此，很有可能想让这种类型检查在初始化时也可以进行。所以，在方法__init__()中设置 self.first_name 时，实际上会调用到 setter()函数（因此就会跳过 self.first_name 而去访问 self._first_name）。执行效果如图 2-55 所示。

```
Traceback (most recent call last):
Li
  File "H:/daima/3/3-19/guanshu.py", line 31, in <module>
Lei
    del a.first_name
出错了！
  File "H:/daima/3/3-19/guanshu.py", line 20, in first_name
    raise AttributeError("不能删除属性！")
AttributeError: 不能删除属性！
```

图 2-55　执行效果

实例 02-53：使用 Python 描述符实现延迟初始化

Python 对象的延迟初始化是指，当它第一次被创建时才进行初始化，或者保存第一次创建的结果，然后每次调用时直接返回该结果。在现实应用中，延迟初始化主要用于提高性能，避免浪费计算，并减少程序的内存需求。

在 Python 程序中，如果想将一个只读属性定义为属性函数，则设置只有在访问它时才参与计算。但是一旦访问了这个属性，我们希望把计算出的值缓存起来，不要在每次访问它时都重新进行计算。Python 实现延迟初始化功能的方法有两种，一种是使用 Python 描述符，另一种是使用@lazy_property标记。下面的实例文件miaoshufu.py演示了使用Python描述符实现延迟初始化功能的过程。

源码路径：daima\2\02-53\miaoshufu.py

```python
class lazy(object):
    def __init__(self, func):
        self.func = func

    def __get__(self, instance, cls):
        val = self.func(instance)
        setattr(instance, self.func.__name__, val)
        return val

class Circle(object):
    def __init__(self, radius):
        self.radius = radius

    @lazy
    def area(self):
        print('evalute')
        return 3.14 * self.radius ** 2

c = Circle(4)
print(c.radius)
print(c.area)
print(c.area)
print(c.area)
```

在上述代码中，在类 lazy 中定义了__get__()方法，这是一个描述符。当第一次执行 c.area 时，Python 解释器会先从 c.__dict__ 中进行查找。如果没有找到，就从 Circle.__dict__ 中进行查找，此时因为 area 被定义为描述符，所以调用__get__()方法。在__get__()方法中，调用实例的 area()方法计算出结果，并动态给实例添加一个同名属性 area，然后将计算出的值赋予它，相当于设置 c.__dict__['area']=val。当再次调用 c.area 时，直接从 c.__dict__ 中进行查找，这时就会直接返回之前计算好的值。执行效果如图 2-56 所示。

```
78.53981633974483
31.41592653589793
```

图 2-56　执行效果

实例 02-54：在宝马汽车中使用继承

下面模拟一个购买汽车的场景：当今市场中的汽车品牌有很多，如宝马、奥迪、奔驰、丰田、比亚迪等。如果想编写一个展示某品牌汽车新款车型的程序，最合理的方法是先定义一个表示汽车的类，然后定义一个表示某个品牌汽车的子类。在下面的实例文件 car_bmw.py 中，首先定义了汽车类 Car（用于表示所有品牌的汽车），然后定义了基于汽车类的子类 Bmw（用于表示宝马汽车）。

源码路径：daima\2\02-54\car_bmw.py

```python
class Car():
    """汽车之家！"""
    def __init__(self, manufacturer, model, year):
        """初始化操作，建立描述汽车的属性"""
        self.manufacturer = manufacturer
        self.model = model
        self.year = year
        self.odometer_reading = 0
    def get_descriptive_name(self):
        """返回描述信息"""
        long_name = str(self.year) + ' ' + self.manufacturer + ' ' + self.model
        return long_name.title()
```

```
        def read_odometer(self):
            """行驶里程"""
            print("这是一辆新车,目前仪表显示行驶里程是" + str(self.odometer_reading) + "公里!")
class Bmw(Car):
    """这是一个子类Bmw,基类是Car"""
    def __init__(self, manufacturer, model, year):
        super().__init__(manufacturer, model, year)
my_tesla = Bmw('宝马', '535Li', '2017款')
print(my_tesla.get_descriptive_name())
```

- 汽车类 Car 是基类（父类），宝马类 Bmw 是派生类（子类）。
- 在创建子类 Bmw 时，父类必须包含在当前文件中，且位于子类前面。
- 上述代码定义了子类 Bmw，在定义子类时，必须在括号内指定父类的名称。方法 __init__()可以接受创建 Car 实例所需的信息。
- super()是一个特殊函数，功能是将父类和子类关联起来。可以让 Python 调用 Car 父类的方法__init__()，可以让 Bmw 的实例包含父类 Car 中的所有属性。父类也称为超类（superclass），名称中的 super 因此而来。
- 为了测试继承是否能够正确地发挥作用，在倒数第 2 行代码中创建了一辆宝马汽车实例，代码提供的信息与创建普通汽车的完全相同。在创建 Bmw 类的一个实例时，将其存储在变量 my_tesla 中。这行代码调用在 Bmw 类中定义的方法__init__()，后者能够让 Python 调用父类 Car 中定义的方法__init__()。在代码中，使用了 3 个实参"宝马"、"535Li" 和 "2017 款" 进行测试。

执行效果如图 2-57 所示。

```
2017款 宝马 535Li
```
图 2-57 执行效果

实例 02-55：在子类中扩展属性

在 Python 程序中，如果在父类中定义了一个属性，则完全可以在子类中扩展这个属性的功能。下面的实例文件 kuo.py 演示了在子类中扩展属性的过程。

源码路径：daima\2\02-55\kuo.py

```python
class Person:
    def __init__(self, name):
        self.name = name

    # Getter function
    @property
    def name(self):
        return self._name

    # Setter function
    @name.setter
    def name(self, value):
        if not isinstance(value, str):
            raise TypeError('不是一个string字符串')
        self._name = value

    @name.deleter
    def name(self):
        raise AttributeError("不能删除attribute属性")

class SubPerson(Person):
    @property
    def name(self):
        print('获取name')
        return super().name

    @name.setter
```

```
        def name(self, value):
            print('设置name为', value)
            super(SubPerson, SubPerson).name.__set__(self, value)

        @name.deleter
        def name(self):
            print('删除name')
            super(SubPerson, SubPerson).name.__delete__(self)
s = SubPerson('AAA')
s.name="BBB"
s.name = 'CCC'
s.name = 42
```

在上述代码中,首先在父类 Person 中定义了属性 name,然后定义了子类 SubPerson,并在子类 SubPerson 中扩展属性 name 的功能。执行效果如图 2-58 所示。

```
设置name为 AAA
设置name为 BBB
设置name为 CCC
设置name为 42
Traceback (most recent call last):
  File "H:/daima/3/3-20/kuo.py", line 39, in <module>
    s.name = 42
  File "H:/daima/3/3-20/kuo.py", line 30, in name
    super(SubPerson, SubPerson).name.__set__(self, value)
  File "H:/daima/3/3-20/kuo.py", line 14, in name
    raise TypeError('不是一个string字符串')
TypeError: 不是一个string字符串
```

图 2-58　执行效果

实例 02-56:模仿 Linux 操作系统中的文件读写接口

在 Python 程序中,接口是一组功能的入口。当要调用某一组功能时,需要通过接口来进行调用,而不需要关注这组功能是如何实现的,要的只是结果。在 Python 类中,接口用于提取一群类中共同的函数,可以把接口当作一个函数的集合。在下面的实例文件 jiekou.py 中,模仿了 Linux 操作系统中的文件读写接口。在 Linux 操作系统中,文本、硬盘和进程都是通过文件实现的,只是具体实现方法不同而已。

源码路径:daima\2\02-56\jiekou.py

```
class File:              #定义一个接口类
    def read(self):      #定义接口函数read()
        pass
    def write(self):     #定义接口函数write()
        pass
#定义子类实现读写功能
#文本文件的读写
class Txt(File):         #文本,具体实现read()和write()
    def du(self):        #注意并不是read()
        print('文本数据的读取方法')
    def xie(self):       #注意并不是write()
        print('文本数据的写入方法')
#硬盘数据的读写
class Sata(File):        #硬盘,具体实现read()和write()
    def read(self):
        print('硬盘数据的读取方法')
    def write(self):
        print('硬盘数据的写入方法')
#进程数据的读写
class Process(File):
    def read(self):
```

```
            print('进程数据的读取方法')
    def write(self):
            print('进程数据的写入方法')

disk=Sata()              #实例化一个硬盘读写对象
disk.read()              #硬盘读
disk.write()             #硬盘写
```

上述代码定义一个接口类 File，其中提供了 read()和 write()接口函数，这两个接口函数一定是没有具体处理功能的，因为具体的处理功能需要交给后面的子类来实现。执行效果如图 2-59 所示。由此可见，通过使用接口，使用者无须关心对象的类是什么，只需要知道这些对象都具备某些功能就可以了，这极大地降低了开发者的使用难度。

硬盘数据的读取方法
硬盘数据的写入方法

图 2-59 执行效果

2.8 迭代器和生成器

迭代是 Python 中最强大的功能之一，是访问集合元素的一种方式。通过使用迭代器，不仅简化了循环程序的代码，而且可以节约内存。迭代器是一种可以连续迭代的一个容器，Python 程序中所有的序列类型都是可迭代的。在 Python 程序中，使用关键字 yield 定义的函数称为生成器（generator）。通过使用生成器，可以生成一个值的序列用于迭代，并且这个值的序列不是一次生成的，而是使用一个，再生成一个，这样最大的好处是可以使程序节约大量内存。

实例 02-57：创建并使用迭代器

在 Python 程序中，迭代器对象是一个可以记住遍历的位置的对象。迭代器对象从集合的第 1 个元素开始访问，直到所有的元素被访问完结束，迭代器只能往前不会后退。在 Python 程序中，要想创建一个自己的迭代器，只需要定义一个实现迭代器协议方法的类即可。下面的实例文件 die.py 演示了创建并使用迭代器的过程。

源码路径：daima\2\02-57\die.py

```
class Use:                                  #定义迭代器类Use
    def __init__(self,x=2,max=50):          #定义构造方法
        self.__mul,self.__x = x,x           #初始化属性，x的初始值是2
        self.__max = max                    #初始化属性
    def __iter__(self):                     #定义迭代器协议方法
        return self                         #返回类的自身
    def __next__(self):                     #定义迭代器协议方法
        if self.__x and self.__x != 1:      #如果x值不是1
            self.__mul *= self.__x          #则设置mul值
            if self.__mul <= self.__max:    #如果mul值小于等于预设的最大值max
                return self.__mul           #则返回mul值
            else:
                raise StopIteration         #超过参数max的值会引发StopIteration异常
        else:
            raise StopIteration
if __name__ == '__main__':
    my = Use()                              #定义类Use的对象实例my
    for i in my:                            #遍历对象实例my
        print('迭代的数据元素为：',i)
```

上述代码的功能是显示迭代器中的数据，首先定义了迭代器类 Use，在其构造方法中，初始化私有的实例属性，功能是生成序列并设置序列中的最大值。这个迭代器总是返回所给整数的幂，当其最大值超过参数 max 值时就会引发 StopIteration 异常，并且马上结束遍历。最后，实

例化迭代器类，并遍历迭代器的值序列，同时输出各个序列值。在本实例中，初始化迭代器时使用了默认参数，遍历得到的序列是 2 的幂的值，最大值不超过 50。执行效果如图 2-60 所示。

```
迭代的数据元素为：   4
迭代的数据元素为：   8
迭代的数据元素为：  16
迭代的数据元素为：  32
```

图 2-60　执行效果

实例 02-58：创建并使用生成器

在 Python 程序中，生成器函数是一个"记住"上一次返回时其在函数体中位置的函数。对生成器函数的第 2 次调用跳转至该函数中间，而上次调用的所有局部变量都保持不变。生成器不仅"记住"了其数据状态，还"记住"了在流控制构造（在命令式编程中，这种构造不只是数据）中的位置。下面的实例文件 sheng1.py 演示了 yield 生成器的执行机制。

源码路径：daima\2\02-58\sheng1.py

```python
def fib(max):                   #定义方法fib()
    a, b = 1, 1                 #为变量a和b赋值为1
    while a < max:              #如果a小于max
        yield a                 #当程序执行到yield这行时就不会继续往下执行
        a, b = b, a+b
for n in fib(15):               #遍历15以内的值
    print (n)
```

在上述实例代码中，程序执行到 yield 这行时就不会继续往下执行，而是返回一个包含当前函数所有参数的状态的 iterator 对象。目的是保证在其第 2 次被调用时，能够访问到的函数所有的参数值都是第一次访问时的值，而不是重新赋值。当程序第 1 次调用时：

yield a #这时a、b值分别为1、1，当然，程序也在执行到这时，返回

从前面可知，当第 1 次调用时，a,b=1,1，那么第 2 次调用时（其实就是调用第 1 次返回的 iterator 对象的 next()函数），程序跳到 yield 语句处，当执行"a,b = b, a+b"语句时，此时 a,b = 1, (1+1) => a,b = 1, 2。然后程序继续执行 while 循环，这样会再一次碰到 yield a 语句，也是像第 1 次那样，保存函数所有参数的状态，返回一个包含这些参数状态的 iterator 对象。然后等待第 3 次调用……执行后会输出：

```
1
1
2
3
5
8
13
```

实例 02-59：实现委托迭代处理

在 Python 程序中构建一个自定义的容器对象后，我们在其内部可以包含一个列表、元组或其他的可迭代对象。如果接下来我们想让这个新容器能够实现迭代操作，应该如何实现呢？这时需要实现委托迭代功能，只需定义一个 __iter__()方法，并将迭代请求委托到对象内部持有的容器上即可。下面的实例文件 weituo.py 演示了实现委托迭代处理的过程。

源码路径：daima\2\02-59\weituo.py

```python
class Node:
    def __init__(self, value):
        self._value = value
        self._children = []

    def __repr__(self):
```

```
        return 'Node({!r})'.format(self._value)

    def add_child(self, node):
        self._children.append(node)

    def __iter__(self):
        return iter(self._children)

#举例
if __name__ == '__main__':
    root = Node(0)
    child1 = Node(1)
    child2 = Node(2)
    root.add_child(child1)
    root.add_child(child2)
    for ch in root:
        print(ch)
```

Python 中的迭代协议要求方法__iter__()返回一个特殊的迭代器对象，通过该对象实现的方法__next__()来实现实际的迭代功能。如果只是迭代另一个容器中的内容，则开发者无须担心底层细节的运作流程，开发者所要做的只是转发迭代请求而已。在上述代码中，方法__iter__()只是简单地将迭代请求转发给对象内部持有的属性_children。方法 iter()对上述代码进行了一定的简化处理，iter(s)通过调用方法 s.__iter__()来返回底层的迭代器，这和 len(s)调用方法 s.__len__()的方式是一样的。

执行后会输出：

```
1
1
2
3
5
8
13
```

实例 02-60：使用生成器创建新迭代模式

在 Python 程序中，可以使用内置函数（如 range()、reversed()等）创建新的迭代模式。除了使用内置函数之外，也可以使用生成器函数来实现。下面的实例文件 xindiedai1.py 演示了使用生成器创建新迭代模式的过程。

源码路径：daima\2\02-60\xindiedai1.py

```
def frange(start, stop, increment):
    x = start
    while x < stop:
        yield x
        x += increment

①for n in frange(0, 4, 0.5):
    print(n)

②print(list(frange(0, 1, 0.125)))
```

在上述代码中，通过一个生成器生成了一系列某个范围内的浮点数。要想使用上面的函数 frange()，可以像①那样使用 for 循环对其迭代，或者像②那样通过其他可以访问、可迭代对象中元素的函数（如 sum()、list()等）来使用。

执行后会输出：

```
0
0.5
1.0
1.5
2.0
2.5
3.0
3.5
[0, 0.125, 0.25, 0.375, 0.5, 0.625, 0.75, 0.875]
```

如果在 Python 函数中出现了 yield 语句，就会将其转换成一个生成器。与普通函数不同，生成器只会在响应迭代操作时才执行。下面的实例文件 xindiedai2.py 演示了生成器函数底层机制的运转过程。

源码路径：daima\2\02-60\xindiedai2.py

```
def countdown(n):
    print('Starting to count from', n)
    while n > 0:
        yield n
        n -= 1
    print('Done!')

# 创建生成器，请注意没有出现输出
c = countdown(3)
print(c)
print(next(c))           #执行4次next()函数
print(next(c))
print(next(c))
print(next(c))
```

在 Python 程序中，生成器函数只会在响应迭代过程中的"next"操作时才会执行。一旦返回生成器函数，迭代操作会立即停止执行。在一般情况下，因为用来处理迭代的 for 循环替开发者处理了这些细节，所以我们一般无须为此操心。执行效果如图 2-61 所示。

```
<generator object countdown at 0x0000020936923990>
Starting to count from 3
3
2
1
Done!
Traceback (most recent call last):
  File "C:\123\人民邮电\2017重点\Python专题\Python开发范例大全\daima\2\02-60\xindiedai2.py", line 14, in <module>
    print(next(c))
StopIteration
>>>
```

图 2-61　执行效果

实例 02-61：使用生成器函数实现一个迭代协议

在 Python 程序中构建一个自定义的对象后，如果希望这个对象支持迭代操作，则最简单的方法是使用生成器函数实现一个迭代协议。下面的实例文件 diedaixie.py 演示了使用生成器函数实现一个迭代协议的过程。

源码路径：daima\2\02-61\diedaixie.py

```
class Node:
    def __init__(self, value):
        self._value = value
        self._children = []

    def __repr__(self):
        return 'Node({!r})'.format(self._value)

    def add_child(self, node):
        self._children.append(node)

    def __iter__(self):
        return iter(self._children)

    def depth_first(self):
        yield self
        for c in self:
            yield from c.depth_first()

if __name__ == '__main__':
    root = Node(0)
    child1 = Node(1)
```

```
        child2 = Node(2)
        root.add_child(child1)
        root.add_child(child2)
        child1.add_child(Node(3))
        child1.add_child(Node(4))
        child2.add_child(Node(5))

        for ch in root.depth_first():
            print(ch)
```

在上述代码中，使用类 Node 来表示树的结构，实现一个能够以深度优先模式遍历树的节点迭代器。上述 depth_first()函数的实现过程并不简单，它首先使用 yield 产生自身，然后迭代每个子节点，利用子节点的 depth_first()函数（通过 yield from 语句）产生其他元素。

执行后会输出：

```
Node(0)
Node(1)
Node(3)
Node(4)
Node(2)
Node(5)
```

实例 02-62：使用函数 reversed()实现反转迭代

在 Python 程序中，可以使用内置函数 reversed()反转迭代序列中的元素，将其元素从后向前颠倒构建成一个新的迭代器。下面的实例文件 fanzhuan1.py 演示了使用函数 reversed()实现反转迭代的过程。

源码路径：daima\2\02-62\fanzhuan1.py

```
a = reversed(range(10))         #传入range对象
print(a)                        #类型变成迭代器
print(list(a))

a = ['a','b','c','d']
print(a)
print(reversed(a))              #传入列表对象
b = reversed(a)
print(b)                        #类型变成迭代器
print(list(b))
```

执行后会输出：

```
<range_iterator object at 0x00000127E75B9F10>
[9, 8, 7, 6, 5, 4, 3, 2, 1, 0]
['a', 'b', 'c', 'd']
<list_reverseiterator object at 0x00000127E76AB2E8>
<list_reverseiterator object at 0x00000127E76AB2E8>
['d', 'c', 'b', 'a']
```

如果函数 reversed()的参数不是一个序列对象，则其必须定义一个__reversed__()方法。下面的实例文件 fanzhuan2.py 演示了使用和不使用__reversed__()方法的对比过程。

源码路径：daima\2\02-62\fanzhuan2.py

```
    #类Student没有定义__reversed__()方法
①class Student:
    def __init__(self, name, *args):
        self.name = name
        self.scores = []
        for value in args:
            self.scores.append(value)

a = Student('Bob', 78, 85, 93, 96)
②#print(reversed(a))      # 实例不能反转
print(type(a.scores))      # 列表类型

    # 重新定义类型，并为其定义__reversed__()方法
③class Student:
    def __init__(self, name, *args):
        self.name = name
        self.scores = []
```

```
            for value in args:
                self.scores.append(value)

    def __reversed__(self):
        self.scores = reversed(self.scores)

a = Student('Bob', 78, 85, 93, 96)
print(a.scores)                    # 列表类型
print(type(a.scores))
print(reversed(a))                 # 实例变得可以反转
print(a.scores)                    # 反转后类型变成迭代器
print(type(a.scores))
print(list(a.scores))
```

在上述代码中，因为①处的类 Student 中没有定义__reversed__()方法，所以②处的代码将无法正确执行。而③处的类 Student 中定义了__reversed__()方法，所以后面的反转操作可以成功实现。执行上述代码后输出：

```
<class 'list'>
[78, 85, 93, 96]
<class 'list'>
None
<list_reverseiterator object at 0x000001ABC6F7D940>
<class 'list_reverseiterator'>
[96, 93, 85, 78]
```

实例 02-63：实现有额外状态的生成器函数

当在 Python 程序中定义一个生成器函数时，如果想让这个函数拥有一些额外的功能，并且能以某种形式将这些功能暴露给用户，那么需要将这些功能定义在一个类中，然后将生成器函数的代码放到__iter__()方法中即可。下面的实例文件 ewai.py 演示了实现有额外状态的生成器函数的过程。

源码路径：daima\2\02-63\ewai.py

```
from collections import deque

class linehistory:
    def __init__(self, lines, histlen=3):
        self.lines = lines
        self.history = deque(maxlen=histlen)

    def __iter__(self):
        for lineno, line in enumerate(self.lines,1):
            self.history.append((lineno, line))
            yield line

    def clear(self):
        self.history.clear()

with open('123.txt') as f:
    lines = linehistory(f)
    for line in lines:
        if 'python' in line:
            for lineno, hline in lines.history:
                print('{}:{}'.format(lineno, hline), end='')
f = open('123.txt')
lines = linehistory(f)
①print(next(lines))
②it = iter(lines)
③print(next(it))
④print(next(it))
```

上述代码定义了额外功能类 linehistory，在使用这个类时需要将其看做一个普通的生成器函数。另外，由于它会创建一个类实例，所以可以访问内部属性（如 history 属性或 clear()函数）。上述代码有一个很明显的缺点，即如果想用除了 for 循环之外的技术来驱动迭代过程，则可能需要额外调用一次 iter()。例如，因为在①中没有额外调用一次 iter()，所以执行后会出错；而在②中额外调用一次 iter()后，后面的③和④处的代码可以执行成功。执行上述代码后，会输出文件 123.txt 中的前两行内容。

实例 02-64：实现迭代器切片处理

在 Python 程序中，内置函数 slice()能够实现迭代器切片处理，主要功能是在切片操作函数中传递参数。下面的实例文件 qiepian.py 演示了使用内置函数 slice()实现迭代器切片处理的过程。

源码路径：daima\2\02-64\qiepian.py

```python
from itertools import islice
la = [x for x in range(20)]
lai = iter(la)
for item in islice(lai,5,9):
    print(item)
for x in lai:
    print(x)
```

在上述代码中，首先 la 通过解析生成一个列表，然后将其迭代器赋值给 lai。lai 赋值给 islice 后会消耗掉前 9 个索引。当在 islice 之后再次遍历 lai 时，会从 index=10 开始迭代。程序执行后会输出 5～19 的连续整数，只是这些连续整数是按照 5～8 和 9～19 两个部分实现的。

实例 02-65：迭代出所有可能的组合

在 Python 程序中，可以迭代出一系列元素中所有可能的组合或排列。下面的实例文件 suoyou.py 演示了使用 3 个函数迭代出列表中所有可能组合的过程。

源码路径：daima\2\02-65\suoyou.py

```python
items = ['a', 'b', 'c']
from itertools import permutations
①for p in permutations(items):
    print(p)

②for p in permutations(items, 2):
    print(p)

from itertools import combinations
③for c in combinations(items, 3):
    print(c)

④for c in combinations(items, 2):
    print(c)

⑤for c in combinations(items, 1):
    print(c)

from itertools import combinations_with_replacement
⑥for c in combinations_with_replacement(items, 3):
    print(c)
```

- 使用函数 combinations()进行迭代操作，itertools 模块中的 permutations()函数能够接受一个元素集合，将其中所有的元素重新排列为可能的情况，并以元组的形式返回，也就是将元素之间的顺序打乱成可能的情形。
- 如果想得到较短长度的所有排列，则可以在 combinations()函数中提供一个可选长度参数。
- ③～⑤处使用 combinations()函数生成输入序列中所有元素的全部组合形式。在使用函数 combinations()时，不用考虑元素之间的实际顺序。也就是说，可以将组合('a', 'b')和 ('b', 'a')看作相同的组合形式（因此只会产生出其中一种）。
- 当生成所有的组合时，会从可能的候选元素中删除已经被选择过的元素。例如，"a" 已经被选过，那么就将它从考虑范围中去掉。此时在⑥中使用的 combinations_with_replacement()函数会取消这一限制，允许多次选择相同的元素。

执行后会输出：
```
('c', 'a', 'b')
('c', 'b', 'a')
('a', 'b')
('a', 'c')
('b', 'a')
('b', 'c')
('c', 'a')
('c', 'b')
('a', 'b', 'c')
('a', 'b')
('a', 'c')
('b', 'c')
('a',)
('b',)
('c',)
('a', 'a', 'a')
('a', 'a', 'b')
('a', 'a', 'c')
('a', 'b', 'b')
('a', 'b', 'c')
('a', 'c', 'c')
('b', 'b', 'b')
('b', 'b', 'c')
('b', 'c', 'c')
('c', 'c', 'c')
```

实例 02-66：使用函数 enumerate() 创建索引迭代序列

在 Python 程序中，函数 enumerate() 通常被用在 for 循环中，功能是将一个可遍历的数据对象（如列表、元组或字符串）组合为一个索引序列，同时列出数据和数据索引。下面的实例文件 suoyin.py 演示了使用函数 enumerate() 创建索引迭代序列的过程。

源码路径：daima\2\02-66\suoyin.py

```
import collections
my_list = ['a', 'b', 'c']
①for idx, val in enumerate(my_list):
    print(idx, val)

my_list = ['a', 'b', 'c']
②for idx, val in enumerate(my_list, 1):
    print(idx, val)

③def parse_data(filename):
    with open(filename, 'rt') as f:
        for lineno, line in enumerate(f, 1):
            fields = line.split()
            try:
                count = int(fields[1])
            except ValueError as e:
④              print('Line {}: Parse error: {}'.format(lineno, e))

⑤word_summary = collections.defaultdict(list)
with open('123.txt', 'r') as f:
    lines = f.readlines()
for idx, line in enumerate(lines):
    # Create a list of words in current line
    words = [w.strip().lower() for w in line.split()]
    for word in words:
⑥       word_summary[word].append(idx)
```

在①中使用函数 enumerate() 创建 "索引,值" 对迭代序列。

在②中为了输出从数字 1 开始的行号，传入一个开始参数作为起始索引。

在③~④中设置一个开始参数作为起始索引，功能是跟踪记录文件中的行号，这样可以及时查看错误信息来自哪一行代码。

在⑤~⑥中，为了在文件中的单词和它们所出现的行之间建立映射关系，使用函数 enumerate() 将每个单词映射到文件行相应的偏移位置。

实例 02-67：使用函数 zip() 或 zip_longest() 迭代多个序列

在 Python 程序中，可以使用函数 zip() 或 zip_longest() 同时迭代多个序列。其中，函数 zip(a, b) 的功能是创建一个迭代器，该迭代器可以产生出元组(x, y)，这里的 x 来自序列 a，而 y 来自序列 b。当其中某个输入序列中没有元素可以继续迭代时，整个迭代过程结束。因此，整个迭代的长度和其中最短的输入序列长度相同。如果没有上述限制要求，则可以使用函数 zip_longest() 来实现迭代。下面的实例文件 duogexulie.py 演示了使用函数 zip() 或 zip_longest() 迭代多个序列的过程。

源码路径：daima\2\02-67\duogexulie.py

```python
xpts = [1, 5, 4, 2, 10, 7]
ypts = [101, 78, 37, 15, 62, 99]
for x, y in zip(xpts, ypts):
    print(x,y)

a = [1, 2, 3]
b = ['w', 'x', 'y', 'z']
for i in zip(a,b):
    print(i)

from itertools import zip_longest
for i in zip_longest(a,b):
    print(i)
for i in zip_longest(a, b, fillvalue=0):
    print(i)
```

有时函数 zip() 可以接受多于的两个序列作为输入，这时得到的元组中的元素数量和输入的序列数量相同。需要注意的是，zip() 函数创建的结果只是一个迭代器。如果需要将配对的数据保存为列表，则建议使用 list() 函数。

第 3 章

标准库函数实战

为了帮助开发者快速实现软件开发功能，Python 提供了大量内置的标准库，如文件操作库、正则表达式库、数学运算库和网络操作库等。这些标准库都提供了大量的内置模块，这些模块中的函数是 Python 程序实现软件项目的最有力的工具。本章将详细讲解 Python 常用标准库函数的演练实例，为读者学习本书后面的知识打下基础。

3.1 字符串处理函数

Python 的内置模块提供了大量的字符串处理函数，通过这些函数，开发者可以快速处理字符串。

实例 03-01：使用函数 split()分割指定的字符串

在内置模块 string 中，函数 split()的功能是通过指定的分隔符对字符串进行分割，如果参数 num 有指定值，则只分隔 num 个子字符串。下面的实例文件 fenge.py 演示了使用函数 split()分割指定的字符串的过程。

源码路径：daima\3\03-01\fenge.py

```
str = "this is string example....wow!!!"
print (str.split( ))
print (str.split('i',1))
print (str.split('w'))
import re
print (re.split('w'))
```

上述代码分别 3 次调用内置函数 str.split()对字符串 str 进行了分割，执行后会输出：

```
['this', 'is', 'string', 'example....wow!!!']
['th', 's is string example....wow!!!']
['this is string example....', 'o', '!!!']
```

实例 03-02：使用函数 re.split()分割指定字符串

在内置模块 re 中，函数 split()的功能是进行字符串分割操作。其语法格式如下：

```
re.split(pattern, string[, maxsplit])
```

上述语法的功能是按照能够匹配的子串对字符串进行分割，然后返回分割列表。参数 maxsplit 用于指定最大的分割次数，不指定则将全部分割。下面的实例文件 refenge.py 演示了使用函数 re.split()分割指定字符串的过程。

源码路径：daima\3\03-02\refenge.py

```
import re
line = 'asdf fjdk; afed, fjek,asdf, foo'
#根据空格符、逗号或分号进行分割
①parts = re.split(r'[;,\s]\s*', line)
print(parts)

#根据捕获组进行分割
②fields = re.split(r'(;|,|\s)\s*', line)
print(fields)
```

在①中，使用的分隔符是空格符、逗号或分号，后面可跟着任意数量的额外空格。

在②中，根据捕获组进行分割，在使用 re.split()时需要注意正则表达式中的捕获组是否包含在括号中。如果用到了捕获组，那么匹配的文本也会包含在最终结果中。

执行后会输出：

```
['asdf', 'fjdk', 'afed', 'fjek', 'asdf', 'foo']
['asdf', ' ', 'fjdk', ';', 'afed', ',', 'fjek', ',', 'asdf', ',', 'foo']
```

实例 03-03：字符串开头和结尾处理

在内置模块 string 中，函数 startswith()的功能是检查字符串是否以指定的子字符串开头，如果是则返回 True，否则返回 False。如果参数 beg 和 end 指定了具体的值，则会在指定的范围内进行检查。

在内置模块 string 中，函数 endswith()的功能是判断字符串是否以指定后缀结尾，如果以指定后缀结尾则返回 True，否则返回 False。其中的可选参数 start 与 end 分别表示检索字符串的开始位置与结束位置。

下面的实例文件 qianhou.py 分别演示了使用函数 startswith()和 endswith()对指定的字符串进行处理的过程。

源码路径：daima\3\03-03\qianhou.py

```
str = "this is string example....wow!!!"
print (str.startswith( 'this' ))
print (str.startswith( 'string', 8 ))
print (str.startswith( 'this', 2, 4 ))

suffix='!!'
print (str.endswith(suffix))
print (str.endswith(suffix,20))
suffix='run'
print (str.endswith(suffix))
print (str.endswith(suffix, 0, 19))
```

由此可见，函数 startswith()和 endswith()提供了一种非常方便的方式，对字符串的前缀和后缀实现了基本的检查。

执行后会输出：

```
True
True
False
True
True
False
False
```

实例 03-04：实现字符串匹配处理

在内置模块 fnmatch 中，函数 fnmatch()的功能是采用大小写区分规则和底层文件相同（根据操作系统而区别）的模式进行匹配。在内置模块 fnmatch 中，函数 fnmatchcase()的功能是根据所提供的字符大小写进行匹配，用法和函数 fnmatch()类似。

函数 fnmatch()和 fnmatchcase()的匹配样式是 UNIX Shell 风格的，其中"*"表示匹配任何单个或多个字符，"?"表示匹配单个字符，[seq]表示匹配单个 seq 中的字符，[!seq]表示匹配单个不是 seq 中的字符。

下面的实例文件 pipeizifu.py 演示了分别使用函数 fnmatch()和 fnmatchcase()实现字符串匹配的过程。

源码路径：daima\3\03-04\pipeizifu.py

```
from fnmatch import fnmatchcase as match
import fnmatch
#匹配以.py 结尾的字符串
①print(fnmatch.fnmatch('py','.py'))

②print(fnmatch.fnmatch('tlie.py','*.py'))

#macOS X
③#print(fnmatch.fnmatch('123.txt', '*.TXT'))
#Windows
④print(fnmatch.fnmatch('123.txt', '*.TXT'))
⑤print(fnmatch.fnmatchcase('123.txt', '*.TXT'))

⑥addresses = [
    '5000 A AAA FF',
    '1000 B BBB',
    '1000 C CCC',
    '2000 D DDD NN',
    '4234 E EEE NN',
]

⑦a = [addr for addr in addresses if match(addr, '* FF')]
print(a)
```

⑧b = [addr for addr in addresses if match(addr, '42[0-9][0-9] *NN*')]
   print(b)

在①~②中演示了函数 fnmatch()的基本用法，可以匹配以.py 结尾的字符串，用法和函数 fnmatchcase()相似。

在③和④中演示了函数 fnmatch()的匹配模式，所采用的大小写区分规则和底层文件系统相同，根据操作系统的不同而有所不同。

在⑤中使用函数 fnmatchcase()以根据提供的大小写方式进行匹配。

在⑥中演示了在处理非文件名式的字符串时的作用，定义了保存一组联系地址的的列表 addresses。

在⑦~⑧中使用 match()进行推导。

由此可见，fnmatch 所实现的匹配操作介于简单的字符串方法和正则表达式之间。如果只想在处理数据时提供一种简单的机制以允许使用通配符，那么通常这是一个合理的解决方案。

执行后会输出：

```
False
True
True
False
['5000 A AAA FF']
['4234 E EEE NN']
```

实例 03-05：文本模式匹配和查找

如果只是想要匹配简单的文字，则只需使用内置模块 string 中的函数 str.find()、str.endswith()、str.startswith()即可实现。下面的实例文件 jdanwenb.py 演示了使用内置模块实现文本模式匹配和查找的过程。

源码路径：daima\3\03-05\jdanwenb.py

```python
text = 'yes, but no, but yes, but no, but yes'
text == 'yeah'

#开头测试匹配
print(text.startswith('yes'))
#结尾测试匹配
print(text.endswith('no'))

#搜索第一次出现的位置
print(text.find('no'))
```

执行后会输出：

```
True
False
9
```

要想在 Python 程序中匹配更复杂的文字模式，则需要使用正则表达式模块 re。下面的实例文件 fuzamoshi.py 演示了使用正则表达式匹配以数字形式构成的日期的过程。

源码路径：daima\3\03-05\fuzamoshi.py

```python
text1 = '11/11/2018'
text2 = 'Nov 11, 2018'
import re

if re.match(r'\d+/\d+/\d+', text1):
    print('yes')
else:
    print('no')

if re.match(r'\d+/\d+/\d+', text2):
    print('yes')
else:
    print('no')
```

执行后会输出：
```
yes
no
```

如果想对同一种模式进行多次匹配工作，则通常会先将正则表达式模式预编译成一个模式对象。函数 match() 总是试图在字符串的开头找到匹配项，如果想针对整个文本搜索出所有的匹配项，那么应该使用 findall() 函数。下面的实例文件 duoci.py 演示了使用 findall() 函数实现多次匹配的过程。

源码路径：daima\3\03-05\duoci.py

```python
import re

#处理字符串
text = '今天是 11/11/2018.  starts 3/13/2018.'

#获取所有匹配日期
①datepat = re.compile(r'\d+/\d+/\d+')
print(datepat.findall(text))

#查找所有与捕获组匹配的日期
datepat = re.compile(r'(\d+)/(\d+)/(\d+)')
②for month, day, year in datepat.findall(text):
    print('{}-{}-{}'.format(year, month, day))

#迭代搜索
for m in datepat.finditer(text):
    print(m.groups())
```

在①中以将部分模式用括号包起来的方式引入捕获组，捕获组可以简化后续对匹配文本的处理，因为每个组的内容都可以单独提取出来。

先在②中使用函数 findall() 搜索整个文本，并找出所有的匹配项，然后将它们以列表的形式返回。当以迭代的方式找出匹配项时，建议使用函数 finditer()。

执行后会输出：
```
['11/11/2018', '3/13/2018']
2018-11-11
2018-3-13
('11', '11', '2018')
('3', '13', '2018')
```

实例 03-06：文本查找和替换

在 Python 程序中，如果只是想实现简单的文本替换功能，则只需使用内置模块 string 中的函数 replace() 即可。函数 replace() 能够把字符串中的 old（旧字符串）替换成 new（新字符串），如果指定第 3 个参数 max，则替换不超过 max 次。下面的实例文件 tihuan.py 演示了使用函数 replace() 实现文本查找和替换的过程。

源码路径：daima\3\03-06\tihuan.py

```python
str = "www.to***.net"
print ("玲珑科技新地址： ", str)
print ("玲珑科技新地址： ", str.replace("chuban***.com", "www.***school"))

str = "this is string example....hehe!!!"
print (str.replace("is", "was", 3))
```

执行后会输出：
```
玲珑科技新地址：    www.to***.net
玲珑科技新地址：    www.to***.net
thwas was string example....hehe!!!
```

要想在 Python 程序中执行更为复杂的替换操作，则需要使用正则表达式模块 re 中的 sub() 函数实现。下面的实例文件 zhengti.py 演示了使用正则表达式函数 sub() 实现文本查找和替换的过程。

源码路径：daima\3\03-06\zhengti.py

```
import re
text = '今天是11/11/2018. Python 3/11/2018.'

①datepat = re.compile(r'(\d+)/(\d+)/(\d+)')

#使用sub()函数
②print(datepat.sub(r'\3-\1-\2', text))

from calendar import month_abbr
③def change_date(m):
    mon_name = month_abbr[int(m.group(1))]
    return '{} {} {}'.format(m.group(2), mon_name, m.group(3))

print(datepat.sub(change_date, text))

④newtext, n = datepat.subn(r'\3-\1-\2', text)
print(newtext)
⑤print(n)
```

①的功能是将日期格式"××/××/×××"修改为"××-××-×××"。

在②中使用函数 sub()，其中，第 1 个参数表示要匹配的模式，第 2 个参数表示要替换的模式。类似"\3"这样的反斜线加数字的模式代表模式中捕获组的数量。为了用相同的模式实现重复替换，需要先把模式编译以获得更好的性能。

在③中定义一个替换回调函数 change_date()，此函数的输入参数是一个匹配对象，由 match() 或 find()返回。使用函数 group()来提取匹配中特定的部分，返回被替换后的文本。

在④中使用函数 subn()获取一共完成了多少次替换，并将次数在⑤中输出。

执行后会输出：

```
今天是2018-11-11. Python 2018-3-11.
今天是11 Nov 2018. Python 11 Mar 2018.
今天是2018-11-11. Python 2018-3-11.
2
```

在 Python 程序中，有时需要以不区分大小写的方式实现文本查找或替换功能，此时可以使用模块 re 实现，并且需要对各种操作都要加上 re.IGNORECASE 标记。下面的实例文件 bufen.py 演示了以不区分大小写的方式实现文本查找和替换的过程。

源码路径：daima\3\03-06\bufen.py

```
①import re
text = '大写 PYTHON, 小写 python, 大小写都有 Python'
print(re.findall('python', text, flags=re.IGNORECASE))
print(re.sub('python', 'snake', text, flags=re.IGNORECASE))

②def matchcase(word):
    def replace(m):
        text = m.group()
        if text.isupper():
            return word.upper()
        elif text.islower():
            return word.lower()
        elif text[0].isupper():
            return word.capitalize()
        else:
            return word
    return replace
print(re.sub('python', matchcase('snake'), text, flags=re.IGNORECASE))
```

在上述代码中，①中的替换操作有一个局限性，待替换的文本与匹配的文本大小写并不吻合。为了解决这个问题，在②中定义了函数 matchcase()，通过 if...elif 语句和内置的大小写函数进行处理。

执行后会输出：

```
['PYTHON', 'python', 'Python']
大写 snake, 小写 snake, 大小写都有 snake
大写 SNAKE, 小写 snake, 大小写都有 Snake
```

> **注意**：在一般情况下，只需加上 re.IGNORECASE 标记即可实现不区分大小写的匹配操作。但是需要注意的是，某些涉及大写转换（case folding）的 Unicode 匹配可能会失效。

实例 03-07：实现最短文本匹配

当在 Python 程序中使用正则表达式对文本模式进行匹配时，被识别出来的匹配结果是最长的可能匹配。要想将匹配结果修改为最短的匹配，此时需要用到正则表达式的知识。下面的实例文件 duan.py 演示了使用正则表达式实现最短文本匹配的过程。

源码路径：daima\3\03-07\duan.py

```
import re
①str_pat = re.compile(r'\"(.*)\"')
text1 = '计算机回复说"no."'
print(str_pat.findall(text1))

text2 = '计算机回复说"no." 手机回复说 "yes."'
②print(str_pat.findall(text2))

③str_pat = re.compile(r'\"(.*?)\"')
print(str_pat.findall(text2))
```

①中的模式 r'\"(.*)\"' 想匹配包含在引号中的文本，但是，星号"*"在正则表达式中采用的是贪心策略，所以匹配过程是基于找出最长的可能匹配来进行的。

在②的 text2 的掩饰代码中，错误地匹配两个被双引号包围的字符串。

在③的模式中的星号"*"后面加上问号"?"修饰符，这样匹配过程就不会以贪心方式进行，从而生成最短的匹配。

执行后会输出：

```
['no.']
['no." 手机回复说  "yes.']
['no.', 'yes.']
```

实例 03-08：处理 Unicode 字符串

当在 Python 程序中处理 Unicode 字符串时，需要确保所有的字符串都拥有相同的底层表示。在 Unicode 字符串中，一些特定的字符可以被表示成多种合法的代码点序列。下面的实例文件 teshu.py 演示了 Unicode 字符串的代码点序列表示方法。

源码路径：daima\3\03-08\teshu.py

```
s1 = 'I Love Python\u00f1o'
s2 = 'I Love Pythonn\u0303o'

print(s1)
print(s2)

print('s1 == s2', s1 == s2)
print(len(s1), len(s2))
```

在上述代码中，字符串"I Love Pythonño"以两种形式显示出来。第 1 种使用的是字符"ñ"的全组成(u+00f1)形式，第 2 种使用的是拉丁字母"n"紧跟着一个"～"组合而成的字符(u+0303)形式。

执行后会输出：

```
I Love Pythonño
I Love Pythonño
s1 == s2 False
15 16
```

在 Python 中，通过使用模块 unicodedata 中的函数 normalize()实现归一化操作。归一化的目标是把需要处理的数据经过处理后（通过某种算法）限制在需要的一定范围内。首先归一化是为了后面数据处理的方便，其次是保证程序执行时收敛加快。归一化的具体作用是归纳统一样本的统计分布性。归一化在 0～1 中是统计的概率分布，在某个区间上是统计的坐标分布。归一化有同一、统一和合一的意思。

简而言之，归一化的目的是使没有可比性的数据变得具有可比性，同时保持相比较的两个数据之间的相对关系，如大小关系；或是为了作图，原来很难在一张图上作出来，归一化后就可以很方便地给出图上的相对位置等。下面的实例文件 guiyihua.py 演示了使用函数 normalize() 归一化 Unicode 字符串的过程。

源码路径：daima\3\03-08\guiyihua.py

```python
import unicodedata
s1 = 'I Love Python\u00f1o'
s2 = 'I Love Pythonn\u0303o'

n_s1 = unicodedata.normalize('NFC', s1)
n_s2 = unicodedata.normalize('NFC', s2)

print('n_s1 == n_s2', n_s1 == n_s2)
print(len(n_s1), len(n_s2))

# 标准化分解和剥离
t1 = unicodedata.normalize('NFD', s1)
print(''.join(c for c in t1 if not unicodedata.combining(c)))
```

在上述代码中，函数 normalize() 的第 1 个参数指定了字符串应该如何规范表示。

执行后会输出：

```
n_s1 == n_s2 True
15 15
I Love Pythonno
```

实例 03-09：删除字符串中的字符

在 Python 程序中，如果想在字符串的开始、结尾或中间删除不需要的字符或空格，则可以使用内置模块 string 中的函数 strip()，它可从字符串的开始和结尾处删除字符。另外，函数 lstrip() 和 rstrip() 可以分别从左侧或右侧开始执行删除字符的操作。下面的实例文件 shanchu.py 演示了使用上述 3 个函数删除字符串字符的过程。

源码路径：daima\3\03-09\shanchu.py

```python
str = "     this is string example....wow!!!     ";
print( str.lstrip() );
str = "88888888this is string example....wow!!!8888888";
print( str.lstrip('8') );


str1 = "     this is string example....wow!!!     "
print (str1.rstrip())
str2 = "*****this is string example....wow!!!*****"
print (str2.rstrip('*'))

str3 = "     this is string example....wow!!!     ";
print( str3.lstrip() );
str3 = "88888888this is string example....wow!!!8888888";
print( str3.lstrip('8') );
```

执行后会输出：

```
this is string example....wow!!!
this is string example....wow!!!8888888
     this is string example....wow!!!
*****this is string example....wow!!!
this is string example....wow!!!
this is string example....wow!!!8888888
```

需要注意的是，上述删除字符的操作函数不会对位于字符串中间的任何文本起作用。例如，下面的演示代码：

```
>>> s = ' hello world \n'
>>> s = s.strip()
>>> s
'hello world'
>>>
```

实例 03-10：字符过滤和清理

在 Python 程序中，有时候想以某种方过滤、清理文本中的某类字符，如用户注册表单中的非法字符。如果想删除整个范围内的字符，则可以使用内置函数 translate()实现。函数 translate()的功能是根据参数 table 给出的表（包含 256 个字符）转换字符串的字符，将要被过滤的字符放到参数 deletechars 中。返回值是翻译后的字符串，如果函数给出了 delete 参数，则删除原来在 bytes 中的属于 delete 的字符，剩下的字符按照 table 给出的映射来进行映射。下面的实例文件 buchangjian.py 演示了使用函数 translate()实现字符过滤的过程。

源码路径：daima\3\03-10\buchangjian.py

```python
intab = "aeiou"
outtab = "12345"
trantab = str.maketrans(intab, outtab)    # 制作翻译表

str = "this is string example....wow!!!"
print(str.translate(trantab))

#制作翻译表
bytes_tabtrans = bytes.maketrans(b'abcdefghijklmnopqrstuvwxyz', b'ABCDEFGHIJKLMNOPQRSTUVWXYZ')

#转换为大写，并删除字母o
print(b'toppr'.translate(bytes_tabtrans, b'o'))
```

执行后会输出：

```
th3s 3s str3ng 2x1mpl2....w4w!!!
b'TPPR'
```

再看下面的实例文件 gaoji.py，其功能是使用函数 translate()删除空格和 Unicode 组合字符。

源码路径：daima\3\03-10\gaoji.py

```python
①s = 'p\xfdt\u0125\xf6\xf1\x0cis\tppppp\r\n'
print(s)

#删除空格
②chuli = {
    ord('\t') : ' ',
    ord('\f') : ' ',
    ord('\r') : None      # 删除
}

a = s.translate(chuli)
③print('whitespace chuliped:', a)

#删除所有的Unicode组合字符标记
④import unicodedata
import sys
cmb_chrs = dict.fromkeys(c for c in range(sys.maxunicode)
                           if unicodedata.combining(chr(c)))

b = unicodedata.normalize('NFD', a)
c = b.translate(cmb_chrs)
⑤print('accents removed:', c)

#使用I/O 解码和编码函数
⑥d = b.encode('ascii','ignore').decode('ascii')
⑦print('accents removed via I/O:', d)
```

在①中输出混乱字符串 s。

在②～③中首先建立一个小型的转换表 chuli，然后使用函数 translate()删除空格。类似 "\t" 和 "\f" 之类的空格字符已经被重新映射成一个单独的空格，回车符 "\r" 已经被完全删除掉。

在④～⑤中删除所有的 Unicode 组合字符，首先使用函数 dict.fromkeys()构建一个将每个 Unicode 组合字符都映射为 None 的字典。原始的输入信息会先通过 unicodedata.normalize()函数被转换为分离的形式，然后通过函数 translate()删除所有的重复符号。同样道理，也可以使用相似的方法来删除其他类型的字符，如控制字符。

在⑥~⑦中首先初步清理文本，然后利用函数 encode()和 decode()修改或清理文本。函数 normalize()先对原始的文本进行操作，后续的 ASCII 编码/解码只是简单地一次性丢弃所有不需要的字符。

执行效果如图 3-1 所示。

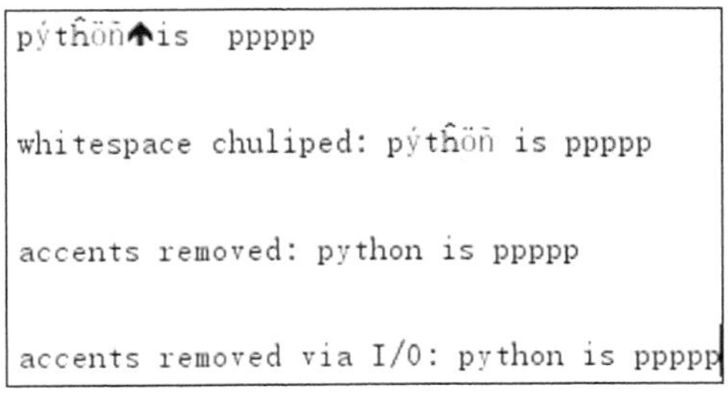

```
pýthōn is  ppppp

whitespace chuliped: pýthōn is ppppp

accents removed: python is ppppp

accents removed via I/O: python is ppppp
```

图 3-1　执行效果

实例 03-11：字符串对齐处理

在 Python 程序中，可以使用如下 3 个内置函数实现字符串对齐处理，这 3 个函数被保存在内置模块 string 中。

（1）函数 ljust()：返回一个左对齐的字符串，并使用空格符填充至指定长度的新字符串。如果指定的长度小于原字符串的长度，则返回原字符串。

（2）函数 rjust()：返回一个右对齐的字符串，并使用空格符填充至长度 width 的新字符串。如果指定的长度小于字符串的长度，则返回原字符串。

（3）函数 center()：返回一个指定宽度 width、居中的字符串，fillchar 为填充的字符，默认认为空格。

下面的实例文件 duiqi.py 演示了使用上述 3 个函数实现字符串对齐处理的过程。

源码路径：daima\3\03-11\duiqi.py

```python
str = "Toppr example....wow!!!"
print (str.ljust(50, '*'))

str1 = "this is string example....wow!!!"
print (str1.rjust(50, '*'))

str3 = "[www.toppr.net]"
print ("str3.center(40, '*') : ", str.center(40, '*'))
```

执行后会输出：

```
Toppr example....wow!!!***************************
******************this is string example....wow!!!
str3.center(40, '*') :  ********Toppr example....wow!!!*********
```

实例 03-12：字符串连接

在 Python 程序中，通常需要将许多个小的字符串合并成一个大的字符串。其中最为简单的方法是使用内置模块 string 中的 join()函数，此函数的功能是将序列中的元素以指定的字符连接生成一个新的字符串。下面的实例文件 lianjie.py 演示了使用函数 join()和其他方式实现字符串连接功能的过程。

源码路径：daima\3\03-12\lianjie.py

```python
① s1 = "-"
s2 = ""
seq = ("t", "o", "p", "p", "r") # 字符串序列
print (s1.join( seq ))
print (s2.join( seq ))
```

```
parts = ['Is', 'Toppr', 'Not', 'Topr?']
print(' '.join(parts))
print(','.join(parts))
②print(''.join(parts))

③a = 'Is Toppr'
  b = 'Not Topr?'
④print(a + ' ' + b)

⑤print('{} {}'.format(a,b))
⑥print(a + ' ' + b)

⑦a = 'Is Toppr" "Not Topr?'
⑧print(a)
```

- 在①~②中使用函数join()实现字符串连接,乍看上去其语法可能显得有些怪异,但是join()函数操作其实是字符串对象的一个方法。因为想要合并在一起的对象可能来自各种不同的数据序列,如可能是列表、元组、字典、文件、集合或生成器,如果每次单独在每一种序列对象中实现一个join()函数,就会显得十分多余。所以,比较好的做法是先指定想要的分隔字符串,然后在字符串对象上使用join()函数将文本片段连接在一起。
- 在③~④中使用加号操作符"+"实现字符串连接。
- 在⑤~⑥中为了实现更加复杂的字符串格式化操作,加号操作符"+"可以作为format()函数的替代者。
- 在⑦~⑧中如果想在源代码中将字符串字面值合并在一起,则可以简单地将它们排列在一起,在中间无须使用加号操作符"+"。

执行后会输出:

```
t-o-p-p-r
toppr
Is Toppr Not Topr?
Is,Toppr,Not,Topr?
IsTopprNotTopr?
Is Toppr Not Topr?
Is Toppr Not Topr?
Is Toppr Not Topr?
Is Toppr Not Topr?
```

需要注意,使用加号"+"连接符做大量的字符串连接是非常低效的,因为这样做会产生内存复制和垃圾收集操作。例如,可能会写出如下字符串连接代码。

```
s = ''
for p in parts:
    s += p
```

上述代码的做法比使用join()函数要慢很多,这是因为每个"+="操作都会创建一个新的字符串对象。最好的做法是先收集所有要连接的部分,然后一次将它们连接起来。

下面的实例文件gaojilian.py演示了使用高效方法实现字符串连接的过程。

源码路径:daima\3\03-12\gaojilian.py.py

```
data = ['Java', 70000, 69.9]
①print(','.join(str(d) for d in data))

a = 'Is Toppr'
b = 'Not Topr?'
c = a + ' ' + b
②print(a + ':' + b + ':' + c) #低效
③print(':'.join([a, b, c])) #低效
④print(a, b, c, sep=':') #高效
```

在①中使用生成器表达式将数据转换为字符串,同时完成连接操作功能。

在②③④中演示了3种连接方案,其中④的效率最佳,很多程序开发者会忘乎所以地使用字符串连接操作,就像②中的代码那样,这是最不值得推荐的。

执行后会输出：
```
Java,70000,69.9
Is Toppr:Not Topr?:Is Toppr Not Topr?
Is Toppr:Not Topr?:Is Toppr Not Topr?
Is Toppr:Not Topr?:Is Toppr Not Topr?
```

在具有字符串连接功能的 I/O 操作程序中，开发者需要对整个程序进行仔细的分析。例如，有如下两段代码。

```
#第1段：字符串连接
f.write(chunk1 + chunk2)
#第2段：单独的I/O操作
f.write(chunk1)
f.write(chunk2)
```

在现实应用中，如果需要编写从多个短字符串中构建输出的程序，建议通过编写生成器函数的方法实现，需要通过 yield 关键字生成字符串片段。下面的实例文件 duogaoji.py 演示了联合使用生成器函数和 I/O 操作实现连接的过程。

源码路径：daima\3\03-12\duogaoji.py

```
①def sample():
②    yield "Is"
      yield "Toppr"
      yield "Not"
      yield "Topr?"

   #简单的字符串连接
③text = ''.join(sample())
   print(text)

   #重定向I/O
④import sys
   for part in sample():
       sys.stdout.write(part)
⑤sys.stdout.write('\n')

   #将字符串片段组合成缓冲区和较大的I/O操作
⑥def combine(source, maxsize):
       parts = []
       size = 0
       for part in source:
           parts.append(part)
           size += len(part)
           if size > maxsize:
               yield ''.join(parts)
               parts = []
               size = 0
       yield ''.join(parts)

   for part in combine(sample(), 70000):
       sys.stdout.write(part)
⑦print(sys.stdout.write('\n'))
```

在①~②中编写生成器函数 sample()，使用 yield 关键字生成字符串片段。

在③中使用函数 join()将字符串简单地连接起来。

在④~⑤中将字符串片段重定向到 I/O。

在⑥~⑦中定义生成器函数 combine()，以混合的方式将 I/O 操作智能化地结合在一起。此生成器函数并不需要知道精确的细节，而只是产生片段而已。

执行效果如图 3-2 所示。

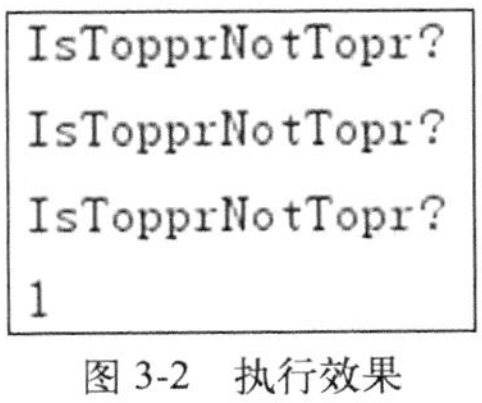

```
IsTopprNotTopr?
IsTopprNotTopr?
IsTopprNotTopr?
1
```

图 3-2 执行效果

实例 03-13：重新格式化字符串

在 Python 程序中，经常需要格式化处理一个字符串，例如，将一个很长的字符串重新格式化，使它们能够按照用户指定的列数显示出来。使用 Python 中的 textwrap 模块，实现重新格式化文本的输出功能。

下面的实例文件 kuan.py 演示了使用函数 wrap() 限制字符串宽度的过程。

源码路径：daima\3\03-13\kuan.py

```
import textwrap
sample_text = '''aaabbbcccdddeeeedddddfffffgggggghhhhhhkkkkkkk'''
sample_text2 = '''aaa bbb ccc ddd eeee ddddd fffff ggggg hhhhhh kkkkkkk'''

print (sample_text)
print (textwrap.wrap(sample_text,width=5))
print (textwrap.wrap(sample_text2,width=5))
```

执行后会输出下面的结果，可以看到在第 3 个输出结果中，并不会保证每个列表元素都是按照 width 切割的，因为不仅要考虑 width，也要考虑空格。

```
aaabbbcccdddeeeedddddfffffgggggghhhhhhkkkkkkk
['aaabb', 'bcccd', 'ddeee', 'edddd', 'dfff', 'fgggg', 'ghhhh', 'hhkkk', 'kkkk']
['aaa', 'bbb', 'ccc', 'ddd', 'eeee', 'ddddd', 'fffff', 'ggggg', 'hhhhh', 'h kkk', 'kkkk']
```

实例 03-14：在字符串中处理 HTML 和 XML 标记

有时字符串中会穿插着 HTML 或 XML 标记，此时需要将这些标记替换为它们相对应的文本，或者对特定的字符（如<、>或&）进行转义处理。如果想在 Python 程序中将 HTML 和 XML 标记生成为文本，可以使用函数 html 模块的 escape() 来替换<or>这样的特殊字符。下面的实例文件 ht1.py 演示了使用函数 escape() 处理 HTML 标记的过程。

源码路径：daima\3\03-14\ht1.py

```
s = '我爱Python"<tag>text</tag>".'
import html
print(s)
print(html.escape(s))
print(html.escape(s, quote=False))
```

上述代码将 quote 设置为 False 后则不会转义单引号和双引号。

执行后会输出：

```
我爱Python"<tag>text</tag>".
我爱Python&quot;&lt;tag&gt;text&lt;/tag&gt;&quot;.
我爱Python"&lt;tag&gt;text&lt;/tag&gt;".
```

在 Python 程序中，如果想在生成 ASCII 文本的同时，将非 ASCII 字符对应的字符编码实体嵌入到文本中，则可以在各种跟 I/O 相关的函数中使用参数 errors='xmlcharrefreplace' 来实现。下面的实例文件 ht2.py 演示了将字符串转换为 ASCII 文本的过程。

源码路径：daima\3\03-14\ht2.py

```
import html
s = 'I Love Python~o'
print(s.encode('ascii', errors='xmlcharrefreplace'))
```

执行后会输出：

```
b'I Love Python~o'
```

实例 03-15：在字节串中实现基本文本处理

在 Python 程序中，有时需要在字节串（byte string）中执行基本的文本处理操作，如拆分、搜索和替换操作。下面的实例文件 zijie1.py 演示了在字节串中实现基本文本处理的过程。

源码路径：daima\3\03-15\zijie1.py

```
①data = b'Hello World'
  print(data[0:5])
  print(data.startswith(b'Hello'))
  print(data.split())
②print(data.replace(b'Hello', b'Hello Cruel'))

③data = bytearray(b'Hello World')
  print(data[0:5])
```

```
    print(data.startswith(b'Hello'))
    print(data.split())
④print(data.replace(b'Hello', b'Hello Cruel'))
```

在①~②中演示了字节串和大多数文本字符串一样的内置操作用法，在③~④中演示了在字节数组中实现文本处理的过程。

执行后会输出：

```
b'Hello'
True
[b'Hello', b'World']
b'Hello Cruel World'
bytearray(b'Hello')
True
[bytearray(b'Hello'), bytearray(b'World')]
bytearray(b'Hello Cruel World')
```

在 Python 程序中，也可以在字节串中实现和正则表达式相关的模式匹配操作，此时模式本身需要以字节串的形式指定。在大多数情况下，几乎所有能在字符串中执行的操作也可以在字节串中进行。下面的实例文件 zijie2.py 演示了在字节串中使用正则表达式实现文本处理的过程。

源码路径：daima\3\03-15\zijie2.py

```
data = b'FOO:BAR,SPAM'
import re
print(re.split(b'[:,]',data))
print(re.split('[:,]',data))
```

执行效果如图 3-3 所示。

```
Traceback (most recent call last):
[b'FOO', b'BAR', b'SPAM']
  File "H:/daima/4/4-1/zijie2.py", line 4, in <module>
    print(re.split('[:,]',data))
  File "C:\Program Files\Anaconda3\lib\re.py", line 212, in split
    return _compile(pattern, flags).split(string, maxsplit)
TypeError: cannot use a string pattern on a bytes-like object
```

图 3-3　执行效果

3.2　文件和 I/O 处理函数

在计算机信息系统中，根据信息的存储时间，信息可以分为临时性信息和永久性信息。简单来说，临时信息存储在计算机系统临时存储设备中（如存储在计算机内存中），这类信息随系统断电而丢失。永久性信息存储在计算机系统的永久性存储设备中（如存储在磁盘中）。永久性信息的最小存储单元为文件，因此文件管理是计算机系统中的一个重要的功能。

实例 03-16：刷新缓冲区

在 Python 程序中，函数 flush() 的功能是刷新缓冲区，即将缓冲区中的数据立刻写入文件，同时清空缓冲区。在一般情况下，缓冲区在文件关闭后会自动刷新，但是有时需要在文件关闭之前刷新，这时就可以使用函数 flush() 实现。下面的实例文件 shua.py 演示了使用 flush() 函数刷新缓冲区的过程。

源码路径：daima\3\03-16\shua.py

```
#以wb的方式打开指定文件
fo = open("456.txt", "wb")
print ("文件名为: ", fo.name)        #显示打开文件的文件名
fo.flush()                          #刷新缓冲区
fo.close()                          #关闭文件
```

在上述代码中，首先使用函数 open() 以 wb 的方式打开了文件 456.txt，然后使用函数 flush()

刷新缓冲区，最后使用函数 close()关闭文件操作。

执行后会输出：
```
文件名为： 456.txt
```

实例 03-17：检测文件是否连接到一个终端设备

在 Python 程序中，函数 isatty()的功能是检测某文件是否连接到一个终端设备，如果是则返回 True，否则返回 False。下面的实例文件 lian.py 演示了使用函数 isatty()检测文件是否连接到一个终端设备的过程。

源码路径：daima\3\03-17\lian.py

```
#以wb的方式打开指定文件
fo = open("456.txt", "wb")
print ("文件名是：", fo.name)       #显示打开文件的文件名
ret = fo.isatty()                    #检测文件是否连接到一个终端设备
print ("返回值是：", ret)           #显示连接检测结果
fo.close()                           #关闭文件
```

在上述代码中，首先使用函数 open()以 wb 的方式打开了文件 456.txt，然后使用函数 isatty()检测这个文件是否连接到一个终端设备，最后使用函数 close()关闭文件操作。

执行后会输出：
```
文件名为是： 456.txt
返回值是： False
```

实例 03-18：返回文件各行内容

在 Python 程序中，File 对象不再支持函数 next()。在 Python 3 程序中，内置函数 next()通过迭代器调用方法 __next__()返回下一项。在循环中，函数 next()会在每次循环中调用，该方法返回文件的下一行。如果到达结尾（EOF），则触发 StopIteration 异常。下面的实例文件 next.py 演示了使用函数 next()返回文件各行内容的过程。

源码路径：daima\3\03-18\next.py

```
#以r的方式打开指定文件
fo = open("456.txt", "r")
print ("文件名为：", fo.name)        #显示打开文件的文件名
for index in range(5):                #遍历文件的内容
    line = next(fo)                   #返回文件各行内容
    print ("第 %d 行 - %s" % (index, line))   #显示5行文件内容
fo.close()                            #关闭文件
```

在上述代码中，首先使用函数 open()以 r 的方式打开了文件 456.txt，然后使用函数 next()返回文件各行内容，最后使用函数 close()关闭文件操作。文件 456.txt 的内容如图 3-4 所示。实例文件 next.py 的执行效果如图 3-5 所示。

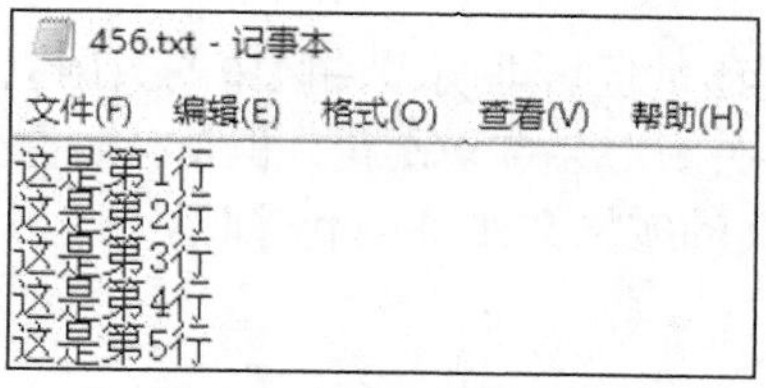

图 3-4 文件 456.txt 的内容

图 3-5 执行效果

> **注意**：本书光盘中的文件 456.txt 的编码格式是 ANSI 编码格式，读者需要确保文件 456.txt 编码格式和当前操作系统编码格式的一致性，否则会出现执行错误。

实例 03-19：返回文件 3 个字节的内容

在 Python 程序中，要想使用某个文本文件中的数据信息，首先需要将这个文件的内容读取到内存中，既可以一次性读取文件的全部内容，也可以按照每次一行的方式进行读取。其中，函数 read()的功能是从目标文件中读取指定的字节数，如果没有给定字节数或字节数为负数，则读取所有内容。下面的实例文件 du.py 演示了使用函数 read()返回文件 3 个字节的内容的过程。

源码路径：daima\3\03-19\du.py

```
#以r+的方式打开指定文件
fo = open("456.txt", "r+")
print ("文件名为: ", fo.name)           #显示打开文件的文件名
line = fo.read(3)                       #读取文件中前3个字节的内容
print ("读取的字符串: %s" % (line))     #显示读取的内容
fo.close()                              #关闭文件
```

在上述代码中，首先使用函数 open()以 r+的方式打开了文件 456.txt，然后使用函数 read()读取了目标文件中前 3 个字节的内容，最后使用函数 close()关闭文件操作。实例文件 du.py 的执行效果如图 3-6 所示。

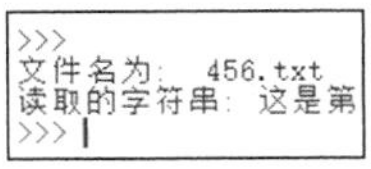

图 3-6 执行效果

在 Python 程序中，使用函数 open()中的 rb 或 wb 方式可以实现对二进制数据的读写操作。下面的实例文件 duer.py 演示了对二进制数据的读写操作的过程。

源码路径：daima\3\03-19\duer.py

```
#以一个单字节字符串来读取整个文件
with open('999.txt', 'rb') as f:
    data = f.read()
#将二进制数据写入文件
with open('999.txt', 'wb') as f:
    f.write(b'Hello World')

with open('999.txt', 'rb') as f:
    data = f.read(16)
①   text = data.decode('utf-8')
with open('999.txt', 'wb') as f:
    text = 'Hello World1'
②   f.write(text.encode('utf-8'))
```

在读取二进制文件时，所有的数据将以字节串的形式返回，而不是文本字符串。同理，当向文件中写入二进制数据时，数据必须以对象的形式来提供，而且该对象可以将数据以字节的形式（如字节串、bytearray 对象等）暴露出来。需要注意的是，在二进制文件中读取或写入文本内容之前，需要确定使用的编码或解码格式，如上述代码中①和②那样。

实例 03-20：返回文件中所有行

在 Python 程序中，函数 readlines()的功能是读取所有行（直到结束符 EOF）并返回列表。如果指定其参数 sizehint 大于 0，则返回大约为 sizehint 个字节的行。实际上读取的值可能比 sizehint 大，因为需要填充缓冲区，如果碰到结束符 EOF 则返回空字符串。下面的实例文件 suo.py 演示了使用函数 readlines()返回文件中所有行的过程。

源码路径：daima\3\03-20\suo.py

```
#以r的方式打开指定文件
fo = open("456.txt", "r")
print ("文件名为: ", fo.name)           #显示打开文件的文件名
line = fo.readlines()                   #读取文件中所有的行
print ("读取的数据为: %s" % (line))     #显示读取的内容
line = fo.readlines(2)
print ("读取的数据为: %s" % (line))
fo.close()                              #关闭文件
```

在上述代码中，首先使用函数 open()以 r 的方式打开了文件 456.txt，然后调用函数 readlines()读取文件中所有的行。实例文件 suo.py 的执行效果如图 3-7 所示。

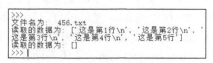

图 3-7　执行效果

实例 03-21：重复读取文件中的第 1 行内容

在 Python 程序中，函数 seek()没有返回值，功能是移动文件读取指针到指定位置。下面的实例文件 zhi.py 演示了使用函数 seek()重复读取文件中第 1 行内容的过程。

源码路径：daima\3\03-21\zhi.py

```
#以r的方式打开指定文件
fo = open("456.txt", "r")
print ("文件名为: ", fo.name)            #显示打开文件的文件名
line = fo.readline()                     #读取文件中的第1行内容
print ("读取的数据为: %s" % (line))      #显示读取的数据
fo.seek(0, 0)                            #重新设置文件读取指针到开头
line = fo.readline()                     #读取文件中的第1行内容
print ("读取的数据为: %s" % (line))      #显示读取的数据
fo.close()                               #关闭文件
```

在上述实例代码中，首先使用函数 open()以 r 的方式打开了文件 456.txt，然后调用函数 readline()读取文件中的第 1 行内容，最后调用函数 seek()设置读取文件指针，重复回到读取文件中第 1 行内容的模式。实例文件 zhi.py 的执行效果如图 3-8 所示。

图 3-8　执行效果

实例 03-22：获取当前文件位置

在 Python 程序中，函数 tell()没有参数，功能是获取文件的当前位置，即文件指针当前位置。下面的实例文件 weizhi.py 演示了使用函数 tell()获取当前文件位置的过程。

源码路径：daima\3\03-22\weizhi.py

```
#以r的方式打开指定文件
fo = open("456.txt", "r")
print ("文件名为: ", fo.name)            #显示打开文件的文件名
line = fo.readline()                     #读取文件中第1行的内容
print ("读取的数据为: %s" % (line))      #显示读取的数据
pos = fo.tell()                          #获取当前文件位置
print ("当前位置: %d" % (pos))           #显示当前文件位置
fo.close()                               #关闭文件
```

在上述实例代码中，首先使用函数 open()以 r 的方式打开了文件 456.txt，然后调用函数 readline()读取文件中第 1 行的内容，最后调用函数 tell()获取文件指针当前位置。实例文件 weizhi.py 的执行效果如图 3-9 所示。在此需要注意，不同执行环境下的效果会有差别。

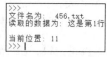

图 3-9　执行效果

实例 03-23：截取文件中前 3 个字符

在 Python 程序中，函数 truncate()的功能是截断文件。如果函数指定了可选参数 size，则截断文件中的 size 个字符；如果没有指定参数 size，则从当前位置起截断文件。在截断文件之后，

size 个字符后面的所有字符将被删除。下面的实例文件 jie.p 演示了使用函数 truncate()截取文件中前 3 个字符的过程。

源码路径：daima\3\03-23\jie.py

```python
#以r+的方式打开指定文件
fo = open("456.txt", "r+")
print ("文件名为: ", fo.name)
# 截取3个字符
fo.truncate(3)
str = fo.read()
print ("读取数据: %s" % (str))
# 关闭文件
fo.close()
```

上述实例代码首先使用函数 open()以 r+的方式打开了文件 456.txt，然后调用函数 truncate()截取文件中的 3 个字符，最后调用函数 read()读取截取后文件的内容。执行效果如图 3-10 所示。

(a) 文件 456.txt 原来的内容　　　(b) 执行效果　　　(c) 截取后文件 456.txt 的内容

图 3-10　执行效果

实例 03-24：向文件中写入多行字符串

在 Python 程序中，函数 writelines()的功能是向文件中写入一序列（多行）的字符串。这一序列字符串可以是由迭代对象产生的，如一个字符串列表，在换行的时候需要指定换行符 "\n"。下面的实例文件 duohang.py 演示了使用函数 writelines()向文件中写入多行字符串的过程。

源码路径：daima\3\03-24\duohang.py

```python
#以w的方式打开指定文件
fo = open("456.txt", "w")
print ("文件名为: ", fo.name)           #显示打开文件的文件名
seq = ["浪潮软件\n", "浪潮信息"]        #设置变量seq的初始文本
fo.writelines( seq )                   #向文件中写入变量seq的文本
fo.close()                             #关闭文件
```

在上述实例代码中，首先使用函数 open()以 w 的方式打开了文件 456.txt，然后调用函数 writelines()向文件中写入了两行文本，最后调用函数 close()关闭文件。执行效果如图 3-11 所示。

(a) 文件 456.txt 原来的内容　　　(b) 执行效果　　　(c) 写入后文件 456.txt 的内容

图 3-11　执行效果

实例 03-25：获取文件操作权限

在 Python 程序中，函数 access()的功能是获取当前文件的操作权限。函数 access()使用当前的 UID/GID 尝试访问指定路径。在下面的实例文件 quan.py 中，演示了使用函数 access()获取文件操作权限的过程。

第 3 章　标准库函数实战

源码路径：daima\3\03-25\quan.py

```
import os, sys
# 假定 123/456.txt 文件存在，并设置有读写权限
ret = os.access("123/456.txt", os.F_OK)
print ("F_OK - 返回值  %s"% ret)          #显示文件是否存在
ret = os.access("123/456.txt", os.R_OK)
print ("R_OK - 返回值  %s"% ret)          #显示文件是否可读
ret = os.access("123/456.txt", os.W_OK)
print ("W_OK - 返回值  %s"% ret)          #显示文件是否可写
ret = os.access("123/456.txt", os.X_OK)
print ("X_OK - 返回值  %s"% ret)          #显示文件是否可执行
```

在执行上述实例代码之前，需要先在实例文件 quan.py 的同目录下创建一个名为"123"的文件夹，然后在里面创建一个文本文件 456.txt。上述代码使用函数 access()获取了文件 123/456.txt 的操作权限。

执行后会输出：

```
F_OK - 返回值  True
R_OK - 返回值  True
W_OK - 返回值  True
X_OK - 返回值  True
```

实例 03-26：修改当前工作目录到指定路径

在 Python 程序中，函数 chdir()的功能是修改当前工作目录到指定路径。下面的实例文件 gai.py 演示了使用函数 chdir()修改当前工作目录到指定路径的过程。

源码路径：daima\3\03-26\gai.py

```
import os, sys
path = "123"                              #设置目录变量的初始值
retval = os.getcwd()                      #获取当前文件的工作目录
print ("当前工作目录为 %s" % retval)       #显示当前文件的工作目录
# 修改当前工作目录
os.chdir( path )
# 查看修改后的工作目录
retval = os.getcwd()                      #再次获取当前文件的工作目录
print ("目录修改成功 %s" % retval)
```

在上述实例代码中，首先使用函数 getcwd()获取了当前文件的工作目录，然后使用函数 chdir()修改当前工作目录到指定路径"123"。

执行后会输出：

```
当前工作目录为  H:\daima\4\4-2
目录修改成功  H:\daima\4\4-2\123
```

实例 03-27：修改文件或目录权限

在 Python 程序中，函数 chmod()的功能是修改文件或目录的权限。下面的实例文件 xiu.py 演示了使用函数 chmod()修改文件或目录权限的过程。

源码路径：daima\3\03-27\xiu.py

```
import os, sys, stat
#假设123/456.txt 文件存在，设置文件可以通过用户组执行
os.chmod("123/456.txt", stat.S_IXGRP)
#设置文件可以被其他用户写入
os.chmod("123/456.txt", stat.S_IWOTH)
print ("修改成功!!")
```

上述实例代码使用函数 chmod()将文件 123/456.txt 的权限修改为 stat.S_IWOTH。执行后会输出"修改成功!!"的提示。

实例 03-28：遍历显示某个目录中所有文件夹和文件列表

在 Python 程序中，函数 walk()的功能是遍历显示目录下所有的文件夹和文件列表，可以向上遍历或者向下遍历。函数 walk()执行后将会得到一个三元组(dirpath, dirnames, filenames)，其中第 1 个参数表示起始路径，第 2 个参数表示起始路径下的文件夹，第 3 个参数表示起始路径

下的文件。下面的实例文件 bian.py 演示了使用函数 walk()遍历显示某个目录中所有文件夹和文件列表的过程。

源码路径：daima\3\03-28\bian.py

```
import os                                          #导入os模块
for root, dirs, files in os.walk(".", topdown=False):   #遍历目录中所有文件夹和文件列表
    for name in files:                             #遍历文件
        print(os.path.join(root, name))
    for name in dirs:                              #遍历目录
        print(os.path.join(root, name))
```

上述实例代码使用函数 walk()遍历显示当前所在目录中所有文件夹和文件列表。执行效果如图 3-12 所示。

图 3-12 执行效果

实例 03-29：修改一个目录名字

在 Python 程序中，函数 rename()的功能是修改目录或文件的名字。如果修改的目录不存在，则会抛出 OSError 异常。下面的实例文件 name.py 演示了使用函数 rename()修改一个目录名字的过程。

源码路径：daima\3\03-29\name.py

```
import os, sys
#列表显示当前目录中所有目录名
print ("目录为: %s"%os.listdir(os.getcwd()))
#将当前目录中的目录"123"重命名为"test2"
os.rename("123","test2")
print ("重命名成功。")
print ("目录为: %s" %os.listdir(os.getcwd()))#列出重命名后的所有目录
```

上述实例代码使用函数 rename()将目录"123"的名字修改为"test2"。执行效果如图 3-13 所示。

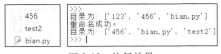

图 3-13 执行效果

实例 03-30：读取两个文本文件内容

在 Python 程序中，fileinput 模块可以对一个或多个文件中的内容实现迭代和遍历等操作，可以对文件进行循环遍历、格式化输出及查找、替换等操作，非常方便。fileinput 模块实现了对文件中行的"懒惰"迭代，读取时不需要把文件内容放入内存，从而提高程序的执行效率。下面的实例文件 lia.py 演示了使用 fileinput 模块读取两个文本文件内容的过程。

源码路径：daima\3\03-30\lia.py

```
import fileinput                    #导入fileinput模块
def demo_fileinput():                #定义函数，用于迭代处理两个文件
    with fileinput.input(['123.txt','456.txt']) as lines:
        for line in lines:           #遍历文件中的各行内容
            print("总第%d行," % fileinput.lineno(), "文件%s中第%d行： " %
                  (fileinput.filename(),fileinput.filelineno()))
            print(line.strip())
if __name__ == '__main__':
    demo_fileinput()                 #输出显示各行的内容
```

第 3 章 标准库函数实战

上述实例代码首先使用 import 语句导入 fileinput 模块,然后使用函数 fileinput.input()来迭代处理两个文件(123.txt 和 456.txt),并将其以列表形式提供给 input()方法作为参数。最后迭代处理每行内容,同时输出每行的行号。程序执行后会显示所有文件中的行号及每一行的内容,执行效果如图 3-14 所示。

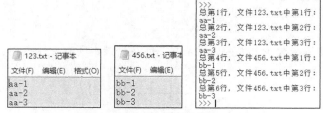

图 3-14 执行效果

实例 03-31:字符串 I/O 操作

在 Python 程序中,如果想将一段文本或二进制字符串写入文件对象中,则可以通过使用类 io.StringIO 和 io.BytesIO 中的函数创建类似于文件对象的方法来实现,这些方法可以操作字符串数据。下面的实例文件 zifuio.py 演示了实现字符串 I/O 操作的过程。

源码路径:daima\3\03-31\zifuio.py

```
import io
s = io.StringIO()
s.write('Hello World\n')
print('This is a test', file=s)
#把所有的数据写入
print(s.getvalue())
#在现有字符串周围包装一个文件接口
s = io.StringIO('Hello\nWorld\n')
print(s.read(4))
print(s.read())

①s = io.BytesIO()
  s.write(b'binary data')
②print(s.getvalue())
```

因为 io.StringIO 类只能用于对文本的处理。为了操作处理二进制数据,在①~②中使用的是 io.BytesIO。

执行后会输出:

```
Hello World
This is a test

Hell
o
World

b'binary data'
```

实例 03-32:读写压缩文件

在现实应用中,有时需要读写 GZIP 或 BZ2 压缩文件中的数据。在 Python 程序中,通过 gzip 和 bz2 这两个模块可以轻松读写压缩文件中的数据信息。在 Python 程序中,模块 gzip 提供了可以读写 GZIP 压缩文件的操作功能,它会自动地压缩写入的数据和对读取出来的数据进行解压缩,使其使用起来就像普通的文件对象一样。

下面的实例文件 xieya.py 演示了使用模块 gzip 读写压缩文件的过程。

源码路径:daima\3\03-32\xieya.py

```
import gzip
test = b'http://www.toppr.net'
```

```
#写入压缩文件
with gzip.open('file.txt.zip', 'wb') as f:
    f.write(test)
#读取压缩文件
with gzip.open('file.txt.zip', 'rb') as f:
    file = f.read()
    print(file)
```

下面的实例文件 yiyouya.py 演示了对已经存在的文件进行压缩的过程。

源码路径：daima\3\03-32\yiyouya.py

```
import gzip
#对已经存在的文件进行压缩
with open('lia.py', 'rb') as fRead:
    with gzip.open('file.txt.zip', 'wb') as f:
        f.writelines(fRead)
#读取压缩的文件
with gzip.open('file.txt.zip', 'rb') as f:
    file = f.read()
    print(file)
```

上述代码对已经存在的文件 lia.py 进行了压缩，并通过函数 open()读取了压缩文件夹 file.txt.zip 的内容。

模块 bz2 提供了和模块 gzip 相同的内置函数，只是模块 bz2 能够处理的压缩文件的格式是 BZ2 而已。下面的实例文件 tongduxie.py 演示了联合使用模块 gzip 和 bz2 读写压缩文件的过程。

源码路径：daima\3\03-32\tongduxie.py

```
import gzip
with gzip.open('somefile.gz', 'rt') as f:
    text = f.read()

with gzip.open('somefile.gz', 'wt') as f:
    f.write(text)
import bz2
with bz2.open('somefile.bz2', 'rt') as f:
    text = f.read()

with bz2.open('somefile.bz2', 'wt') as f:
    f.write(text)
```

在上述代码中，所有的 I/O 操作都会采用文本的形式并执行 Unicode 编码或解码操作。如果想处理二进制数据，则需要使用 rb 或 wb 方式实现。

实例 03-33：对二进制文件做内存映射

在 Python 程序中，可以使用模块 mmap 实现对文本文件的内存映射操作。通过内存映射的方式，可将一个二进制文件加载到可变的字节数组中，这样可以随机访问或随时修改其内容。mmap 是一种虚拟内存映射文件的方法，能够将一个文件或其它对象映射到进程的地址空间中，实现文件磁盘地址和进程虚拟地址空间中虚拟地址的一一映射关系。下面的实例文件 neicun.py 演示了使用模块 mmap 对二进制文件做内存映射的过程。

源码路径：daima\3\03-33\neicun.py

```
  import os
  import mmap
① def memory_map(filename, access=mmap.ACCESS_WRITE):
      size = os.path.getsize(filename)
      fd = os.open(filename, os.O_RDWR)
② return mmap.mmap(fd, size, access=access)

③ size = 1000000
  with open('data', 'wb') as f:
      f.seek(size-1)
④ f.write(b'\x00')
```

```
⑤m = memory_map('data')
  print(len(m))
  print(m[0:10])
  print(m[0])
  # 指派一个切片
  m[0:11] = b'Hello World'
  m.close()
  # 确认已做出更改
  with open('data', 'rb') as f:
⑥    print(f.read(11))

⑦with memory_map('data') as m:
      print(len(m))
      print(m[0:10])
⑧print(m.closed)
```

①～②中定义函数 memory_map()，功能是打开一个文件并对它进行内存映射操作。

为了使用函数 memory_map()，需要准备一个已经创建好的文件，并为之填充一些数据。为此在③～④中先创建了一个初始文件，然后将其扩展为我们需要的大小。

⑤～⑥中使用函数 memory_map()对文件内容进行内存映射操作。

⑦～⑧中将函数 mmap()返回的 mmap 对象当作上下文管理器来使用，在这种情况下，底层的文件会自动关闭。

执行后会输出：

```
1000000
b'\x00\x00\x00\x00\x00\x00\x00\x00\x00\x00'
0
b'Hello World'
1000000
b'Hello Worl'
True
```

实例 03-34：检测某个文件或目录是否存在

在现实应用中，模块 os.path 经常用于检测某个文件或目录是否存在，也可以用于检查某个文件的类型。下面的实例文件 cunzai.py 演示了使用 os.path 模块检测某个文件或目录是否存在的过程。

源码路径：daima\3\03-34\cunzai.py

```
import os
print(os.path.exists('/etc/aaa'))
print(os.path.exists('/tmp/bbb'))

print(os.path.isfile('/etc/passwd'))
#是一个目录
print(os.path.isdir('/etc/passwd'))
#是符号链接
print(os.path.islink('/usr/local/bin/python3'))
#获取链接到的文件
print(os.path.realpath('/usr/local/bin/python3'))
```

上述代码不但检测了某个文件或目录是否存在，而且进行了进一步的测试来验证这个文件的类型。

执行后会输出：

```
False
False
False
False
False
H:\usr\local\bin\python3
```

实例 03-35：获取某个目录中的文件列表信息

其实模块 os.path 的功能还不止如此，它还可以获取某个目录中的文件列表信息。下面的实例文件 liebiao.py 演示了使用 os.path 模块获取某个目录中文件列表信息的过程。

源码路径：daima\3\03-35\liebiao.py

```
import os
① path = "test2/"  # 打开文件
   dirs = os.listdir( path )
   # 输出所有文件和目录
   for file in dirs:
②     print (file)

   import os.path
   #获取所有的常规文件
③ names = [name for name in os.listdir('test2')
           if os.path.isfile(os.path.join('test2', name))]
④ print(names)
   #获取所有的目录
⑤ dirnames = [name for name in os.listdir('test2')
              if os.path.isdir(os.path.join('test2', name))]
⑥ print(dirnames)
⑦ import glob
   pyfiles = glob.glob('somedir/*.py')
   from fnmatch import fnmatch
   pyfiles = [name for name in os.listdir('test2') #获取所有的Python文件
              if fnmatch(name, '*.py')]
⑧ print(pyfiles)
```

在①～②中使用 os.listdir()函数获取目录 test2 中的文件列表信息。

在③～④中使用 os.path 模块获取目录 test2 中的常规文件信息。

在⑤～⑥中使用 os.path 模块获取目录 test2 中目录文件信息。

在⑦～⑧中使用模块 glob 和 fnmatch 获取目录 test2 中 Python 文件的信息。

执行后后输出：

```
456
789
bucunzai.py
chongding.py
['bucunzai.py', 'chongding.py']
['456', '789']
['bucunzai.py', 'chongding.py']
```

实例 03-36：获取目录的详细信息

在 Python 程序中，os.path 模块不但可以获取目录中的文件列表信息，而且可以获取更加详细的附加信息，如文件大小、修改日期等。使用 os.path 模块中的 os.stat()函数即可实现上述功能。下面的实例文件 xiangxi.py 演示了使用 os.path 模块获取目录的详细信息的过程。

源码路径：daima\3\03-36\xiangxi.py

```
import os
import os.path
import glob

pyfiles = glob.glob('*.py')

#获取文件大小和修改日期
name_sz_date = [(name, os.path.getsize(name), os.path.getmtime(name))
                for name in pyfiles]

for r in name_sz_date:
    print(r)

#获取文件的元数据
file_metadata = [(name, os.stat(name)) for name in pyfiles]
for name, meta in file_metadata:
    print(name, meta.st_size, meta.st_mtime)
```

上述代码执行后会输出目录中每一个文件的大小和修改日期。

实例 03-37：绕过文件名编码设置编码格式

在默认情况下，Python 程序中的所有文件名都会根据 sys.getfilesystemencoding()返回的文本编码格式进行编码和解码。但是如果基于某些原因想忽略这种编码，则可以使用原始字节串来指定文件名。下面的实例文件 raoguo.py 演示了绕过文件名编码设置编码格式的过程。

源码路径：daima\3\03-37\raoguo.py

```
import sys
print(sys.getfilesystemencoding())

#使用Unicode写入文件Name
with open('123\123.txt', 'w') as f:
    print(f.write('Python!'))
#目录列表（解码）
import os
print(os.listdir('.'))

#目录列表（原始文件）
print(os.listdir(b'.')) # string

#使用原始文件名打开文件
with open(b'123\678\123.txt') as f:
    print(f.read())
```

通过上述两个操作过程可以看到，当文件处理函数（如 open()和 os.listdir()）提供字节串参数时，其对文件名的处理会发生很小的改变。一般情况下，开发者无须担心文件名编码和解码的问题。但是有时一些操作系统可能会允许用户通过意外或恶意的方式创建出文件名不遵守期望的编码规则的文件，这样的文件名可能会导致处理大量文件的 Python 程序崩溃。

实例 03-38：创建并读取临时文件信息

应用程序经常要保存一些临时的信息，这些信息不是特别重要，没有必要写在配置文件中，但又不能没有，此时就可以把这些信息写到临时文件中。其实很多程序在执行时都会产生一大堆临时文件，有些用于保存日志，有些用于保存一些临时数据，还有一些用于保存无关紧要的设置。在 Windows 操作系统中，临时文件一般被保存在这个文件夹下：C:/Documents and Settings/User/Local Settings/Temp。其实我们最常用的 IE 在浏览网页时，会产生大量的临时文件，这些临时文件一般是我们浏览过的网页的本地副本。Python 提供了一个 tempfile 模块，用来对临时数据进行操作。下面的实例文件 linshi.py 演示了创建并读取临时文件信息的过程。

源码路径：daima\3\03-38\linshi.py

```
from tempfile import TemporaryFile
①with TemporaryFile('w+t') as f:
    #读写文件操作
    f.write('Hello World\n')
    f.write('Testing\n')
    #开始寻找并读取数据
    f.seek(0)
②data = f.read()
③f = TemporaryFile('w+t')           #使用临时文件
④f.close()                          #销毁临时文件
    import tempfile
⑤print(tempfile.mkstemp())
⑥print(tempfile.mkdtemp())
⑦print(tempfile.gettempdir())

from tempfile import NamedTemporaryFile
⑧with NamedTemporaryFile('w+t') as f:
⑨print('文件名是:', f.name)

⑩with NamedTemporaryFile('w+t', delete=False) as f:
⑪print('文件名是:', f.name)

from tempfile import TemporaryDirectory
⑫with TemporaryDirectory() as dirname:
⑬print('目录是:', dirname)
```

3.2 文件和 I/O 处理函数

在①~②中使用 TemporaryFile() 创建一个未命名的临时文件。

在③~④中简单使用临时文件，没有编写具体的功能代码。

创建并使用临时目录的较方便的方式就是使用 TemporaryFile()、NamedTemporaryFile()及 TemporaryDirectory()这 3 个函数，它们能自动处理创建和清除有关的所有步骤。从较低的层次来看，也可以使用函数 mkstemp()和 mkdtemp()来创建临时文件和目录。例如，上面代码中的⑤和⑥分别使用函数 mkstemp()和 mkdtemp()创建了临时文件和目录。

⑤和⑥的这些函数并不会进一步处理文件管理任务。例如，函数 mkstemp()只是简单地返回一个原始的操作系统文件描述符，然后由开发者自行将其转换为一个合适的文件。同样，删除临时文件任务也需要由开发者自己完成。在一般情况下，临时文件都是在操作系统默认的区域中创建的，如"/tmp"等类似的地方。要想找出实际的位置，可以像⑦中那样使用 tempfile.gettempdir()函数实现。

在大多数 UNIX 操作系统中，由函数 TemporaryFile()创建的文件都是未命名的，而且在目录中也没有对应的条目。要想解决这种限制问题，可以使用 NamedTemporaryFile()函数。在⑧~⑨中，在已打开文件的属性 f.name 中包含了临时文件的文件名。当需要将它传给其他需要打开这个文件的代码时，这个属性就显得非常重要了。

对于 TemporaryFile()函数来说，临时文件会在程序关闭时自动删除。如果不想使用这种功能，可以像⑩~⑪中那样使用关键字参数 "delete=False" 关闭这个功能。

在⑫~⑬中使用函数 tempfile.TemporaryDirectory()创建一个临时目录。

执行后会输出：

```
(3, 'C:\\Users\\apple\\AppData\\Local\\Temp\\tmpyvhggvjd')
C:\Users\apple\AppData\Local\Temp\tmpgncoefue
C:\Users\apple\AppData\Local\Temp
文件名是：C:\Users\apple\AppData\Local\Temp\tmpx5jva2h2
文件名是：C:\Users\apple\AppData\Local\Temp\tmpvbrim5iz
目录是：C:\Users\apple\AppData\Local\Temp\tmpe6uk_l9i
```

实例 03-39：实现数据序列化

如果想透明地存储 Python 对象，而不丢失其身份和类型等信息，则需要某种形式的对象序列化处理，这是一个将任意复杂的对象转成对象的文本或二进制表示的过程。同样，必须能够将对象经过序列化后的形式恢复到原有的对象。在 Python 程序中，这种序列化称为 pickle，通过内置的模块 pickle 实现这一功能。可以将对象 pickle 成字符串、磁盘上的文件或任何类似于文件的对象，也可以将字符串、文件或任何类似于文件的对象 unpickle 成原来的对象。

在模块 pickle 中，函数 dumps()和函数 dump()使用 ASCII 表示来创建 pickle。两者都有一个 final 参数（可选），如果其为 True，则使用更快及更小的二进制表示来创建 pickle。函数 loads()和函数 load()会自动检测 pickle 是二进制格式还是文本格式。下面的实例文件 xuliehua.py 演示了使用 pickle 模块实现数据序列化的过程。

源码路径：daima\3\03-39\xuliehua.py

```python
import pickle
dataList = [[1, 1, 'yes'],
            [1, 1, 'yes'],
            [1, 0, 'no'],
            [0, 1, 'no'],
            [0, 1, 'no']]
dataDic = { 0: [1, 2, 3, 4],
            1: ('a', 'b'),
            2: {'c':'yes','d':'no'}}

#使用dump()将数据序列化到文件中
fw = open('123.txt','wb')
pickle.dump(dataList, fw, -1)
```

```
pickle.dump(dataDic, fw)
fw.close()

#使用load()将数据从文件中序列化读出
fr = open('123.txt','rb')
data1 = pickle.load(fr)
print(data1)
data2 = pickle.load(fr)
print(data2)
fr.close()

#使用dumps()和loads()举例
p = pickle.dumps(dataList)
print( pickle.loads(p) )
p = pickle.dumps(dataDic)
print( pickle.loads(p) )
```

执行后会输出：

```
[[1, 1, 'yes'], [1, 1, 'yes'], [1, 0, 'no'], [0, 1, 'no'], [0, 1, 'no']]
{0: [1, 2, 3, 4], 1: ('a', 'b'), 2: {'c': 'yes', 'd': 'no'}}
[[1, 1, 'yes'], [1, 1, 'yes'], [1, 0, 'no'], [0, 1, 'no'], [0, 1, 'no']]
{0: [1, 2, 3, 4], 1: ('a', 'b'), 2: {'c': 'yes', 'd': 'no'}}
```

3.3 数字处理函数

Python 的内置模块提供了大量的数字处理函数。通过这些函数，开发者可以灵活高效地处理数字。

实例 03-40：使用函数 abs()返回绝对值

在 Python 中，模块 math 提供了一些实现基本数学运算功能的函数，如求弦、求根、求对等。其中，函数 abs()的功能是计算一个数字的绝对值。下面的实例文件 juedui.py 演示了使用函数 abs()返回数字绝对值的过程。

源码路径：daima\3\03-40\juedui.py

```
print ("abs(-40) : ", abs(-40))
print ("abs(100.10) : ", abs(100.10))
```

执行后会输出：

```
abs(-40) :  40
abs(100.10) :  100.1
```

实例 03-41：返回最小整数

函数 ceil()的功能是返回一个大于或等于参数 x 的的最小整数。在 Python 程序中，函数 ceil()是不能被直接访问的，在使用时需要导入 math 模块，通过静态对象调用该函数。下面的实例文件 zuixiaozheng.py 演示了使用函数 ceil()返回最小整数的过程。

源码路径：daima\3\03-41\zuixiaozheng.py

```
import math         #导入math模块

print ("math.ceil(-45.17) : ", math.ceil(-45.17))
print ("math.ceil(100.12) : ", math.ceil(100.12))
print ("math.ceil(100.72) : ", math.ceil(100.72))
print ("math.ceil(math.pi) : ", math.ceil(math.pi))
```

执行后会输出：

```
math.ceil(-45.17) :   -45
math.ceil(100.12) :   101
math.ceil(100.72) :   101
math.ceil(math.pi) :   4
```

上述代码如果删除了导入语句 import math，则后面的代码会提示错误。

实例 03-42：返回参数指数值

函数 exp() 的功能是返回参数 x 的指数：e^x。在 Python 程序中，函数 exp() 是不能被直接访问的，在使用时需要导入 math 模块，通过静态对象调用该函数。下面的实例文件 zhishu.py 演示了使用函数 exp() 返回参数指数值的过程。

源码路径：daima\3\03-42\zhishu.py

```python
import math        #导入 math 模块
print ("math.exp(-45.17) : ", math.exp(-45.17))
print ("math.exp(100.12) : ", math.exp(100.12))
print ("math.exp(100.72) : ", math.exp(100.72))
print ("math.exp(math.pi) : ", math.exp(math.pi))
```

执行后会输出：

```
math.exp(-45.17) :  2.4150062132629406e-20
math.exp(100.12) :  3.0308436140742566e+43
math.exp(100.72) :  5.522557130248187e+43
math.exp(math.pi) :  23.140692632779267
```

实例 03-43：返回参数的绝对值

函数 fabs() 的功能是返回参数 x 的绝对值，如 math.fabs(-10) 返回 10.0。在 Python 程序中，函数 fabs() 是不能被直接访问的，在使用时需要导入 math 模块，通过静态对象调用该函数。下面的实例文件 juedui1.py 演示了使用函数 fabs() 返回参数绝对值的过程。

源码路径：daima\3\03-43\juedui1.py

```python
import math        #导入 math 模块

print ("math.fabs(-45.17) : ", math.fabs(-45.17))
print ("math.fabs(100.12) : ", math.fabs(100.12))
print ("math.fabs(100.72) : ", math.fabs(100.72))
print ("math.fabs(math.pi) : ", math.fabs(math.pi))
```

执行后会输出：

```
math.fabs(-45.17) :   45.17
math.fabs(100.12) :   100.12
math.fabs(100.72) :   100.72
math.fabs(math.pi) :  3.141592653589793
```

实例 03-44：返回指定数字的下舍整数

函数 floor() 的功能是返回参数 x 的下舍整数，返回值小于或等于 x。在 Python 程序中，函数 floor() 是不能被直接访问的，在使用时需要导入 math 模块，通过静态对象调用该函数。下面的实例文件 xiaode.py 演示了使用函数 floor() 返回指定数字的下舍整数的过程。

源码路径：daima\3\03-44\xiaode.py

```python
import math        #导入math模块

print ("math.floor(-45.17) : ", math.floor(-45.17))
print ("math.floor(100.12) : ", math.floor(100.12))
print ("math.floor(100.72) : ", math.floor(100.72))
print ("math.floor(math.pi) : ", math.floor(math.pi))
```

执行后会输出：

```
math.floor(-45.17) :   -46
math.floor(100.12) :   100
math.floor(100.72) :   100
math.floor(math.pi) :  3
```

实例 03-45：计算指定数字自然对数

函数 log() 的功能是返回参数 x(x > 0) 的自然对数。在 Python 程序中，函数 log() 是不能被直接访问的，在使用时需要导入 math 模块，通过静态对象调用该函数。下面的实例文件 zirandui.py 演示了使用函数 log() 计算指定数字自然对数的过程。

源码路径：daima\3\03-45\zirandui.py

```
import math    #导入math模块

print ("math.log(100.12) : ", math.log(100.12))
print ("math.log(100.72) : ", math.log(100.72))
print ("math.log(math.pi) : ", math.log(math.pi))
```

执行后会输出：

```
math.log(100.12) :  4.6063694665635735
math.log(100.72) :  4.612344389736092
math.log(math.pi) :  1.1447298858494002
```

实例 03-46：计算指定数字以 10 为基数的对数

函数 log10()的功能是返回参数 x(x>0)的以 10 为基数的对数。在 Python 程序中，函数 log10()是不能被直接访问的，在使用时需要导入 math 模块，通过静态对象调用该函数。下面的实例文件 shidedui.py 演示了使用函数 log10()计算指定数字以 10 为基数的对数的过程。

源码路径：daima\3\03-46\shidedui.py

```
import math    # 导入math模块

print ("math.log10(100.12) : ", math.log10(100.12))
print ("math.log10(100.72) : ", math.log10(100.72))
print ("math.log10(119) : ", math.log10(119))
print ("math.log10(math.pi) : ", math.log10(math.pi))
```

执行后会输出：

```
math.log10(100.12) :  2.0005208409361854
math.log10(100.72) :  2.003115717099806
math.log10(119) :  2.0755469613925306
math.log10(math.pi) :  0.49714987269413385
```

实例 03-47：获取参数最大值

函数 max()的功能是返回指定参数 x 的最大值，参数可以是序列。下面的实例文件 zuidazhi.py 演示了使用函数 max()获取参数最大值的过程。

源码路径：daima\3\03-47\zuidazhi.py

```
print ("max(80, 100, 1000) : ", max(80, 100, 1000))
print ("max(-20, 100, 400) : ", max(-20, 100, 400))
print ("max(-80, -20, -10) : ", max(-80, -20, -10))
print ("max(0, 100, -400) : ", max(0, 100, -400))
```

执行后会输出：

```
max(80, 100, 1000) :   1000
max(-20, 100, 400) :   400
max(-80, -20, -10) :   -10
max(0, 100, -400) :    100
```

实例 03-48：获取参数最小值

函数 min()的功能是返回参数 x 的最小值，参数可以是序列。下面的实例文件 zuixiaozhi.py 演示了使用函数 min()获取参数最小值的过程。

源码路径：daima\3\03-48\zuixiaozhi.py

```
print ("min(80, 100, 1000) : ", min(80, 100, 1000))
print ("min(-20, 100, 400) : ", min(-20, 100, 400))
print ("min(-80, -20, -10) : ", min(-80, -20, -10))
print ("min(0, 100, -400) : ", min(0, 100, -400))
```

执行后会输出：

```
min(80, 100, 1000) :   80
min(-20, 100, 400) :   -20
min(-80, -20, -10) :   -80
min(0, 100, -400) :    -400
```

3.3 数字处理函数

实例 03-49：获取参数的整数部分和小数部分

函数 modf() 的功能是分别返回参数 x 的整数部分和小数部分，两部分的数值符号与参数 x 相同，整数部分以浮点数表示。在 Python 程序中，函数 modf() 是不能被直接访问的，在使用时需要导入 math 模块，通过静态对象调用该函数。下面的实例文件 zhengxiao.py 演示了使用函数 modf() 获取参数的整数部分和小数部分的过程。

源码路径：daima\3\03-49\zhengxiao.py

```
import math         #导入math模块
print ("math.modf(100.12) : ", math.modf(100.12))
print ("math.modf(100.72) : ", math.modf(100.72))
print ("math.modf(119) : ", math.modf(119))
print ("math.modf(math.pi) : ", math.modf(math.pi))
```

执行后会输出：

```
math.modf(100.12) :  (0.12000000000000455, 100.0)
math.modf(100.72) :  (0.7199999999999989, 100.0)
math.modf(119) :  (0.0, 119.0)
math.modf(math.pi) :  (0.14159265358979312, 3.0)
```

实例 03-50：计算 x 的 y 次方的结果

函数 pow() 的功能是返回 x 的 y 次方的结果。在 Python 程序中，pow() 函数有两种语法格式。在 math 模块中，函数 pow() 的语法格式如下。

math.pow(x, y)

Python 内置的标准函数 pow() 的语法格式如下。

pow(x, y[, z])

如果 z 存在，则再对结果进行取模，其结果等效于 pow(x,y)%z。

如果通过 Python 内置函数的方式直接调用 pow()，则内置函数 pow() 会把其本身的参数作为整数。而在 math 模块中，则内置函数 pow() 会把参数转换为浮点数。下面的实例文件 cifang.py 演示了使用两种 pow() 函数的过程。

源码路径：daima\3\03-50\cifang.py

```
import math         #导入math模块
print ("math.pow(100, 2) : ", math.pow(100, 2))
#使用内置函数，查看输出结果的区别
print ("pow(100, 2) : ", pow(100, 2))
print ("math.pow(100, -2) : ", math.pow(100, -2))
print ("math.pow(2, 4) : ", math.pow(2, 4))
print ("math.pow(3, 0) : ", math.pow(3, 0))
```

执行后会输出：

```
math.pow(100, 2) :   10000.0
pow(100, 2) :   10000
math.pow(100, -2) :   0.0001
math.pow(2, 4) :   16.0
math.pow(3, 0) :   1.0
```

实例 03-51：计算指定数字的四舍五入值

函数 round() 的功能是返回参数 x 的四舍五入值。下面的实例文件 sishe.py 演示了使用函数 round() 计算指定数字的四舍五入值的过程。

源码路径：daima\3\03-51\sishe.py

```
print ("round(70.23456) : ", round(70.23456))
print ("round(56.659,1) : ", round(56.659,1))
print ("round(80.264, 2) : ", round(80.264, 2))
print ("round(100.000056, 3) : ", round(100.000056, 3))
print ("round(-100.000056, 3) : ", round(-100.000056, 3))
```

执行后会输出：

```
round(70.23456) :   70
round(56.659,1) :   56.7
round(80.264, 2) :   80.26
```

```
round(100.000056, 3)：   100.0
round(-100.000056, 3)：  -100.0
```

实例 03-52：使用格式化方式设置数字精度

在 Python 程序中，输出数字时不能将取整操作和格式化操作混为一谈。如果只是将数字以固定的位数输出，则无须使用 round()函数，只需在格式化时指定所需要的精度即可。下面的实例文件 buyonground.py 演示了使用格式化方式设置数字精度的过程。

源码路径：daima\3\03-52\buyonground.py

```python
x = 1.2345678
print(format(x, '0.2f'))
print(format(x, '0.3f'))
print(format(x, '0.6f'))
print('值是：{:0.3f}'.format(x))
```

执行后会输出：

```
1.23
1.235
1.234568
值是：1.235
```

实例 03-53：计算指定数字的平方根

函数 sqrt()的功能是返回参数 x 的平方根。在 Python 程序中，函数 sqrt()是不能被直接访问的，在使用时需要导入 math 模块，通过静态对象调用该函数。下面的实例文件 pingfanggen.py 演示了使用函数 sqrt()计算指定数字平方根的过程。

源码路径：daima\3\03-53\pingfanggen.py

```python
import math    #导入math模块

print ("math.sqrt(100) : ", math.sqrt(100))
print ("math.sqrt(7) : ", math.sqrt(7))
print ("math.sqrt(math.pi) : ", math.sqrt(math.pi))
```

执行后会输出：

```
math.sqrt(100) :   10.0
math.sqrt(7) :   2.6457513110645907
math.sqrt(math.pi) :   1.7724538509055159
```

实例 03-54：分别实现无穷大数和 NaN 验证处理

函数 isinf()的功能是如果 x 为无穷大数则返回 True，否则返回 False。函数 isnan()的功能是如果 x 不是数字则返回 True，否则返回 False。下面的实例文件 wuqiong.py 演示了分别实现无穷大数和 NaN（不是数字）验证处理的过程。

源码路径：daima\3\03-54\wuqiong.py

```python
①a = float('inf')
  b = float('-inf')
②c = float('nan')
  print(a)
  print(b)
  print(c)
  import math    #导入 math 模块
③print(math.isinf(a))
  print(math.isnan(c))
④print(math.isinf(1.0e+308))

⑤a = float('inf')
  print(a + 45)
  print(a * 10)
⑥print(10 / a)

⑦a = float('inf')
  print(a/a)
  b = float('-inf')
⑧print(a + b)
```

在①~②中使用函数 float()创建特殊的浮点数。
在③~④中分别使用函数 math.isinf()和 math.isnan()验证是否出现无穷大数和 NaN。
在⑤~⑥中无穷大数在数学计算中进行传播。
在⑦~⑧中一些特定的操作会产生未定义的行为并产生 NaN 的结果。
执行后会输出：

```
inf
-inf
nan
True
True
False
inf
inf
0.0
nan
nan
```

实例 03-55：实现误差运算和精确运算

在 Python 程序中，模块 decimal 的功能是实现定点数和浮点数的数学运算。decimal 实例可以准确地表示任何数字，对其上取整或下取整，还可以对有效数字的个数加以限制。当在程序中需要对小数进行精确计算，不希望因为浮点数本身存在的误差带来影响时，decimal 模块是开发者的最佳选择。下面的实例文件 wucha.py 演示了实现误差运算和精确运算的过程。

源码路径：daima\3\03-55\wucha.py

```
①a = 4.2
  b = 2.1
  print(a + b)
②print((a + b) == 6.3)

  from decimal import Decimal
③a = Decimal('4.2')
  b = Decimal('2.1')
  print(a + b)
  print(Decimal('6.3'))
  print(a + b)
④print((a + b) == Decimal('6.3'))
  from decimal import localcontext
⑤a = Decimal('1.3')
  b = Decimal('1.7')
  print(a / b)

  with localcontext() as ctx:
      ctx.prec = 3              #设置3位精度
      print(a / b)
  with localcontext() as ctx:
      ctx.prec = 50             #设置50位精度
⑥print(a / b)
```

在①~②中展示浮点数的一个尽人皆知的问题——无法精确表达出所有的十进制小数位。从原理上讲，这些误差是底层 CPU 的浮点运算单元和 IEEE 754 浮点数算术标准的一种"特性"。因为 Python 使用原始表示形式保存浮点数，所以如果编写的代码用到了浮点数，那么就无法避免类似的误差。

在③~④中使用 decimal 模块解决浮点数误差，将数字以字符串的形式进行指定。Decimal 对象能以任何期望的方式来工作，能够支持所有常见的数学操作。如果将它们输出或在字符串格式化函数中使用它们，那么它们看起来就和普通的数字一样。

在⑤~⑥中使用 decimal 模块设置运算数字的小数位数，在实现时需要创建一个本地的上下文环境，然后修改其设定。

执行后会输出：
```
6.300000000000001
False
6.3
6.3
6.3
True
0.7647058823529411764705882353
0.765
0.7647058823529411764705882352941176470588235294117 6
```

实例 03-56：将整数转换为二进制、八进制或十六进制数据

在 Python 程序中，当需要对以二进制、八进制或十六进制表示的数值进行转换或输出操作时，通常可以使用内置函数 bin()、oct()和 hex()来实现，这 3 个函数可以将整数转换为二进制、八进制或十六进制表示的文本字符串。如果不想在程序中出现 0b、0o 或 0x 之类的进制前缀符，可以使用 format()函数来处理。如果需要将字符串形式的整数转换为不同的进制，可以使用函数 int()来实现。下面的实例文件 erbashiliu.py 演示了将整数转换为二进制、八进制或十六进制数据的过程。

源码路径：daima\3\03-56\erbashiliu.py

```
①x = 123
  print(bin(x))
  print(oct(x))
②print(hex(x))

③print(format(x, 'b'))
  print(format(x, 'o'))
④print(format(x, 'x'))

⑤x = -123
  print(format(x, 'b'))
⑥print(format(x, 'x'))

⑦x = -123
  print(format(2**32 + x, 'b'))
⑧print(format(2**32 + x, 'x'))

⑨print(int('4d2', 16))
⑩print(int('10011010010', 2))
```

①～②中使用内置函数 bin()、oct()和 hex()实现进制转换。

③～④中使用函数 format()取消进制的前缀符。

⑤～⑥中转换处理负整数。

⑦～⑧中添加最大值来设置位的长度，这样可以生成一个无符号的数值。

⑨～⑩中使用函数 int()设置进制，将字符串形式的整数转换为不同的进制。

执行后会输出：
```
0b1111011
0o173
0x7b
1111011
173
7b
-1111011
-7b
11111111111111111111111110000101
ffffff85
1234
1234
```

实例 03-57：实现复数运算

在 Python 程序中，实现复数运算有如下两种方式。

❑ 使用"浮点数+后缀j"的格式进行指定。

❑ 使用函数 complex(real,imag)实现复数运算功能，函数 complex()的功能是创建一个值为"real + imag * j"的复数或转化一个字符串或数为复数。

下面的实例文件 complexYONG.py 演示了实现复数运算的过程。

源码路径：daima\3\03-57\complexYONG.py

```
a = complex(2, 4)
b = 3 - 5j
①print(a)
②print(b)

③print(a.real)
  print(a.imag)
④print(a.conjugate())

⑤print(a + b)
  print(a * b)
  print(a / b)
⑥print(abs(a))

⑦import cmath
  print(cmath.sin(a))
  print(cmath.cos(a))
⑧print(cmath.exp(a))
```

①中使用函数 complex()处理复数，②中使用浮点数加后缀 j 的格式来处理复数。

③～④中提取复数的实部、虚部和共轭值。

⑤～⑥中使用常见的算术运算来处理复数。

⑦～⑧中使用 cmath 模块执行和复数有关的求正弦、余弦或平方根函数。

执行后会输出：

```
(2+4j)
(3-5j)
2.0
4.0
(2-4j)
(5-1j)
(26+2j)
(-0.4117647058823529+0.6470588235294118j)
4.47213595499958
(24.83130584894638-11.356612711218174j)
(-11.36423470640106-24.814651485634187j)
(-4.829809383269385-5.5920560936409816j)
```

实例 03-58：使用 fractions 模块处理分数

在 Python 程序中，内置模块 fractions 用于处理分数。类 Fraction 是 fractions 模块的核心，它继承了 numbers.Rational 类并且实现了该类所有的方法。下面的实例文件 fenshu.py 演示了使用 fractions 模块处理分数的过程。

源码路径：daima\3\03-58\fenshu.py

```
from fractions import Fraction
print(Fraction(16, -10))
print(Fraction(123))
print(Fraction())
print(Fraction('3/7'))
print(Fraction(' -3/7 '))
print(Fraction('1.414213 \t\n'))
print(Fraction('-.125'))
print(Fraction('7e-6'))
print(Fraction(2.25))
print(Fraction(1.1))
from decimal import Decimal
print(Fraction(Decimal('1.1')))
```

执行后会输出：

```
-8/5
123
```

```
0
3/7
-3/7
1414213/1000000
-1/8
7/1000000
9/4
2476979795053773/2251799813685248
11/10
```

实例 03-59：使用 NumPy 模块分别创建一维数组和二维数组

在 Python 程序中，模块 NumPy（Numerical Python）提供了对多维数组对象的支持，不但具有矢量运算能力，而且支持大量的高级维度数组与矩阵运算，也针对数组运算提供大量的数学函数。下面的实例文件 daxing.py 演示了使用 NumPy 模块分别创建一维数组和二维数组的过程。

源码路径：daima\3\03-59\daxing.py

```python
import numpy as np
data= np.array([2,5 ,6 ,8 ,3 ])              #创建一个一维数组
print(data)
data1=np.array([[2,5,6,8,3],np.arange(5)])   #创建一个二维数组
print(data1)
```

执行后会输出：

```
[2 5 6 8 3]
[[2 5 6 8 3]
 [0 1 2 3 4]]
```

实例 03-60：使用函数 choice()创建随机数

在 Python 程序中，如果想生成随机数或从序列中随机挑选元素，则可以通过 random 模块实现。随机数可以用于数学、游戏和算法等领域中，以提高算法效率，并增强程序的安全性。下面的实例文件 suiji01.py 演示了使用函数 choice()创建随机数的过程。

源码路径：daima\3\03-60\suiji01.py

```python
import random

print ("从 range(100) 返回一个随机数 : ",random.choice(range(100)))
print ("从列表中 [1, 2, 3, 5, 9]) 返回一个随机元素 : ", random.choice([1, 2, 3, 5, 9]))
print ("从字符串中 'toppr' 返回一个随机字符 : ", random.choice('Runoob'))
```

因为是随机的，所以上述代码每次执行后的效果都会不同，例如，在作者的计算机上执行后会输出：

```
从 range(100) 返回一个随机数 :  88
从列表中 [1, 2, 3, 5, 9]) 返回一个随机元素 :  3
从字符串中 'toppr' 返回一个随机字符 :  R
```

3.4 日期和时间函数

Python 的内置模块提供了大量的日期和时间函数。通过这些函数，开发者可以快速实现日期和时间功能。

实例 03-61：使用函数 time.clock()处理时间

在 Python 程序中，时间模块 time 的常用内置函数有如下几种。

（1）函数 time.asctime([tupletime])：接受时间元组并返回一个可读形式为 "Tue Dec 11 18:07:14 2018"（2018 年 12 月 11 日周二 18 时 07 分 14 秒）的 24 个字符的字符串。

（2）函数 time.clock()：以浮点数计算的秒数返回当前的 CPU 时间，用于衡量不同程序的耗时，比 time.time()函数更有用。需要注意，函数 time.clock()在不同的操作系统中的含义不同。在

UNIX 操作系统中，它返回的是"进程时间"，是用秒表示的浮点数（时间戳）。当在 Windows 操作系统中第一次调用该函数时，它返回的是进程执行的实际时间，而在第二次之后调用它时，它返回的是自第一次调用以后到现在的执行时间（实际上是以 WIN32 上 QueryPerformanceCounter() 为基础，比用毫秒表示更为精确）。下面的实例文件 shijian01.py 演示了使用函数 time.clock() 处理时间的过程。

源码路径：daima\3\03-61\shijian01.py

```python
import time

def procedure():
    time.sleep(2.5)

# time.clock
t0 = time.clock()
procedure()
print (time.clock() - t0)

# time.time
t0 = time.time()
procedure()
print (time.time() - t0)
```

实例 03-62：使用函数 time.tzset() 操作时间

函数 time.tzset() 的功能是根据环境变量 TZ 重新初始化时间的相关设置。注意，不能在 Windows 操作系统下使用此函数。下面的实例文件 shijian03.py 演示了使用函数 time.tzset() 操作时间的过程。

源码路径：daima\3\03-62\shijian03.py

```python
import time
import os

os.environ['TZ'] = 'EST+05EDT,M4.1.0,M10.5.0'
time.tzset()
print(time.strftime('%X %x %Z'))

os.environ['TZ'] = 'AEST-10AEDT-11,M10.5.0,M3.5.0'
time.tzset()
print(time.strftime('%X %x %Z'))
```

执行后会输出：

```
23:25:45 04/06/18 EDT
13:25:45 04/07/18 AEST
```

实例 03-63：使用 calendar 模块函数操作日期

在 Python 程序中，日历模块 calendar 的常用内置函数有如下几种。

（1）函数 calendar.calendar(year,w=2,l=1,c=6)：返回一个多行字符串格式的 year 年年历，3 个月一行，间隔距离为 c。每日宽度间隔为 w。每行长度为 21w+18+2c。l 代表每星期行数。

（2）函数 calendar.firstweekday( )：返回当前每周起始日期的设置。在默认情况下，首次载入 calendar 模块时返回 0，即表示星期一。

（3）函数 calendar.isleap(year)：如果 year 是闰年则返回 True，否则返回 False。

（4）函数 calendar.leapdays(y1,y2)：返回 y1 和 y2 两年之间的闰年总数。

（5）函数 calendar.month(year,month,w=2,l=1)：返回一个多行字符串格式的 year 年 month 月日历，两行标题，一周一行。每日宽度间隔为 w，每行的长度为 7w+6。l 表示每星期的行数。

（6）函数 calendar.monthcalendar(year,month)：返回一个整数的单层嵌套列表，每个子列表装载代表一个星期的整数，year 年 month 月外的日期都设为 0。范围内的日子都由该月第几日表示，从 1 开始。

(7) 函数 calendar.monthrange(year,month)：返回两个整数，第 1 个整数表示该月的首日是星期几，第 2 个整数表示该月的天数（28～31）。

(8) 函数 calendar.prcal(year,w=2,l=1,c=6)：相当于 print calendar.calendar(year,w,l,c)。

(9) calendar.prmonth(year,month,w=2,l=1)：相当于 print calendar.calendar（year,w,l,c）。

(10) 函数 calendar.setfirstweekday(weekday)：设置每周的起始日期码，取 0（星期一）到 6（星期日）。

(11) 函数 calendar.timegm(tupletime)：和函数 time.gmtime()相反，功能是接受一个时间元组，返回该时刻的时间辍（1970 年后经过的浮点秒数）。很多 Python 程序用一个元组装起来的 9 组数字处理时间，具体说明如表 3-1 所示。

表 3-1　　　　　　　　　　9 组数字处理时间举例

序号	字　　段	值（举例）
1	4 位数年	2018
2	月	1～12
3	日	1～31
4	小时	0～23
5	分钟	0～59
6	秒	0～61（60 或 61 是闰秒）
7	一周的第几日	0～6（0 是周一）
8	一年的第几日	1～366
9	夏令时	-1，0，1，-1（决定是否为夏令时的标志）

这样我们可以定义一个元组，在元组中设置 9 个属性分别来表示表 3-1 所示的 9 组数字。

(12) 函数 calendar.weekday(year,month,day)：返回给定日期的日期码，day 为 0（星期一）～6（星期日），month 为 1（1 月）～12（12 月）。

下面的实例文件 rili.py 演示了使用上述 calendar 模块函数操作日期的过程。

源码路径：daima\3\03-63\rili.py

```
import calendar

calendar.setfirstweekday(calendar.SUNDAY)
print(calendar.firstweekday())
c = calendar.calendar(2018)
# c = calendar.TextCalendar()
# c = calendar.HTMLCalendar()
print(c)
print(calendar.isleap(2018))
print(calendar.leapdays(2010, 2018))
m = calendar.month(2018, 7)
print(m)
print(calendar.monthcalendar(2018, 7))
print(calendar.monthrange(2018, 7))
print(calendar.timegm((2018, 7, 24, 11, 19, 0, 0, 0, 0)))     #定义有9组数字的元组
print(calendar.weekday(2018, 7, 23))
```

执行后会输出：

```
6
                                  2018

        January                   February                    March
Su Mo Tu We Th Fr Sa      Su Mo Tu We Th Fr Sa      Su Mo Tu We Th Fr Sa
    1  2  3  4  5  6                   1  2  3                    1  2  3
 7  8  9 10 11 12 13       4  5  6  7  8  9 10       4  5  6  7  8  9 10
14 15 16 17 18 19 20      11 12 13 14 15 16 17      11 12 13 14 15 16 17
21 22 23 24 25 26 27      18 19 20 21 22 23 24      18 19 20 21 22 23 24
28 29 30 31               25 26 27 28               25 26 27 28 29 30 31
```

```
         April                    May                     June
Su Mo Tu We Th Fr Sa    Su Mo Tu We Th Fr Sa    Su Mo Tu We Th Fr Sa
 1  2  3  4  5  6  7           1  2  3  4  5                    1  2
 8  9 10 11 12 13 14     6  7  8  9 10 11 12     3  4  5  6  7  8  9
15 16 17 18 19 20 21    13 14 15 16 17 18 19    10 11 12 13 14 15 16
22 23 24 25 26 27 28    20 21 22 23 24 25 26    17 18 19 20 21 22 23
29 30                   27 28 29 30 31          24 25 26 27 28 29 30

         July                   August                 September
Su Mo Tu We Th Fr Sa    Su Mo Tu We Th Fr Sa    Su Mo Tu We Th Fr Sa
 1  2  3  4  5  6  7              1  2  3  4                       1
 8  9 10 11 12 13 14     5  6  7  8  9 10 11     2  3  4  5  6  7  8
15 16 17 18 19 20 21    12 13 14 15 16 17 18     9 10 11 12 13 14 15
22 23 24 25 26 27 28    19 20 21 22 23 24 25    16 17 18 19 20 21 22
29 30 31                26 27 28 29 30 31       23 24 25 26 27 28 29
                                                30

        October                November                 December
Su Mo Tu We Th Fr Sa    Su Mo Tu We Th Fr Sa    Su Mo Tu We Th Fr Sa
    1  2  3  4  5  6                 1  2  3                       1
 7  8  9 10 11 12 13     4  5  6  7  8  9 10     2  3  4  5  6  7  8
14 15 16 17 18 19 20    11 12 13 14 15 16 17     9 10 11 12 13 14 15
21 22 23 24 25 26 27    18 19 20 21 22 23 24    16 17 18 19 20 21 22
28 29 30 31             25 26 27 28 29 30       23 24 25 26 27 28 29
                                                30 31

False
2
      July 2018
Su Mo Tu We Th Fr Sa
 1  2  3  4  5  6  7
 8  9 10 11 12 13 14
15 16 17 18 19 20 21
22 23 24 25 26 27 28
29 30 31

[[1, 2, 3, 4, 5, 6, 7], [8, 9, 10, 11, 12, 13, 14], [15, 16, 17, 18, 19, 20, 21], [22, 23, 24, 25, 26, 27, 28], [29, 30, 31, 0, 0, 0, 0]]
(6, 31)
1532431140
0
```

实例 03-64：使用类 date 的实例方法和属性实现日期操作

在 Python 程序中，datetime 是一个使用面向对象编程设计的模块，可以在 Python 程序中使用日期和时间。相比于 time 模块，datetime 模块的接口更加直观、更加容易被调用。模块 datetime 中的类 date 表示一个日期，日期由年、月、日组成。下面的实例文件 datetime02.py 演示了使用类 date 的实例方法和属性实现日期操作的过程。

源码路径：daima\3\03-64\datetime02.py

```python
from datetime import *
import time
now = date(2018,4,6 )
tomorrow = now.replace(day = 7 )
print('now:' , now,    ', tomorrow:' , tomorrow)
print( 'timetuple():' , now.timetuple())
print('weekday():' , now.weekday())
print('isoweekday():' , now.isoweekday())
print('isocalendar():' , now.isocalendar())
print('isoformat():' , now.isoformat())
```

执行后会输出：

```
now: 2018-04-06 , tomorrow: 2018-04-07
timetuple(): time.struct_time(tm_year=2018, tm_mon=4, tm_mday=6, tm_hour=0, tm_min=0, tm_sec=0, tm_wday=4, tm_yday=96, tm_isdst=-1)
weekday(): 4
isoweekday(): 5
isocalendar(): (2018, 14, 5)
isoformat(): 2018-04-06
```

实例 03-65：使用类 time 实现日期操作

在 Python 程序中，类 time 表示时间，由时、分、秒及微秒组成。下面的实例文件 datetime04.py 演示了使用类 time 实现日期操作的过程。

源码路径：daima\3\03-65\datetime04.py

```python
from datetime import *
tm = time(23, 46, 10)
print('tm:', tm)
print('hour: %d, minute: %d, second: %d, microsecond: %d'% (tm.hour, tm.minute, tm.second, tm.microsecond))
tm1 = tm.replace(hour = 20)
print('tm1:', tm1)
print('isoformat():', tm.isoformat())
```

执行后会输出：

```
tm: 23:46:10
hour: 23, minute: 46, second: 10, microsecond: 0
tm1: 20:46:10
isoformat(): 23:46:10
```

实例 03-66：使用类 datetime 实现日期操作

在 Python 程序中，类 datetime 是 date 与 time 的结合体，包括了 date 与 time 的所有功能。下面的实例文件 datetime05.py 演示了使用类 datetime 实现日期操作的过程。

源码路径：daima\3\03-66\datetime05.py

```python
from datetime import *
import time
print('datetime.max:', datetime.max)
print('datetime.min:', datetime.min)
print('datetime.resolution:', datetime.resolution)
print('today():', datetime.today())
print('now():', datetime.now())
print('utcnow():', datetime.utcnow())
print('fromtimestamp(tmstmp):', datetime.fromtimestamp(time.time()))
print('utcfromtimestamp(tmstmp):', datetime.utcfromtimestamp(time.time()))
```

执行后会输出：

```
datetime.max: 9999-12-31 23:59:59.999999
datetime.min: 0001-01-01 00:00:00
datetime.resolution: 0:00:00.000001
today(): 2017-11-21 23:44:22.366920
now(): 2017-11-21 23:44:22.366919
utcnow(): 2017-11-21 15:44:22.366919
fromtimestamp(tmstmp): 2017-11-21 23:44:22.366920
utcfromtimestamp(tmstmp): 2017-11-21 15:44:22.366920
```

类 datetime 提供的实例方法与属性跟 date 和 time 的相似，在此不再讲解这些相似的方法与属性。

实例 03-67：使用类 datetime 格式化日期

在 Python 程序中，类 datetime、date 和 time 都提供了 strftime()函数，该函数能够接收一个格式字符串，输出日期时间的字符串表示。各个格式字符串的具体含义如下。

- ❑ %a：星期几的简写，例如，星期三为 Web。
- ❑ %A：星期几的全写，例如，星期三为 Wednesday。
- ❑ %b：月份的简写，例如，4 月份为 Apr。
- ❑ %B：月份的全写，例如，4 月份为 April。
- ❑ %c：日期时间的字符串表示，如 04/07/10 10:43:39。
- ❑ %d：这日在这个月中的天数（是这个月的第几天）。
- ❑ %f：微秒（范围为[0,999999]）。
- ❑ %H：小时（24 小时制，范围为[0, 23]）。

- %I：小时（12 小时制，范围为[0, 11]）。
- %j：今天在今年中的天数（当年的第几天，范围为[001,366]）。
- %m：月份（范围为[01,12]）。
- %M：分钟（范围为[00,59]）。
- %p：AM 或 PM。
- %S：秒（范围为[00,61]）。
- %U：本周在今年的周数（当年的第几周），星期天作为周的第一天。
- %w：今天在这周的天数（范围为[0, 6]，6 表示星期天）。
- %W：周在当年的周数（当年的第几周），星期一作为周的第一天。
- %x：日期字符串，如 04/07/10。
- %X：时间字符串，如 10:43:39。
- %y：2 个数字表示的年份。
- %Y：4 个数字表示的年份。
- %z：与 UTC 时间的间隔（如果是本地时间，则返回空字符串）。
- %Z：时区名称（如果是本地时间，则返回空字符串）。
- %%：%% => %。

下面的实例文件 datetime06.py 演示了使用类 datetime 格式化日期的过程。

源码路径：daima\3\03-67\datetime06.py

```
from datetime import *
import time
dt = datetime.now()
print('(%Y-%m-%d %H:%M:%S %f):' , dt.strftime( '%Y-%m-%d %H:%M:%S %f' ) )
print('(%Y-%m-%d %H:%M:%S %p): ' , dt.strftime( '%y-%m-%d %I:%M:%S %p' )  )
print('%%a: %s '      % dt.strftime( '%a' ) )
print('%%A: %s '      % dt.strftime( '%A' ) )
print('%%b: %s '      % dt.strftime( '%b' ) )
print('%%B: %s '      % dt.strftime( '%B' ) )
print('日期时间%%c: %s '  % dt.strftime( '%c' ) )
print('日期%%x：%s '    % dt.strftime( '%x' ) )
print('时间%%X：%s '    % dt.strftime( '%X' ) )
print('今天是这周的第%s天 '  % dt.strftime( '%w' ) )
print('今天是今年的第%s天 '  % dt.strftime( '%j' ) )
print('本周是今年的第%s周 '  % dt.strftime( '%U' ) )
```

执行后会输出：

```
(%Y-%m-%d %H:%M:%S %f):  2017-11-21 23:56:09 580715
(%Y-%m-%d %H:%M:%S %p):  17-11-21 11:56:09 PM
%a: Tue
%A: Tuesday
%b: Nov
%B: November
日期时间%c: Tue Nov 21 23:56:09 2017
日期%x: 11/21/17
时间%X: 23:56:09
今天是这周的第2天
今天是今年的第325天
本周是今年的第47周
```

实例 03-68：使用类 datetime 实现时间换算

在 Python 程序中，有时需要实现简单的时间换算，例如，将日转换为秒，将小时转换为分钟等，此时只需要使用 datetime 模块来实现这个功能。下面的实例文件 datetime07.py 演示了使用类 datetime 实现时间换算的过程。

源码路径：daima\3\03-68\datetime07.py

```
from datetime import timedelta
①a = timedelta(days=2, hours=6)
```

```
    b = timedelta(hours=4.5)
    c = a + b
    print(c.days)
    print(c.seconds)
    print(c.seconds / 3600)
②print(c.total_seconds() / 3600)

    from datetime import datetime
③a = datetime(2012, 9, 23)
    print(a + timedelta(days=10))
    b = datetime(2012, 12, 21)
    d = b - a
    print(d.days)
    now = datetime.today()
    print(now)
④print(now + timedelta(minutes=10))

⑤a = datetime(2012, 3, 1)
    b = datetime(2012, 2, 28)
    a - b
    print((a - b).days)
    c = datetime(2013, 3, 1)
    d = datetime(2013, 2, 28)
⑥print((c - d).days)
```

①～②中创建一个 timedelta 实例来表示一个时间间隔。

③～④中创建 datetime 实例对象，使用标准的数学运算来表示特定的日期和时间。

⑤～⑥中使用模块 datetime 正确地处理闰年。

执行后会输出：

```
2
37800
10.5
58.5
2012-10-03 00:00:00
89
2017-11-22 16:27:31.049802
2017-11-22 16:37:31.049802
2
1
```

实例 03-69：获取某一周中某一天的日期

在 Python 程序中，有时需要获取某一周中某一天的日期，例如，获取上周三是几月几号，此时可以使用模块 datetime 中的函数和类实现。下面的实例文件 datetime08.py 演示了使用 datetime 获取某一周中某一天的日期的方法。

源码路径：daima\3\03-69\datetime08.py

```
from datetime import datetime, timedelta
weekdays = ['Monday', 'Tuesday', 'Wednesday', 'Thursday', 'Friday', 'Saturday', 'Sunday']

def get_previous_byday(dayname, start_date=None):
    if start_date is None:
        start_date = datetime.today()
    day_num = start_date.weekday()
    day_num_target = weekdays.index(dayname)
    days_ago = (7 + day_num - day_num_target) % 7
    if days_ago == 0:
        days_ago = 7
    target_date = start_date - timedelta(days=days_ago)
    return target_date
print(datetime.today())                          #获取今天的具体日期
print(get_previous_byday('Monday'))              #获取上周一的具体日期
print(get_previous_byday('Tuesday'))             #获取上周二的具体日期
print(get_previous_byday('Friday'))              #获取上周五的具体日期
```

上述代码定义了函数 get_previous_byday()，功能是获取上周某一天（周几）的具体日期。首先将起始日期和目标日期映射到它们在一周之中的位置上（周一为第 0 天，依此类推），然后用取模运算计算上一次目标日期出现时到起始日期为止一共经过了多少天，最后从起始日期中减去一个合适的 timedelta 实例就得到了我们所要的日期。

执行后会输出：
```
2017-11-22 16:58:19.571494
2017-11-20 16:58:19.571494
2017-11-21 16:58:19.571494
2017-11-17 16:58:19.571494
```

实例 03-70：输出当月每一天的日期

在 Python 程序中，有时需要循环输出当月每一天的日期，此时需要先计算出当月开始和结束的具体范围，然后使用 datetime.timedelta 循环递增并输出每一个日期即可。下面的实例文件 datetime09.py 演示了使用 datetime 循环输出当月每一天的日期的方法。

源码路径：daima\3\03-70\datetime09.py

```python
from datetime import datetime, date, timedelta
import calendar

def get_month_range(start_date=None):
    if start_date is None:
        start_date = date.today().replace(day=1)
    days_in_month = calendar.monthrange(start_date.year, start_date.month)[1]
    end_date = start_date + timedelta(days=days_in_month)
    return (start_date, end_date)

first_day, last_day = get_month_range()
a_day = timedelta(days=1)
while first_day < last_day:
    print(first_day)
    first_day += a_day
```

上述代码定义了函数 get_month_range()，此函数可接受任意的 datetime 对象，返回一个包含本月第一天和下个月第一天日期的元组。

执行后会输出：
```
2017-11-01
2017-11-02
2017-11-03
//为了节省本书篇幅，在此省略中间其他天数的日期
2017-11-28
2017-11-29
2017-11-30
```

实例 03-71：循环输出当月每一天的日期

我们不但可以输出当月每一天的日期，而且可以循环输出指定起始时间后的每一天的日期和时间（小时、分钟、秒）间隔。下面的实例文件 datetime10.py 演示了使用 datetime 循环输出当月每一天的日期的方法。

源码路径：daima\3\03-71\datetime10.py

```python
from datetime import datetime, date, timedelta
import calendar

def daterange(start, stop, step):
    while start < stop:
        yield start
        start += step

for d in daterange(date(2018, 8, 1), date(2018, 8, 11), timedelta(days=1)):
    print(d)

for d in daterange(datetime(2018, 8, 1), datetime(2018, 8, 3), timedelta(minutes=30)):
    print(d)
```

上述代码创建了一个专门处理日期的函数 daterange()，其用法和普通的 Python 内置函数 range() 一样。上述实现方法比实例文件 datetime9.py 简单很多，原因是日期和时间可以通过标准的算术和比较操作符进行操作。

执行后会输出：
```
2018-08-01
2018-08-02
//为了节省本书篇幅，在此省略中间其他天数的日期
2018-08-02 23:00:00
2018-08-02 23:30:00
```

实例 03-72：将字符串转换为日期

在 Python 程序中，可以将字符串形式的数据转换为日期，具体方法是先将字符串转换为 datetime 对象，然后对它们执行一些非字符串操作。下面的实例文件 datetime11.py 演示了使用 datetime 将字符串转换为日期的方法。

源码路径：daima\3\03-72\datetime11.py

```python
from datetime import datetime
text = '2018-09-20'
y = datetime.strptime(text, '%Y-%m-%d')
z = datetime.now()
diff = z - y
print(z)
print(diff)
nice_z = datetime.strftime(z, '%A %B %d, %Y')
print(nice_z)
```

上述代码中用到了函数 datetime.strptime()，此函数支持多种格式化操作功能，例如，"%Y"代表以 4 位数字表示的年份，而"%m"代表以 2 位数字表示的月份。值得注意的是，这些格式化占位符也可以反过来用于将 datetime 对象转换为字符串。当需要以字符串形式来表示 datetime 对象并且让输出格式变得美观时，建议使用函数 datetime.strptime()。

执行后会输出：
```
2017-11-22 21:00:42.197880
-302 days, 21:00:42.197880
Wednesday November 22, 2017
```

第 4 章
进程通信和联网操作实战

在计算机信息系统中,我们经常需要面临在不同进程之间及不同网络之间进行通信的问题。Python 提供了内置的相关库,本章将通过具体实例的实现过程,详细讲解使用 Python 实现进程通信和联网操作的知识,为读者学习本书后面的知识打下基础。

4.1 使用 Socket 网络接口库

Python 提供了两种访问网络服务的功能,其中低级别的网络服务通过 Socket(套接字)实现,它提供了标准的 BSD Sockets API,可以访问底层操作系统 Socket 接口的全部方法。而高级别的网络服务通过模块 socketserver 实现,它提供了服务器中心类,可以简化网络服务器的开发。

实例 04-01:分别创建简单 Socket 服务器和客户端

在 Python 程序中,模块 socket 针对服务器和客户端进行打开、读写和关闭的操作。下面的实例文件 jiandanfuwu.py 演示了创建一个简单 Socket 服务器的过程。

源码路径:daima\4\04-01\jiandanfuwu.py

```python
import socket
sk = socket.socket()
sk.bind(("127.0.0.1",8080))
sk.listen(5)
conn,address = sk.accept()
sk.sendall(bytes("Hello world",encoding="utf-8"))
```

下面的实例文件 jiandankehu.py 演示了创建一个简单 Socket 客户端的过程。

源码路径:daima\4\04-01\jiandankehu.py

```python
import socket

obj = socket.socket()
obj.connect(("127.0.0.1",8080))

ret = str(obj.recv(1024),encoding="utf-8")
print(ret)
```

实例 04-02:使用 Socket 建立 TCP"客户端/服务器"连接

在 Python 程序中,所有 Socket 都是通过 socket.socket()函数创建的。因为服务器需要占用一个端口并等待客户端的请求,所以它们必须绑定一个本地地址。因为 TCP 是一种面向连接的通信协议,所以在 TCP 服务器开始操作之前,必须安装一些基础设施。特别地,TCP 服务器必须监听传入的连接。一旦这个安装过程完成后,服务器就可以开始它的无限循环。在调用 accept() 函数之后,一个简单的(单线程)服务器就开启了,它会等待客户端的连接。在默认情况下,accept()函数是阻塞的,这说明执行操作会被暂停,直到一个连接到达为止。一旦服务器接受了一个连接,就会利用 accept()函数返回一个独立的客户端 Socket,用来与即将到来的消息进行交换。在下面的实例代码中,演示了使用 Socket 建立 TCP"客户端/服务器"连接的过程,这是一个可靠的、相互通信的"客户端/服务器"。

(1)实例文件 ser.py 的功能是以 TCP 连接方式建立一个服务器程序,能够将收到的信息直接发回客户端。实例文件 ser.py 的具体实现代码如下。

源码路径:daima\4\04-02\ser.py

```python
import socket                                       #导入socket模块
HOST = ''                                           #定义变量HOST的初始值
PORT = 10000                                        #定义变量PORT的初始值
#创建Socket对象s,参数分别表示地址和协议类型
s = socket.socket(socket.AF_INET, socket.SOCK_STREAM)
s.bind((HOST, PORT))                                #将Socket与地址绑定
s.listen(1)                                         #监听连接
conn, addr = s.accept()                             #接受客户端连接
print('客户端地址:', addr)                           #显示客户端地址
while True:                                         #连接成功后
    data = conn.recv(1024)                          #实现对话操作(接收/发送)
    print("获取信息: ",data.decode('utf-8'))        #显示获取的信息
    if not data:                                    #如果没有信息
```

```
        break                          #则退出循环
        conn.sendall(data)             #发送信息
conn.close()                           #关闭连接
```

在上述实例代码中，建立 TCP 连接之后使用 while 循环多次与客户端进行信息交换，直到收到信息为空时，服务器终止执行。因为这只是一个服务器程序，所以执行之后程序不会立即返回交互信息，还需要等待和客户端建立连接，和客户端建立连接后才能看到具体的交互效果。

（2）实例文件 cli.py 的功能是建立客户端程序，在此需要创建一个 Socket 对象，然后调用这个 Socket 对象的 connect()函数来连接服务器。实例文件 cli.py 的具体实现代码如下。

源码路径：daima\4\04-02\cli.py

```
import socket                                    #导入socket模块
HOST = 'localhost'                               #定义变量HOST的初始值
PORT = 10000                                     #定义变量PORT的初始值
#创建Socket对象s，参数分别表示地址和协议类型
s = socket.socket(socket.AF_INET, socket.SOCK_STREAM)
s.connect((HOST, PORT))                          #建立和服务器的连接
data = "你好！"                                   #设置变量
while data:
    s.sendall(data.encode('utf-8'))              #发送信息"你好"
    data = s.recv(512)                           #实现对话操作（接收/发送）
    print("获取服务器信息：\n",data.decode('utf-8'))  #显示接收到的服务器信息
    data = input('请输入信息：\n')                 #信息输入
s.close()                                        #关闭连接
```

上述代码使用 Socket 以 TCP 连接方式建立了一个简单的客户端程序，基本功能是将从键盘输入的信息发送给服务器，并从服务器接收信息。因为服务器建立在 localhost 的 10000 端口上，所以上述代码作为客户端程序，连接的就是 localhost 的 10000 端口。当连接成功之后，程序向服务器发送了一个默认的信息"你好！"之后，便将从键盘输入的信息发送给服务器，直到输入空信息（按 Enter 键）时退出 while 循环，关闭连接。首先执行 ser.py 服务器程序，然后执行 cli.py 客户端程序，除了发送一个默认的信息外，从键盘输入的信息都会被发送给服务器，服务器收到后显示它并再次将它转发回客户端进行显示。执行效果如图 4-1 所示。

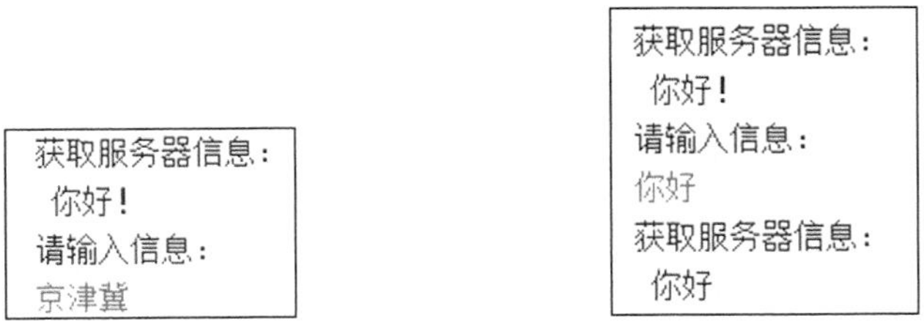

图 4-1 执行效果

实例 04-03：TCP"客户端/服务器"模式的机器人聊天程序

再看下面的这个实例，其功能是建立一个 TCP"客户端/服务器"模式的机器人聊天程序。

（1）服务器的实例文件 jiqirenser.py 的具体实现代码如下。

源码路径：daima\4\04-03\jiqirenser.py

```
import  socketserver
class Myserver(socketserver.BaseRequestHandler):
    def handle(self):

        conn = self.request
        conn.sendall(bytes("你好，我是机器人",encoding="utf-8"))
        while True:
            ret_bytes = conn.recv(1024)
            ret_str = str(ret_bytes,encoding="utf-8")
            if ret_str == "q":
                break
```

```
            conn.sendall(bytes(ret_str+"你好我好大家好",encoding="utf-8"))

if __name__ == "__main__":
    server = socketserver.ThreadingTCPServer(("127.0.0.1",8000),Myserver)
    server.serve_forever()
```

（2）客户端的实例文件 jiqirencli.py 的具体实现代码如下。

源码路径：daima\4\04-03\jiqirencli.py

```
import socket

obj = socket.socket()

obj.connect(("127.0.0.1",8000))

ret_bytes = obj.recv(1024)
ret_str = str(ret_bytes,encoding="utf-8")
print(ret_str)

while True:
    inp = input("你好请问您有什么问题？ \n >>>")
    if inp == "q":
        obj.sendall(bytes(inp,encoding="utf-8"))
        break
    else:
        obj.sendall(bytes(inp, encoding="utf-8"))
        ret_bytes = obj.recv(1024)
        ret_str = str(ret_bytes,encoding="utf-8")
        print(ret_str)
```

执行效果如图 4-2 所示。

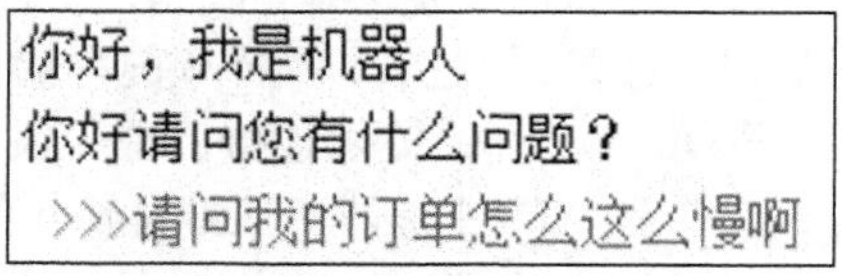

```
你好，我是机器人
你好请问您有什么问题？
    >>>请问我的订单怎么这么慢啊
```

图 4-2　执行效果

实例 04-04：实现一个文件上传系统

实现的文件上传系统分为服务端程序和客户端程序，具体实现如下。
（1）服务器的实例文件 shangchuanfuwu.py 的具体实现代码如下。

源码路径：daima\4\04-04\shangchuanfuwu.py

```
import socket
sk = socket.socket()
sk.bind(("127.0.0.1",8000))
sk.listen(5)

while True:
    conn,address = sk.accept()
    conn.sendall(bytes("欢迎光临我爱我家",encoding="utf-8"))

    size = conn.recv(1024)
    size_str = str(size,encoding="utf-8")
    file_size = int(size_str)

    conn.sendall(bytes("开始传送", encoding="utf-8"))

    has_size = 0
    f = open("123.jpg","wb")
    while True:
        if file_size == has_size:
            break
        date = conn.recv(1024)
        f.write(date)
        has_size += len(date)
    f.close()
```

(2) 客户端的实例文件 shangchuancli.py 的具体实现代码如下。

源码路径：daima\4\04-04\shangchuancli.py

```python
import socket
import os

obj = socket.socket()

obj.connect(("127.0.0.1",8000))

ret_bytes = obj.recv(1024)
ret_str = str(ret_bytes,encoding="utf-8")
print(ret_str)

size = os.stat("yan.jpg").st_size
obj.sendall(bytes(str(size),encoding="utf-8"))

obj.recv(1024)

with open("yan.jpg","rb") as f:
    for line in f:
        obj.sendall(line)
```

实例 04-05：使用 Socket 建立 UDP "客户端/服务器"连接

在 Python 程序中，当使用 Socket 应用传输层的 UDP 建立客户端与服务器的连接时，整个实现过程要比使用 TCP 简单一点儿。基于 UDP 的客户端与服务器在进行数据传送时，不是先建立连接，而是直接进行数据传送。下面的实例代码演示了使用 Socket 建立 UDP "客户端/服务器"连接的过程，这是一个不可靠的、相互通信的"客户端/服务器"。其中实例文件 serudp.py 的功能是使用 UDP 连接方式建立一个服务器程序，将收到的信息直接发回客户端。实例文件 serudp.py 的具体实现代码如下。

源码路径：daima\4\04-05\serudp.py

```python
import socket                                      #导入socket模块
HOST = ''                                          #定义变量HOST的初始值
PORT = 10000                                       #定义变量PORT的初始值
#创建Socket对象s，参数分别表示地址和协议类型
s = socket.socket(socket.AF_INET, socket.SOCK_DGRAM)
s.bind((HOST, PORT))                               #将Socket与地址绑定
data = True                                        #设置变量data的初始值
while data:                                        #如果有信息
    data,address = s.recvfrom(1024)                #则实现对话操作（接收/发送）
    if data=='zaijian':                            #当接收的信息是zaijian时
        break                                      #退出循环
    print('接收信息：',data.decode('utf-8'))       #显示接收到的信息
    s.sendto(data,address)                         #发送信息
s.close()                                          #关闭连接
```

在上述实例代码中，建立 UDP 连接之后使用 while 循环多次与客户端进行信息交换。上述服务器程序建立在 localhost 的 10000 端口，当收到"zaijian"信息时先退出 while 循环，然后关闭连接。

实例文件 cliudp.py 的具体实现代码如下。

源码路径：daima\4\04-05\cliudp.py

```python
import socket                                      #导入socket模块
HOST = 'localhost'                                 #定义变量HOST的初始值
PORT = 10000                                       #定义变量PORT的初始值
#创建Socket对象s，参数分别表示地址和协议类型
s = socket.socket(socket.AF_INET, socket.SOCK_DGRAM)
data = "你好！"                                    #定义变量data的初始值
while data:                                        #如果有data数据
    s.sendto(data.encode('utf-8'),(HOST,PORT))     #则发送信息
    if data=='zaijian':                            #如果data的值是zaijian
        break                                      #则退出循环
    data,addr = s.recvfrom(512)                    #读取信息
```

```
        print("从服务器接收信息：\n",data.decode('utf-8'))    #显示从服务器接收的信息
        data = input('输入信息：\n')                          #信息输入
    s.close()                                                 #关闭连接
```

上述代码使用 Socket 以 UDP 连接方式建立了一个简单的客户端程序，客户端创建 Socket 后，会直接向服务器（localhost 的 10000 端口）发送数据，而没有进行连接。当用户输入"zaijian"时退出 while 循环，关闭连接。本实例执行效果与 TCP"客户端/服务器"连接实例的基本相同。执行效果如图 4-3 所示。

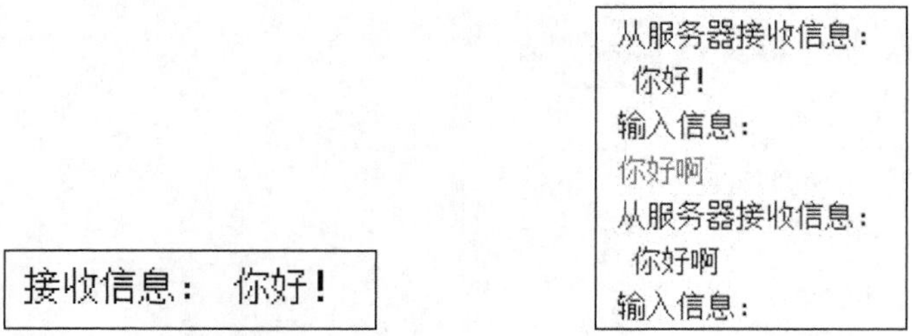

```
接收信息：  你好！

    服务器
```

```
从服务器接收信息：
  你好！
输入信息：
  你好啊
从服务器接收信息：
  你好啊
输入信息：

    客户端
```

图 4-3 执行效果

4.2 实现安全 Socket 编程

在 Python 中，模块 ssl 为客户端和服务器的网络 Socket 提供了安全加密处理机制，可以实现传输层安全（通常称为"安全 Socket"）加密和对等体认证功能的访问。模块 ssl 是通过使用第三方库 OpenSSL 实现的。

实例 04-06：创建 SSL Socket 连接

下面的实例演示了创建 SSL Socket 连接的过程，具体实现代码如下。

（1）SSL 的通信与 TCP 相似，不同之处是首先要生成证书。可以通过 openssl 命令生成 CA 证书，例如：

```
openssl req -new -x509 -days 362 -nodes -out cert.pem -keyout key.pem
```

（2）服务器文件 ssl.py 的具体实现代码如下。

源码路径：daima\4\04-06\ssl.py

```
import socket, ssl, time

bindsocket = socket.socket()
print("socket create success")
bindsocket.bind(('127.0.0.1', 10023))
print("socket bind success")
bindsocket.listen(5)
print("socket listen success")

def do_something(connstream, data):
    print("data length:", len(data))
    return True

def deal_with_client(connstream):
    t_recv = 0
    t_send = 0
    n = 0
    t1 = time.clock()
    data = connstream.recv(1024)
    t2 = time.clock()
    print("receive time:", t2 - t1)
    while data:
```

```python
            if not do_something(connstream, data):
                break
            n = n + 1
            t1 = time.clock()
            connstream.send(b'b' * 1024)
            t2 = time.clock()
            t_send += t2 - t1
            print("send time:", t2 - t1)
            t1 = time.clock()
            data = connstream.recv(1024)
            t2 = time.clock()
            t_recv += t2 - t1
            print("receive time:", t2 - t1)
        print("avg send time:", t_send / n, "avg receive time:", t_recv / n)


while True:
    newsocket, fromaddr = bindsocket.accept()
    print("socket accept one client")
    connstream = ssl.wrap_socket(newsocket, "key.pem", "cert.pem", server_side=True, ssl_version=ssl.PROTOCOL_TLSv1)
    try:
        deal_with_client(connstream)
    finally:
        connstream.shutdown(socket.SHUT_RDWR)
        connstream.close()
```

（3）对应的客户端文件 kehu.py 的具体实现代码如下。

源码路径：daima\4\04-06\kehu.py

```python
import socket, ssl, pprint,time

s = socket.socket(socket.AF_INET, socket.SOCK_STREAM)
print( "socket create success" )
#服务端认证
ssl_sock = ssl.wrap_socket(s,
                           ca_certs="cert.pem",
                           cert_reqs=ssl.CERT_REQUIRED)
ssl_sock.connect(('10.0.0.9', 10023))
print( "socket connect success" )

pprint.pprint(ssl_sock.getpeercert())
n=0
t_send=0
t_recv=0
while n <10:
    n = n+1
    t1=time.clock()
    ssl_sock.send(b'a'*100)
    t2=time.clock()
    t_send += t2-t1
    print("send time:",t2-t1)
    t1=time.clock()
    data=ssl_sock.recv(1024)
    t2=time.clock()
    t_recv += t2-t1
    print("receive time:",t2-t1)
print("avg send time:",t_send/n,"avg receive time:",t_recv/n)
ssl_sock.close()
```

实例 04-07：实现客户端和服务器 SSL 安全交互

实现客户端和服务器 SSL 安全交互的步骤如下。

（1）通过 openssl 命令生成 CA 证书 key.pem，证书生成后会产生两个文件，用文本编辑器打开文件，其中文件 cert.pem 的内容如下。

```
-----BEGIN CERTIFICATE-----
MIIDlTCCAn2gAwIBAgIJAPQtvkMnVSOsMA0GCSqGSIb3DQEBCwUAMGExCzAJBgNV
BAYTAndwMQswCQYDVQQIDAJ3cDELMAkGA1UEBwwCd3AxCzAJBgNVBAoMAndwMQsw
CQYDVQQLDAJ3cDELMAkGA1UEAwwCd3AxETAPBgkqhkiG9w0BCQEWAndwMB4XDTE1
MTAxMDAxNTY0MloXDTE2MTAwOTAxNTY0MlowYTELMAkGA1UEBhMCd3AxCzAJBgNV
BAgMAndwMQswCQYDVQQHDAJ3cDELMAkGA1UECgwCd3AxCzAJBgNVBAsMAndwMQsw
CQYDVQQDDAJ3cDERMA8GCSqGSIb3DQEJARYCd3AwggEiMA0GCSqGSIb3DQEBAQUA
```

```
A4IBDwAwggEKAoIBAQDvUQ5jIeIZaFXWYKR4AUIIyrbkMKj8Xcjvfy4KMB0qXy7V
YuP98Jy/hp5g/YzANq1G+a8p04P9F8HwFO3Z4jYYitqCKuXzos9zGILXarjwFTTA
eiI0Sc2xzYCkD4p8HYKSvwUbVWRYmI3+JIZQgQykT+07AMPbV8/y/rpZRi7OZdb5
XSBXwxvOGD+mZ4xBNurUjX4LXlQaQf4+uXpR9W/TI/6ZF6tl0F4QqnM8IWNO+bq6
PsLNdHWt9+0TjD/iB3offA5bdIqmlReBE5jXcka1g6lIPJfBUdlnylYRX8lvY/EV
YRQcG/IzNsPIsTdhUcDW/I2qMokz98uPG0d8ElYbAgMBAAGjUDBOMB0GA1UdDgQW
BBQ8XDm7KHptC2duHeWZCVslpSogvTAfBgNVHSMEGDAWgBQ8XDm7KHptC2duHeWZ
CVslpSogvTAMBgNVHRMEBTADAQH/MA0GCSqGSIb3DQEBCwUAA4IBAQDRjGB98wwS
aPFIZ2NB2LqoQ95oOndSQQvJJPreJBnMJdgpo7D/C3A8CuuqoN8zP4SWhh4Z6yUj
YEGZ/wZDciUhy7wO9XIdFyv7V87EcUcH6XQqr8co2xz6Ss77mUWBI6SbNdwRdoOr
vM4MLYgl8G0aPIwUb5GZqaNYGs3jRjVNk9xaRLK9MT18noGG3+ULcf9eIgN28aIM
knqStE1w6cZrK8CaqX/8hIxi2031+apDibTsxQprRIJpVnCMjnUYR7xhWKVox7kf
8uSZAKbONspeGNrDhcqEszvRntNgiR9n4s9Hxa9VfSKWEQs9VodioI6p36aNp+th
5fmpo2/rr8wT
-----END CERTIFICATE-----
```

其中文件 key.pem 的内容如下。

```
-----BEGIN PRIVATE KEY-----
MIIEvgIBADANBgkqhkiG9w0BAQEFAASCBKgwggSkAgEAAoIBAQDvUQ5jIeIZaFXW
YKR4AUIIyrbkMKj8Xcjvfy4KMB0qXy7VYuP98Jy/hp5g/YzANq1G+a8p04P9F8Hw
FO3Z4jYYitqCKuXzos9zGILXarjwFTTAeiI0Sc2xzYCkD4p8HYKSvwUbVWRYmI3+
JIZQgQykT+07AMPbV8/y/rpZRi7OZdb5XSBXwxvOGD+mZ4xBNurUjX4LXlQaQf4+
uXpR9W/TI/6ZF6tl0F4QqnM8IWNO+bq6PsLNdHWt9+0TjD/iB3offA5bdIqmlReB
E5jXcka1g6lIPJfBUdlnylYRX8lvY/EVYRQcG/IzNsPIsTdhUcDW/I2qMokz98uP
G0d8ElYbAgMBAAECggEBALTFJXj8Py2yAiTNG28KhDbf6Qa1OvBmZ0GBb+WCMoWv
IFFPQaiv97c0mK3q2EbZBkp2kDmn7Cthpr0TEhdjKDmhqSxp+wjuEoV+HldZ5hhz
7ET3/J5CoX2NHX7PvmvCXr86S0E6X3IMyjUOoeZtlH8JYMkQ6uDkk8+ZWmnU+cYs
8Re5puhqYOJSTQ5bllPmjQKXSUde86Z6JroFFFBT603Q60PCwK3lz4uUv30DRIi1
k4qNNAbfIwQya5hgSWl2+a+Fwo+4+EFcCa/9rOC4MWEfQ+W+X3e7umQa98avBmyb
t2bDLiOrGnEt1kCgGAs+EuHNnbTtmrqR5fBYyHiumwECgYEA+9q/vTvRM1/Rdk5y
iCpyhB4CH72ql9wlsaAYpKKTGMbDbupXm7x1WzlricRCntPHJI8luV0aqXQzGQyy
zijTCh6Uu9zQzGhV8pob+GPb0tsvcDDumg2MHwaatHGerr2jD0xOAs4pUgm6WCi/
V+dJWV1tLUrR5j4wUzbvSxiK+i0CgYEA80F50NAQo7CzbanTZiOw3qPSoSo/GLf/
xKerL5irq8QUznVrzt9E2YvDkiroMSQE7t2mCNavTPP9gboBJotdS5BhBe5c4DI6
H2EN+6Ph0YzMK9OcZv/rMtngjW0/TIbh9SJMDPdvkvFYTpc5Zh689EzgoWTHpo7z
J5Pq0sM5JmcCgYBzkoxOUDbN9nhua81PAvuN+R9MZYL1MQbzHd5xHlGWGw/vxAXz
52WLOSxKCg/wRoqqEi5jg4AKptlT+qnKxR0jFal3E/uU8YQPUfzn3RsxBXgdabb0
ZqcYTbWMfF8vHMLI8GEUFHsNtH0Ym4pC0lXsdlz1XdI4b+1JInpo4ZhU+QKBgQCP
kOorMlFPup77LwrEMnZVwEw0sDFTDm1WjDp9Oda/5lb9TtHU43LIDjPblZU6Q27h
51Dx0HrfqPTKVpQNQp1KVqjcjkSkUSB6mpZdGOjS+w0ZJKgfJhWTCoH8FikQql67
hYVq9bIVAHnE0H2g5q1QZfsBZfBrPd0GU8d4op2dKwKBgG1SQM06q+4jiXzypyok
wcFfk35X2sCuXjxmnKMC24a8BOg3sAK/qg7GF+Q2nUD0FrW/xZ2HD7sQ4+3zLqcz
Ai5aJMer2n28Qn1S0T9cPeO+o5YWNCe+vKbZHuxtLJMCAvf9cGPYCKg19pKVY6cT
AAiCY3adHJHDjU2LxfaG2vr+
-----END PRIVATE KEY-----
```

(2) 服务器文件 ssl1.py 的具体实现代码如下。

源码路径：daima\4\04-07\ssl1.py

```python
import socket, ssl, time

HOST = ''
PORT = 10023
BUFSIZE = 1024
ADDR = (HOST,PORT)
#创建socket成功
bindsocket = socket.socket(socket.AF_INET, socket.SOCK_STREAM)
bindsocket.bind(ADDR)
bindsocket.listen(5)

def do_something(connstream, data):
    return len(data)

def deal_with_client(connstream):
    data = connstream.recv(BUFSIZE)
    while data:
        backdata = do_something(connstream,data)
        if not backdata:
            break
        connstream.send(str(backdata))
        data = connstream.recv(BUFSIZE)

while True:
```

```
        newsocket, fromaddr = bindsocket.accept()
        print("socket accept one client from ",fromaddr)

        connstream = ssl.wrap_socket(newsocket, "key.pem", "cert.pem", server_side=True, ssl_version = ssl.PROTOCOL_TLSv1)

        try:
            deal_with_client(connstream)
        finally:
            connstream.shutdown(socket.SHUT_RDWR)
            connstream.close()
```

（3）对应的客户端文件 kehu1.py 的具体实现代码如下。

源码路径：daima\4\04-07\kehu1.py

```
import socket, ssl, pprint,time

HOST = '10.0.3.83'
PORT = 10023
BUFSIZE = 1024
ADDR = (HOST,PORT)

#创建socket成功
s = socket.socket(socket.AF_INET, socket.SOCK_STREAM)
#服务器认证
ssl_sock = ssl.wrap_socket(s,ca_certs="cert.pem",cert_reqs=ssl.CERT_REQUIRED)
#socket连接成功
ssl_sock.connect(ADDR)

pprint.pprint(ssl_sock.getpeercert())

while True:
    data = input('> ')
    if not data:
        break
    ssl_sock.send(data)
    data=ssl_sock.recv(BUFSIZE)
    if not data:
        break
    print(data)
ssl_sock.close()
```

执行时先启动服务器，然后连接客户端。服务器的执行效果如图 4-4 所示。

```
socket accept one client from ('172.30.1.55', 32984)
```

图 4-4　服务器的执行效果

在客户端和服务器的交互过程中，首先将服务器生成的证书复制到客户端的同路径下，然后执行客户端。客户端的执行效果如图 4-5 所示。

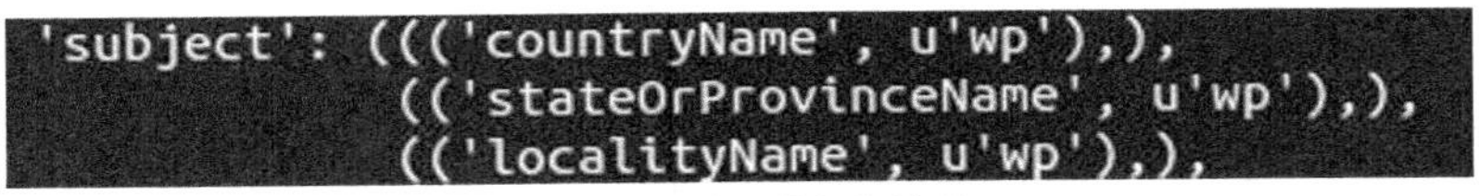

```
'subject': (((('countryName', u'wp'),),
             (('stateOrProvinceName', u'wp'),),
             (('localityName', u'wp'),),
```

图 4-5　客户端的执行效果

4.3　实现 I/O 多路复用

I/O 多路复用指通过一种机制可以监视多个描述符，一旦某个描述符就绪（一般是读就绪或写就绪），程序就进行相应的读写操作。

实例 04-08：使用 select 同时监听多个端口

在 Python 中，select 模块专注于实现 I/O 多路复用功能，提供了 select()、poll()和 epoll() 3 个功能函数，其中，后两个函数在 Linux 操作系统中可用，Windows 仅支持 select()函数。另外，

select 还提供了 kqueue()函数供 FreeBSD 操作系统使用。下面的实例代码演示了使用 select 同时监听多个端口的过程。

(1) 文件 duoser.py 实现了服务器的功能，具体实现代码如下。

源码路径：daima\4\04-08\duoser.py

```python
import socket
import select

sk1 = socket.socket()
sk1.bind(("127.0.0.1",8000))
sk1.listen()

sk2 = socket.socket()
sk2.bind(("127.0.0.1",8002))
sk2.listen()

sk3 = socket.socket()
sk3.bind(("127.0.0.1",8003))
sk3.listen()

li = [sk1,sk2,sk3]

while True:
    r_list,w_list,e_list = select.select(li,[],[],1) # r_list是可变化的
    for line in r_list:
        conn,address = line.accept()
        conn.sendall(bytes("Hello World !",encoding="utf-8"))
```

- select 内部会自动监听 sk1、sk2 和 sk3 这 3 个对象，监听 3 个句柄是否发生变化，把发生变化的元素放入 r_list。
- 如果有人连接 sk1，则 r_list = [sk1]；如果有人连接 sk1 和 sk2，则 r_list = [sk1,sk2]。
- select()中的第 1 个参数表示将 inputs 中发生变化的句柄放入 r_list；select()中的第 2 个参数表示[]中的值原封不动地传递给 w_list；select()中的第 3 个参数表示将 inputs 中发生错误的句柄放入 e_list；参数 1 表示每隔 1s 监听 1 次。
- 当有用户连接时，r_list 的内容为[<socket.socket fd=220, family=AddressFamily.AF_INET, type=SocketKind.SOCK_STREAM, proto=0, laddr=('0.0.0.0', 8001)>]。

(2) 文件 duocli.py 实现了客户端的功能，实现代码非常简单，例如，可通过如下相似的代码建立客户端和两个端口的通信。

源码路径：daima\4\04-08\duocli.py

```python
import socket

obj = socket.socket()
obj.connect(('127.0.0.1', 8001))

content = str(obj.recv(1024), encoding='utf-8')
print(content)

obj.close()

#客户端c2.py
import socket

obj = socket.socket()
obj.connect(('127.0.0.1', 8002))

content = str(obj.recv(1024), encoding='utf-8')
print(content)

obj.close()
```

实例 04-09：模拟多线程并实现读写分离

下面的实例代码演示了使用 select 模拟多线程并实现读写分离的过程。
(1) 文件 fenliser.py 实现了服务器的功能，具体实现代码如下。

源码路径：daima\4\04-09\fenliser.py

```python
#使用select模拟多线程，使多用户可以同时连接
import socket
import select

sk1 = socket.socket()
sk1.bind(('0.0.0.0', 8000))
sk1.listen()

inputs = [sk1, ]
outputs = []
message_dict = {}

while True:
    r_list, w_list, e_list = select.select(inputs, outputs, inputs, 1)
    print('正在监听的Socket对象%d' % len(inputs))
    print(r_list)
    for sk1_or_conn in r_list:
        #每一个连接对象
        if sk1_or_conn == sk1:
            #表示有新用户来连接
            conn, address = sk1_or_conn.accept()
            inputs.append(conn)
            message_dict[conn] = []
        else:
            #有老用户发消息了
            try:
                data_bytes = sk1_or_conn.recv(1024)
            except Exception as ex:
                #如果用户终止连接
                inputs.remove(sk1_or_conn)
            else:
                data_str = str(data_bytes, encoding='utf-8')
                message_dict[sk1_or_conn].append(data_str)
                outputs.append(sk1_or_conn)

    #w_list中仅仅保存了谁给我发过消息
    for conn in w_list:
        recv_str = message_dict[conn][0]
        del message_dict[conn][0]
        conn.sendall(bytes(recv_str+'好', encoding='utf-8'))
        outputs.remove(conn)

    for sk in e_list:

        inputs.remove(sk)
```

(2) 文件 fenlicli.py 实现了客户端的功能，具体实现代码如下。

源码路径：daima\4\04-09\fenlicli.py

```python
import socket
obj = socket.socket()
obj.connect(('127.0.0.1', 8000))

while True:
    inp = input('>>>')
    obj.sendall(bytes(inp, encoding='utf-8'))
    ret = str(obj.recv(1024),encoding='utf-8')
    print(ret)

obj.close()
```

实例 04-10：使用 select 实现一个可并发的服务器

（1）文件 bingser.py 实现了服务器的功能，具体实现代码如下。

源码路径：daima\4\04-10\bingser.py

```python
import socket
import select

s = socket.socket()
s.bind(('127.0.0.1', 8888))
s.listen(5)
r_list = [s, ]
num = 0
while True:
    rl, wl, error = select.select(r_list, [], [], 10)
    num += 1
    print('counts is %s' % num)
    print("rl's length is %s" % len(rl))
    for fd in rl:
        if fd == s:
            conn, addr = fd.accept()
            r_list.append(conn)
            msg = conn.recv(200)
            conn.sendall(('first----%s' % conn.fileno()).encode())
        else:
            try:
                msg = fd.recv(200)
                fd.sendall('second'.encode())
            except ConnectionAbortedError:
                r_list.remove(fd)
s.close()
```

（2）文件 bingcli.py 实现了客户端的功能，具体实现代码如下。

源码路径：daima\4\04-10\bingcli.py

```python
import socket

flag = 1
s = socket.socket()
s.connect(('127.0.0.1', 8888))
while flag:
    input_msg = input('input>>>')
    if input_msg == '0':
        break
    s.sendall(input_msg.encode())
    msg = s.recv(1024)
    print(msg.decode())

s.close()
```

从服务器可以看到，我们需要不停地调用 select()，这充分说明了如下 3 点。
- ❑ 当文件描述符过多时，文件描述符在用户空间与内核空间内进行复制会很费时。
- ❑ 当文件描述符过多时，内核对文件描述符的遍历也很费时。
- ❑ select 最多支持 1024 个文件描述符。

实例 04-11：实现一个可并发的服务器

下面的实例代码演示了使用 epoll() 函数实现一个可并发的服务器的过程。

（1）文件 epollser.py 实现了服务器的功能，具体实现代码如下。

源码路径：daima\4\04-11\epollser.py

```python
import socket
import select

s = socket.socket()
s.bind(('127.0.0.1', 8888))
s.listen(5)
```

```python
epoll_obj = select.epoll()
epoll_obj.register(s, select.EPOLLIN)
connections = {}
while True:
    events = epoll_obj.poll()
    for fd, event in events:
        print(fd, event)
        if fd == s.fileno():
            conn, addr = s.accept()
            connections[conn.fileno()] = conn
            epoll_obj.register(conn, select.EPOLLIN)
            msg = conn.recv(200)
            conn.sendall('ok'.encode())
        else:
            try:
                fd_obj = connections[fd]
                msg = fd_obj.recv(200)
                fd_obj.sendall('ok'.encode())
            except BrokenPipeError:
                epoll_obj.unregister(fd)
                connections[fd].close()
                del connections[fd]

s.close()
epoll_obj.close()
```

（2）文件 epollcli.py 实现了客户端的功能，具体实现代码如下。

源码路径：daima\4\04-11\epollcli.py

```python
import socket

flag = 1
s = socket.socket()
s.connect(('127.0.0.1', 8888))
while flag:
    input_msg = input('input>>>')
    if input_msg == '0':
        break
    s.sendall(input_msg.encode())
    msg = s.recv(1024)
    print(msg.decode())

s.close()
```

实例 04-12：实现高级 I/O 多路复用

在 Python 中，模块 selectors 允许高效地实现 I/O 多路复用，这是建立在普通 select 模块原函数基础之上的。Python 官方鼓励开发者使用 selectors 模块。下面的实例代码演示了使用 selector 实现高级 I/O 多路复用的过程。

（1）文件 ser.py 实现了服务器的功能，具体实现代码如下。

源码路径：daima\4\04-12\ser.py

```python
import selectors  #基于select模块实现的I/O多路复用，建议大家使用
import socket
sock=socket.socket()
sock.bind(('127.0.0.1',8800))
sock.listen(5)
sock.setblocking(False)
sel=selectors.DefaultSelector() #根据操作系统选择最佳的I/O多路机制，例如，Linux就会选择epoll

def read(conn,mask):
    try:
        data=conn.recv(1024)
        print(data.decode('utf8'))
        data2=input('>>>>')
        conn.send(data2.encode('utf8'))
    except Exception:
        sel.unregister(conn)

def accept(sock,mask):
```

```
            conn,addr=sock.accept()
            print('-------',conn)
            sel.register(conn,selectors.EVENT_READ,read)
        sel.register(sock, selectors.EVENT_READ, accept)      #注册功能
        while True:
            print('wating....')
            events=sel.select()       #[(sock), (), ()]       #监听

            for key,mask in events:

                func=key.data
                obj=key.fileobj

                func(obj,mask)
```

（2）文件 cli.py 实现了客户端的功能，具体实现代码如下。

源码路径：daima\4\04-12\cli.py

```
import socket
tin=socket.socket()
tin.connect(('127.0.0.1',8800))
while True:
    inp=input('>>>>')
    tin.send(inp.encode('utf8'))
    data=tin.recv(1024)
    print(data.decode('utf8'))
```

4.4 实现异步 I/O 处理

在 Python 中，内置模块 asyncio 实现了异步 I/O 功能。模块 asyncio 的编程模型是一个消息循环。开发者先从 asyncio 模块中直接获取一个 EventLoop 的引用，然后把需要执行的协程放到 EventLoop 中执行，这样就实现了异步 I/O 功能。

实例 04-13：使用 asyncio 实现 Hello world 代码

下面的实例文件 hello.py 演示了使用 asyncio 实现 Hello world 代码的过程。

源码路径：daima\4\04-13\hello.py

```
import asyncio

@asyncio.coroutine
def hello():
    print("Hello world!")
    #异步调用asyncio.sleep(1):
    r = yield from asyncio.sleep(1)
    print("Hello again!")

#获取EventLoop
loop = asyncio.get_event_loop()
#执行coroutine
loop.run_until_complete(hello())
loop.close()
```

在上述代码中，@asyncio.coroutine 先把一个生成器标记为 coroutine 类型，然后把这个 coroutine 放到 EventLoop 中执行。hello()会首先输出 Hello world!，然后通过 yield from 语法方便地调用另一个生成器。由于 asyncio.sleep()也是一个 coroutine，所以线程不会等待 asyncio.sleep()，而是直接中断并执行下一个消息循环。当 asyncio.sleep()返回时，线程就可以从通过 yield from 得到返回值（此处是 None），然后接着执行下一行语句。把 asyncio.sleep(1)看作一个耗时 1s 的 I/O 操作，在其执行期间，主线程并未等待，而是执行 EventLoop 中其他可以执行的 coroutine，因此可以实现并发执行。

上述实例文件 hello.py 执行后会每隔 1s 输出如下两段文本。

```
Hello world!
Hello again!
```

下面的实例文件 yibu01.py 演示了使用 Task 封装两个 coroutine 的过程。

源码路径：daima\4\04-13\yibu01.py

```python
import threading
import asyncio

@asyncio.coroutine
def hello():
    print('Hello world! (%s)' % threading.currentThread())
    yield from asyncio.sleep(1)
    print('Hello again! (%s)' % threading.currentThread())

loop = asyncio.get_event_loop()
tasks = [hello(), hello()]
loop.run_until_complete(asyncio.wait(tasks))
loop.close()
```

执行后会输出：

```
Hello world! (<_MainThread(MainThread, started 1328)>)
Hello world! (<_MainThread(MainThread, started 1328)>)
(暂停约1s)
Hello again! (<_MainThread(MainThread, started 1328)>)
Hello again! (<_MainThread(MainThread, started 1328)>)
```

由输出的当前线程名称可以看出，两个 coroutine 是由同一个线程并发执行的。如果把函数 asyncio.sleep()换成真正的 I/O 操作，则多个 coroutine 可以由一个线程并发执行。

实例 04-14：使用 asyncio 获取网站首页信息

下面的实例文件 yibu02.py 演示了使用 asyncio 获取网站首页信息的过程。

源码路径：daima\4\04-14\yibu02.py

```python
import asyncio

@asyncio.coroutine
def wget(host):
    print('wget %s...' % host)
    connect = asyncio.open_connection(host, 80)
    reader, writer = yield from connect
    header = 'GET / HTTP/1.0\r\nHost: %s\r\n\r\n' % host
    writer.write(header.encode('utf-8'))
    yield from writer.drain()
    while True:
        line = yield from reader.readline()
        if line == b'\r\n':
            break
        print('%s header > %s' % (host, line.decode('utf-8').rstrip()))
    writer.close()

loop = asyncio.get_event_loop()
tasks = [wget(host) for host in ['www.sina.com.cn', 'www.sohu.com', 'www.163.com']]
loop.run_until_complete(asyncio.wait(tasks))
loop.close()
```

执行后会输出：

```
wget www.sohu.com...
wget www.sina.com.cn...
wget www.163.com...
(等待一段时间)
(输出sohu的header)
www.sohu.com header > HTTP/1.1 200 OK
www.sohu.com header > Content-Type: text/html
...
(输出sina的header)
www.sina.com.cn header > HTTP/1.1 200 OK
www.sina.com.cn header > Date: Wed, 20 May 2015 04:56:33 GMT
...
(输出163的header)
www.163.com header > HTTP/1.0 302 Moved Temporarily
www.163.com header > Server: Cdn Cache Server V2.0
...
```

实例 04-15：以动画的方式显示文本式旋转指针

Python 并没有提供终止线程的 API，所以要想关闭线程，必须给线程发送消息。下面的实例文件 yibu03.py 演示了通过协程以动画的方式显示文本式旋转指针的过程。

源码路径：daima\4\04-15\yibu03.py

```python
import asyncio
import itertools
import sys

@asyncio.coroutine                          #打算交给asyncio处理的协程要使用@asyncio.coroutine装饰
def spin(msg):
    write, flush = sys.stdout.write, sys.stdout.flush
    for char in itertools.cycle('|/-\\'):    # itertools.cycle()函数从指定的序列中反复不断地生成元素
        status = char + ' ' + msg
        write(status)
        flush()
        write('\x08' * len(status))          #使用退格符把光标移回行首
        try:
            yield from asyncio.sleep(0.1)
            #使用yield from asyncio.sleep(0.1)代替time.sleep(.1)，这样的休眠不会阻塞事件循环
        except asyncio.CancelledError:
            #spin()函数苏醒后抛出asyncio.CancelledError异常
            break
    write(' ' * len(status) + '\x08' * len(status))  #使用空格清除状态消息，把光标移回开头

@asyncio.coroutine
def slow_function():    #现在此函数是协程，当用休眠假装进行I/O操作时，使用yield from继续执行事件循环
    # 假装等待I/O一段时间
    yield from asyncio.sleep(3)    #此表达式用于把控制权交给主循环，在休眠结束后恢复这个协程
    return 42


@asyncio.coroutine
def supervisor():     #这个函数也是协程，因此可以使用yield from驱动slow_function()
    spinner = asyncio.async(spin('thinking!'))
    # asyncio.async()函数排定协程的执行时间，使用一个Task对象包装spin()协程，并立即返回
    print('spinner object:', spinner)
    # Task对象，输出类似spinner object: <Task pending coro=<spin() running at spinner_asyncio.py:6>>
    #驱动slow_function()函数，结束后，获取返回值。同时事件循环继续执行
    #因为slow_function()函数最后使用yield from asyncio.sleep(3)表达式把控制权交给主循环
    result = yield from slow_function()
    #Task对象可以取消；取消后会在协程当前暂停的yield处抛出asyncio.CancelledError异常
    #协程可以捕获这个异常，也可以延迟取消，甚至拒绝取消
    spinner.cancel()
    return result


def main():
    loop = asyncio.get_event_loop()    #获取事件循环引用
    #驱动supervisor()协程，让它执行完毕；这个协程的返回值是这次调用的返回值
    result = loop.run_until_complete(supervisor())
    loop.close()
    print('Answer', result)


if __name__ == '__main__':
    main()
```

执行后会输出：

```
| thinking!
/ thinking!
- thinking!
\ thinking!
| thinking!
/ thinking!
- thinking!
\ thinking!
| thinking!
```

```
/ thinking!
- thinking!
\ thinking!
| thinking!
/ thinking!
- thinking!
\ thinking!
| thinking!
/ thinking!
- thinking!
\ thinking!
| thinking!
/ thinking!
- thinking!
\ thinking!
| thinking!
/ thinking!
- thinking!
\ thinking!
| thinking!
/ thinking!
Answer 42
```

4.5 实现异步 Socket 处理

在 Python 中，内置模块 asyncore 提供了以异步的方式写入 Socket 服务的客户端和服务器的基础结构。

实例 04-16：使用模块 asyncore 实现一个基本的 HTTP 客户端

在 Python 程序中，其内置模块 asyncore 和 asynchat 的基本思路是创建一个或多个网络通道，即 asyncore.dispatcher 和 asynchat.async_chat 的实例，然后将其添加到全局映射，如果没有创建自己的映射，则可以直接使用 loop()函数。函数 loop()能够激活所有的通道服务，直到执行到最后一个通道关闭为止。例如，下面的实例文件 cli.py 使用模块 asyncore 实现了一个基本的 HTTP 客户端，演示了使用类 dispatcher 实现了其 Socket 处理的过程。

源码路径：daima\4\04-16\cli.py

```python
import asyncore
class HTTPClient(asyncore.dispatcher):

    def __init__(self, host, path):
        asyncore.dispatcher.__init__(self)
        self.create_socket()
        self.connect( (host, 80) )
        self.buffer = bytes('GET %s HTTP/1.0\r\nHost: %s\r\n\r\n' %(path, host), 'ascii')

    def handle_connect(self):
        pass

    def handle_close(self):
        self.close()

    def handle_read(self):
        print(self.recv(8192))

    def writable(self):
        return (len(self.buffer) > 0)

    def handle_write(self):
        sent = self.send(self.buffer)
        self.buffer = self.buffer[sent:]

client = HTTPClient('www.python.org', '/')
asyncore.loop()
```

执行后会输出：

```
b'HTTP/1.1 301 Moved Permanently\r\nServer: Varnish\r\nRetry-After: 0\r\nLocation: https://www.python.org/\r\nContent-Length: 0\r\nAccept-Ranges: bytes\r\nDate: Tue, 06 Mar 2018 02:21:22 GMT\r\nVia: 1.1 varnish\r\nConnection: close\r\nX-Served-By: cache-hnd18734-HND\r\nX-Cache: HIT\r\nX-Cache-Hits: 0\r\nX-Timer: S1520302883.680251,VS0,VE0\r\nStrict-Transport-Security: max-age=63072000; includeSubDomains\r\n\r\nb"
```

实例 04-17：使用模块 asyncore 响应客户端发送数据

下面的实例文件 asy01.py 演示了使用模块 asyncore 响应客户端发送数据的过程。

源码路径：daima\4\04-17\asy01.py

```python
class EchoHandler(asyncore.dispatcher_with_send):

    def handle_read(self):
        data = self.recv(1024)
        if data:
            self.send(data)

class EchoServer(asyncore.dispatcher):

    def __init__(self, host, port):
        asyncore.dispatcher.__init__(self)
        self.create_socket(socket.AF_INET, socket.SOCK_STREAM)
        self.set_reuse_addr()
        self.bind((host, port))
        self.listen(5)

    def handle_accept(self):
        conn, addr = self.accept()
        print('Incoming connection from %s' % repr(addr))
        self.handler = EchoHandler(conn)

class EchoClient(asyncore.dispatcher):

    def __init__(self, host, port):
        asyncore.dispatcher.__init__(self)
        self.messages = ['1', '2', '3', '4', '5', '6', '7', '8', '9', '10']
        self.create_socket(socket.AF_INET, socket.SOCK_STREAM)
        self.connect((host, port))

    def handle_connect(self):
        pass

    def handle_close(self):
        self.close()

    def handle_read(self):
        print(self.recv(1024))

    def writable(self):
        return (len(self.messages) > 0)

    def handle_write(self):
        if len(self.messages) > 0:
            self.send(self.messages.pop(0))

class EchoServerThread(threading.Thread):
    def __init__(self):
        threading.Thread.__init__(self)

    def run(self):
        server = EchoServer('localhost', 8888)
        asyncore.loop()

class EchoClientThread(threading.Thread):
    def __init__(self):
        threading.Thread.__init__(self)

    def run(self):
        client = EchoClient('localhost', 8888)
        asyncore.loop()
```

```
EchoServerThread().start()
time.sleep(2)
EchoClientThread().start()
```

(1) EchoServer：响应服务器程序，负责监听一个端口，并响应客户端发送的数据，然后将其原样返回给客户端。其中，handle_accept()函数定义当一个连接到来的时候要执行的操作，这里指定了使用一个 Handler 来发送数据。

(2) EchoHandler：服务器数据响应类，接收数据并把数据原样发回。

(3) EchoClient：响应客户端程序，负责连接响应服务器，包含如下 3 个函数。

- messages：定义了一个要发送的数据列表，每次发送一个数据，直到列表为空为止。
- handle_read()：处理接收到的数据，这里把收到的数据输出到终端上。
- writable()：判断是否有数据可以向服务器发送。
- handle_write()：当 writable()函数返回 True 时，写入数据。

(4) EchoServerThread：用于启动服务器程序的线程。

(5) EchoClientThread：用于启动客户端程序的线程。

执行以上代码后可以看到如下输出，当服务器和客户端建立了连接后，会响应客户端发送来的 10 个数字，然后关闭连接。

```
Incoming connection from ('127.0.0.1', 51424)
1
2
3
4
5
6
7
8
9
10
```

4.6 实现内存映射

在 Python 中，内置模块 mmap 用于实现对大文件的处理。mmap 是一种虚拟内存映射文件的方法，即将一个文件或者其它对象映射到进程的地址空间，实现文件磁盘地址和进程虚拟地址空间中虚拟地址的一一对映关系。

实例 04-18：读取文件 test.txt 的内容

下面的实例文件 mm01.py 演示了使用模块 mmap 的内置成员读取文件 test.txt 的内容的过程。

源码路径：daima\4\04-18\mm01.py

```
import os,mmap
m=mmap.mmap(os.open('test.txt',os.O_RDWR),0)   #创建内存映射对象
print(m.read(10))                               #可以使用方法
print(m.read(10))                               #读取10字节的字符串
print(m.read(10))                               #读取上面10字节再往后的10字节的字符串
print(m.read_byte())                            #读取第1个字节
print(m.read_byte())                            #读取第2个字节
print(m.read_byte())                            #读取第3个字节
print(m.readline())                             #读取一整行
print(m.readline())                             #读取下一整行
print(m.readline())                             #读取下一整行
print(m.readline())                             #读取下一整行
print(m.readline())                             #读取下一整行
print(m.readline())                             #读取下一整行
m.close()                                       #关闭对象
```

执行后会输出：
```
b'\xef\xbb\xbf-- MySQ'
b'L dump 10.'
b'13    Distri'
```

```
98
32
53
b'.6.19, for osx10.7 (x86_64)\r\n'
b'--\r\n'
b'-- Host: localhost    Database: test\r\n'
b'-- -------------------------------------------------------\r\n'
b'-- Server version       4.6.19\r\n'
b'\r\n'
```

实例 04-19：读取整个文件 test.txt 的内容

下面的实例文件 mm02.py 演示了使用模块 mmap 读取整个文件 test.txt 内容的过程。

源码路径：daima\4\04-19\mm02.py

```python
import mmap
import contextlib

f = open('test.txt', 'r')
with contextlib.closing(mmap.mmap(f.fileno(), 0,access=mmap.ACCESS_READ)) as m:
    #readline()需要循环才能读取整个文件
    while True:
        line = m.readline().strip()
        print(line)
        #光标移到最后位置（即读完），就退出
        if m.tell()==m.size():
            break
```

执行后会输出：

```
b'\xef\xbb\xbf-- MySQL dump 10.13   Distrib 4.6.19, for osx10.7 (x86_64)'
b'--'
b'-- Host: localhost    Database: test'
b'-- -------------------------------------------------------'
b'-- Server version       4.6.19'
b''
b'/*!40101 SET @OLD_CHARACTER_SET_CLIENT=@@CHARACTER_SET_CLIENT */;'
b'/*!40101 SET @OLD_CHARACTER_SET_RESULTS=@@CHARACTER_SET_RESULTS */;'
b'/*!40101 SET @OLD_COLLATION_CONNECTION=@@COLLATION_CONNECTION */;'
b'/*!40101 SET NAMES utf8 */;'
b'/*!40103 SET @OLD_TIME_ZONE=@@TIME_ZONE */;'
b"/*!40103 SET TIME_ZONE='+00:00' */;"
b'/*!40014 SET @OLD_UNIQUE_CHECKS=@@UNIQUE_CHECKS, UNIQUE_CHECKS=0 */;'
b'/*!40014 SET @OLD_FOREIGN_KEY_CHECKS=@@FOREIGN_KEY_CHECKS, FOREIGN_KEY_CHECKS=0 */;'
b"/*!40101 SET @OLD_SQL_MODE=@@SQL_MODE, SQL_MODE='NO_AUTO_VALUE_ON_ZERO' */;"
b'/*!40111 SET @OLD_SQL_NOTES=@@SQL_NOTES, SQL_NOTES=0 */;'
```

实例 04-20：逐步读取文件 test.txt 中的指定字节数内容

下面的实例文件 mm03.py 演示了使用模块 mmap 逐步读取文件 test.txt 中的指定字节数内容的过程。

源码路径：daima\4\04-20\mm03.py

```python
import mmap
import contextlib

with open('test.txt', 'r') as f:
    with contextlib.closing(mmap.mmap(f.fileno(), 0,access=mmap.ACCESS_READ)) as m:
        print('读取10字节的字符串:', m.read(10))
        print('支持切片，对读取到的字符串进行切片操作:', m[2:10])
        print('读取之前光标后的10字节的字符串:', m.read(10))
```

执行后会输出：

```
读取10字节的字符串: b'\xef\xbb\xbf-- MySQ'
支持切片，对读取到的字符串进行切片操作: b'\xbf-- MySQ'
读取之前光标后的10字节的字符串:b'L dump 10.'
```

4.7 socketserver 编程

Python 提供了高级别的网络服务模块 socketserver，该模块提供了服务器类，可以简化网络服务器的开发步骤。

实例 04-21：使用 socketserver 创建 TCP "客户端/服务器"程序

socketserver 模块使用的服务器类主要有 TCPServer、UDPServer、ThreadingTCPServer、ThreadingUDPServer、ForkingTCPServer、ForkingUDPServer 等。其中，有 TCP 字符的是使用 TCP 的服务器类，有 UDP 字符的是使用 UDP 的服务器类，有 Threading 字符的是多线程服务器类，有 Forking 字符的是多进程服务器类。要创建不同类型的服务器程序，只需继承其中之一或直接实例化，然后调用服务器类方法 serve_forever() 即可。

在下面的实例代码中，演示了使用 socketserver 创建 TCP "客户端/服务器"程序的过程，本实例使用 socketserver 创建了一个可靠的、相互通信的"客户端/服务器"。

实例文件 ser.py 的功能是使用 socketserver 模块创建基于 TCP 的服务器程序，能够将收到的信息直接发回客户端。文件 socketserverser.py 的具体实现代码如下。

源码路径：daima\4\04-21\socketserverser.py

```
#定义类StreamRequestHandler的子类MyTcpHandler
class MyTcpHandler(socketserver.StreamRequestHandler):
    def handle(self):                          #定义函数handle()
        while True:
            data = self.request.recv(1024)     #返回接收到的信息
            if not data:
                Server.shutdown()              #关闭连接
                break                          #退出循环
            print('接收信息：',data.decode('utf-8')) #显示接收信息
            self.request.send(data)            #发送信息
        return
#定义类TCPServer的对象实例
Server = socketserver.TCPServer((HOST,PORT),MyTcpHandler)
Server.serve_forever()                         #循环并等待其停止
```

上述实例代码自定义了一个继承 StreamRequestHandler 的子类，并覆盖了函数 handle() 以实现信息处理，然后直接实例化类 TCPServer，调用方法 serve_forever() 启动服务器。

客户端实例文件 socketservercli.py 的和前面的实例文件 socketservercli.py 完全相同，本实例的执行效果如图 4-6 所示。

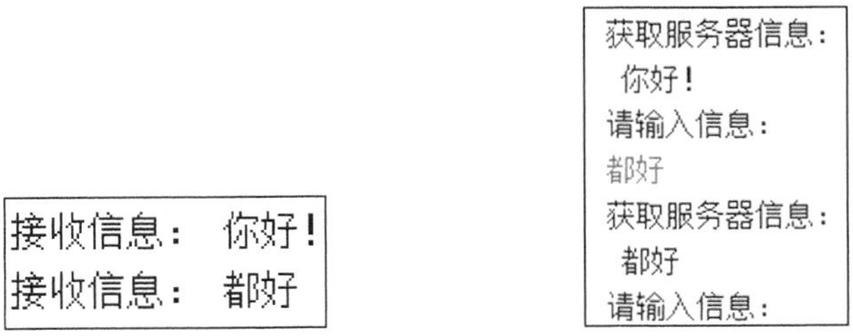

```
接收信息： 你好！          获取服务器信息：
接收信息： 都好            你好！
                        请输入信息：
                        都好
                        获取服务器信息：
                        都好
                        请输入信息：
      服务器                    客户端
```

图 4-6 执行效果

实例 04-22：使用 ThreadingTCPServer 创建 "客户端/服务器" 通信程序

在 ThreadingTCPServer 实现的 Soket 服务器内部，会为每一个客户端创建一个"线程"，该线程用于和客户端进行交互。在 Python 程序中使用 ThreadingTCPServer 的步骤如下。

第 4 章 进程通信和联网操作实战

（1）创建一个继承 SocketServer.BaseRequestHandler 的类。
（2）必须在类中定义一个名为 handle 的方法。
（3）启动 ThreadingTCPServer。

下面的实例代码演示了使用 ThreadingTCPServer 创建"客户端/服务器"通信程序的过程。

（1）实例文件 ser.py 的功能是使用 socketserver 模块创建服务器程序，能够将收到的信息直接发回客户端。文件 ser.py 的具体实现代码如下。

源码路径：daima\4\04-22\ser.py

```python
import socketserver

class Myserver(socketserver.BaseRequestHandler):

    def handle(self):

        conn = self.request
        conn.sendall(bytes("你好，我是机器人",encoding="utf-8"))
        while True:
            ret_bytes = conn.recv(1024)
            ret_str = str(ret_bytes,encoding="utf-8")
            if ret_str == "q":
                break
            conn.sendall(bytes(ret_str+"你好我好大家好",encoding="utf-8"))

if __name__ == "__main__":
    server = socketserver.ThreadingTCPServer(("127.0.0.1",8000),Myserver)
    server.serve_forever()
```

（2）实例文件 cli.py 的功能是使用 socketserver 模块创建客户端程序，能够接收服务器发回的信息。文件 cli.py 的具体实现代码如下。

源码路径：daima\4\04-22\cli.py

```python
import socket
obj = socket.socket()
obj.connect(("127.0.0.1",8000))
ret_bytes = obj.recv(1024)
ret_str = str(ret_bytes,encoding="utf-8")
print(ret_str)

while True:
    inp = input("您好，请问您有什么问题？ \n >>>")
    if inp == "q":
        obj.sendall(bytes(inp,encoding="utf-8"))
        break
    else:
        obj.sendall(bytes(inp, encoding="utf-8"))
        ret_bytes = obj.recv(1024)
        ret_str = str(ret_bytes,encoding="utf-8")
        print(ret_str)
```

第 5 章

结构化标记处理实战

互联网技术应用中存在着大量的标记语言，如 HTML 和 XML。Python 提供了内置模块来处理各种形式的结构化标记，包括使用标准通用标记语言（Standard Generalized Markup Language，SGML）、超文本标记语言（Hyper Text Markup Language，HTML）的模块，以及使用可扩展标记语言（Extensible Markup Language，XML）的几个接口。本章将通过具体实例的实现过程，详细讲解使用 Python 处理结构化标记的知识，为读者学习本书后面的知识打下基础。

5.1 使用内置模块 html

在 Python 中，内置模块 html 可以操作 HTML。模块 html 包含如下两个子模块。
- html.parser：具有宽松解析模式的 HTML/XHTML 解析器。
- html.entities：HTML 实体的定义。

实例 05-01：使用 html.parser 创建 HTML 解析器

在 Python 中，模块 html.parser 定义了一个类 HTMLParser，作为解析以 HTML 和 XHTML 格式化的文本文件的基础。下面的实例文件 html01.py 演示了使用 html.parser 创建 HTML 解析器，并解析不同类型 HTML 元素的过程。

源码路径：daima\5\05-01\html01.py

```python
from html.parser import HTMLParser
from html.entities import name2codepoint

class MyHTMLParser(HTMLParser):
    def handle_starttag(self, tag, attrs):
        print("Start tag:", tag)
        for attr in attrs:
            print("     attr:", attr)

    def handle_endtag(self, tag):
        print("End tag   :", tag)

    def handle_data(self, data):
        print("Data     :", data)

    def handle_comment(self, data):
        print("Comment  :", data)

    def handle_entityref(self, name):
        c = chr(name2codepoint[name])
        print("Named ent:", c)

    def handle_charref(self, name):
        if name.startswith('x'):
            c = chr(int(name[1:], 16))
        else:
            c = chr(int(name))
        print("Num ent  :", c)

    def handle_decl(self, data):
        print("Decl     :", data)

parser = MyHTMLParser()
#解析DOCTYPE
parser.feed('<!DOCTYPE HTML PUBLIC "-//W3C//DTD HTML 4.01//EN" '
            '"http://www.w3.org/TR/html4/strict.dtd">')
#解析具有几个属性的元素
parser.feed('<img src="python-logo.png" alt="The Python logo">')
parser.feed('<script type="text/javascript">'
            'alert("<strong>hello!</strong>");</script>')
#script和style元素的内容按原样返回，无须进一步解析
parser.feed('<style type="text/css">#python { color: green }</style>')
parser.feed('<script type="text/javascript">'
            'alert("<strong>hello!</strong>");</script>')
#解析注释
parser.feed('<!-- a comment -->'
            '<!--[if IE 9]>IE-specific content<![endif]-->')
#解析命名和数字字符引用并将它们转换为正确的字符（注意：这3个引用都等效于'>'）
parser.feed('&gt;&#62;&#x3E;')
#向feed()提供不完整的块，但handle_data()可能会被调用多次（除非convert_charrefs设置为True）
for chunk in ['<sp', 'an>buff', 'ered ', 'text</s', 'pan>']:
    parser.feed(chunk)
```

```
#解析无效的HTML标记（如无属性）时也工作
parser.feed('<p><a class=link href=#main>tag soup</p ></a>')
```

执行后会输出：

```
Decl       : DOCTYPE HTML PUBLIC "-//W3C//DTD HTML 4.01//EN" "http://www.w3.org/TR/html4/strict.dtd"
Start tag: img
     attr: ('src', 'python-logo.png')
     attr: ('alt', 'The Python logo')
Start tag: script
     attr: ('type', 'text/javascript')
Data       : alert("<strong>hello!</strong>");
End tag    : script
Start tag: style
     attr: ('type', 'text/css')
Data       : #python { color: green }
End tag    : style
Start tag: script
     attr: ('type', 'text/javascript')
Data       : alert("<strong>hello!</strong>");
End tag    : script
Comment    :      a comment
Comment    : [if IE 9]>IE-specific content<![endif]
Data       : >>>
Start tag: span
Data       : buff
Data       : ered
Data       : text
End tag    : span
Start tag: p
Start tag: a
     attr: ('class', 'link')
     attr: ('href', '#main')
Data       : tag soup
End tag    : p
End tag    : a
```

实例 05-02：使用 html.entities 解析 HTML

在 Python 中，子模块 html.entities 定义了 HTML 的实体，其内置成员由如下 4 个字典构成。

- html.entities.html5：将 HTML5 命名字符引用映射到等效的 Unicode 字符的字典，如 html5 ['gt;'] == '>>'。注意，尾部分号包括在名称中（如'gt;'），但是有些名称在没有分号的情况下也被标准接受；在这种情况下，名称存在于没有';'。
- html.entities.entitydefs：将 XHTML 1.0 实体定义映射到 ISO Latin-1 中的替换文本的字典。
- html.entities.name2codepoint：将 HTML 实体名称映射到 Unicode 代码点的字典。
- html.entities.codepoint2name：将 Unicode 代码点映射到 HTML 实体名称的字典。

下面的实例文件 html02.py 演示了使用 html.entities 解析指定 HTML 内容的过程。

源码路径：daima\5\05-02\html02.py

```python
from html.parser import HTMLParser
import html.entities

class BaseHTMLProcessor(HTMLParser):
    def reset(self):
        self.pieces = []
        HTMLParser.reset(self)
    #调用每一个开始标记
    def handle_starttag(self, tag, attrs):
        strattrs = "".join([' %s="%s"' % (key, value) for key, value in attrs])
        self.pieces.append("<%(tag)s%(strattrs)s>" % locals())

    def handle_endtag(self, tag):
        #调用每一个结束标记
        self.pieces.append("</%(tag)s>" % locals())
```

```python
def handle_charref(self, ref):
    #调用每一个字符引用.
    self.pieces.append("&#%(ref)s;" % locals())

def handle_entityref(self, ref):
    #调用每一个实体引用
    self.pieces.append("&%(ref)s" % locals())
    #使用分号关闭标准的HTML实体
    if html.entities.entitydefs.has_key(ref):
        self.pieces.append(";")

def handle_data(self, text):
    #调用每一个纯文本块，即在任何标记之外的文本
    #不包含任何字符或实体引用
    self.pieces.append(text)

def handle_comment(self, text):
    #调用每一个HTML注释
    self.pieces.append("<!--%(text)s-->" % locals())

def handle_pi(self, text):
    #调用每一个处理指令
    self.pieces.append("<?%(text)s>" % locals())

def handle_decl(self, text):
    #调用DOCTYPE
    #重建原始文档
    self.pieces.append("<!%(text)s>" % locals())

def output(self):
    """将处理后的HTML标记作为单个字符串返回"""
    return "".join(self.pieces)

if __name__ == '__main__':
    a = '<!DOCTYPE html PUBLIC "-//W3C//DTD HTML 4.01 Transitional//EN" "http://www.w3.org/TR/html4/loose.dtd"> \
        <html><head><!--insert javaScript here!--><title>test</title><body><a href="http: //www.163.com">链接到 \
        163</a></body></html>'
    bhp = BaseHTMLProcessor()
    bhp.feed(a)
    print(bhp.output())
```

执行后会输出：

<!DOCTYPE html PUBLIC "-//W3C//DTD HTML 4.01 Transitional//EN" "http://www.w3.org/TR/html4/loose.dtd"> <html><head><!--insert javaScript here!--><title>test</title><body><a href="http: //www.163.com">链接到163</a></body></html>

5.2 使用内置模块解析 XML

在 Python 应用程序中，常见的 XML 编程接口有两种，分别是 SAX 和 DOM。所以与之对应的是，Python 有两种解析 XML 文件的方法，分别是 SAX 方法和 DOM 方法。Python 通过库 xml 实现 XML 处理功能，库 xml 由如下核心模块构成。

- ❑ xml.etree.ElementTree：ElementTree API，一个简单、轻量的 XML 处理器。
- ❑ xml.dom：DOM API。
- ❑ xml.dom.minidom：最小的 DOM 实现。
- ❑ xml.dom.pulldom：支持构建部分 DOM 树。
- ❑ xml.sax：SAX2 基类和便利函数。
- ❑ xml.parsers.expat：Expat 解析器绑定。

实例 05-03：使用模块 **xml.etree.ElementTree** 读取 XML 文件

在 Python 程序中，Element 是一种灵活的容器对象，用于在内存中存储结构化数据。需要注意的是，模块 xml.etree.ElementTree 在应对恶意结构的数据时并不安全。下面的实例代码演

5.2 使用内置模块解析 XML

示了使用模块 xml.etree.ElementTree 读取 XML 文件的过程。

（1）XML 文件 test.xml 的具体实现代码如下。

源码路径：daima\5\05-03\test.xml

```xml
<students>
    <student name='刘玉' sex='男' age='35'/>
    <student name='吕仁' sex='男' age='38'/>
    <student name='刘文' sex='女' age='22'/>
</students>
```

（2）文件 ElementTreeuse.py 的功能是获取上述 XML 文件中的节点元素，具体实现代码如下。

源码路径：daima\5\05-03\ElementTreeuse.py

```python
#从文件中读取数据
import xml.etree.ElementTree as ET

#全局唯一标识
unique_id = 1

#遍历所有的节点
def walkData(root_node, level, result_list):
    global unique_id
    temp_list = [unique_id, level, root_node.tag, root_node.attrib]
    result_list.append(temp_list)
    unique_id += 1

    #遍历每个子节点
    children_node = root_node.getchildren()
    if len(children_node) == 0:
        return
    for child in children_node:
        walkData(child, level + 1, result_list)
    return

#获得原始数据
def getXmlData(file_name):
    level = 1    #节点的深度从1开始
    result_list = []
    root = ET.parse(file_name).getroot()
    walkData(root, level, result_list)

    return result_list


if __name__ == '__main__':
    file_name = 'test.xml'
    R = getXmlData(file_name)
    for x in R:
        print(x)
        pass
```

执行后会输出：

```
[1, 1, 'students', {}]
[2, 2, 'student', {'name': '刘玉', 'sex': '男', 'age': '35'}]
[3, 2, 'student', {'name': '吕仁', 'sex': '男', 'age': '38'}]
[4, 2, 'student', {'name': '刘文', 'sex': '女', 'age': '22'}]
```

实例 05-04：使用 SAX 方法解析 XML 文件

Python 的标准库包含了 SAX 解析器，SAX 通过使用事件驱动模型，在解析 XML 的过程中触发事件并调用用户定义的回调函数来处理 XML 文件。下面的实例代码演示了使用 SAX 方法解析 XML 文件的过程。

（1）实例文件 movies.xml 是一个基本的 XML 文件，保存了一些和电影有关的资料信息。文件 movies.xml 的具体实现代码如下。

源码路径：daima\5\05-04\movies.xml

```xml
<collection shelf="New Arrivals">
<movie title="AAA">
   <type>AAAA</type>
   <format>DVD</format>
   <year>2003</year>
   <rating>PG</rating>
   <stars>10</stars>
   <description>AAAA</description>
</movie>
<movie title="BBB">
   <type>BBB</type>
   <format>DVD</format>
   <year>1989</year>
   <rating>R</rating>
   <stars>8</stars>
   <description>BBB</description>
</movie>
<movie title="CCC">
   <type>CCC</type>
   <format>DVD</format>
   <episodes>4</episodes>
   <rating>PG</rating>
   <stars>10</stars>
   <description>CCC</description>
</movie>
<movie title="DDD">
   <type>DDD</type>
   <format>VHS</format>
   <rating>PG</rating>
   <stars>2</stars>
   <description>DDD</description>
</movie>
</collection>
```

(2) 实例文件 sax.py 的功能是解析文件 movies.xml 的内容，具体实现代码如下。

源码路径：daima\5\05-04\sax.py

```python
import xml.sax
class MovieHandler( xml.sax.ContentHandler ):
   def __init__(self):
      self.CurrentData = ""
      self.type = ""
      self.format = ""
      self.year = ""
      self.rating = ""
      self.stars = ""
      self.description = ""
   #元素开始调用
   def startElement(self, tag, attributes):
      self.CurrentData = tag
      if tag == "movie":
         print ("*****Movie*****")
         title = attributes["title"]
         print ("Title:", title)
   #元素结束调用
   def endElement(self, tag):
      if self.CurrentData == "type":          #处理XML中的type元素
         print ("Type:", self.type)
      elif self.CurrentData == "format":      #处理XML中的format元素
         print ("Format:", self.format)
      elif self.CurrentData == "year":        #处理XML中的year元素
         print ("Year:", self.year)
      elif self.CurrentData == "rating":      #处理XML中的rating元素
         print ("Rating:", self.rating)
      elif self.CurrentData == "stars":       #处理XML中的stars元素
         print ("Stars:", self.stars)
      elif self.CurrentData == "description": #处理XML中的description元素
         print ("Description:", self.description)
      self.CurrentData = ""
   #读取字符时调用
   def characters(self, content):
```

```python
        if self.CurrentData == "type":
            self.type = content
        elif self.CurrentData == "format":
            self.format = content
        elif self.CurrentData == "year":
            self.year = content
        elif self.CurrentData == "rating":
            self.rating = content
        elif self.CurrentData == "stars":
            self.stars = content
        elif self.CurrentData == "description":
            self.description = content
if ( __name__ == "__main__"):
    #创建一个XMLReader
    parser = xml.sax.make_parser()
    parser.setFeature(xml.sax.handler.feature_namespaces, 0)
    #重写ContentHandler
    Handler = MovieHandler()
    parser.setContentHandler( Handler )
    parser.parse("movies.xml")
```

执行效果如图 5-1 所示。

```
*****Movie*****
Title: AAA
Type: AAAA
Format: DVD
Year: 2003
Rating: PG
Stars: 10
Description: AAAA
*****Movie*****
Title: BBB
Type: BBB
Format: DVD
Year: 1989
Rating: R
Stars: 8
Description: BBB
*****Movie*****
Title: CCC
Type: CCC
Format: DVD
Rating: PG
Stars: 10
Description: CCC
*****Movie*****
Title: DDD
Type: DDD
Format: VHS
Rating: PG
Stars: 2
Description: DDD
```

图 5-1 执行效果

实例 05-05：使用 DOM 解析 XML 文件

一个 DOM 解析器在解析一个 XML 文件时，可以一次性读取整个文件。将文件中的所有元素保存在内存的一个树结构中后，可以利用 DOM 提供的不同函数来读取或修改文件的内容和结构，也可以把修改过的内容写入到 XML 文件中。下面的实例文件 dom.py 演示了使用 DOM 解析 XML 文件的过程。

实例文件 dom.py 的功能是解析文件 movies.xml 的内容，具体实现代码如下。

源码路径：daima\5\05-05\dom.py

```python
from xml.dom.minidom import parse
import xml.dom.minidom
#使用minidom解析器打开XML文件
DOMTree = xml.dom.minidom.parse("movies.xml")
collection = DOMTree.documentElement
if collection.hasAttribute("shelf"):
    print ("Root element : %s" % collection.getAttribute("shelf"))
```

```python
#在集合中获取所有电影
movies = collection.getElementsByTagName("movie")
#输出每部电影的详细信息
for movie in movies:
    print ("*****Movie*****")
    if movie.hasAttribute("title"):
        print ("Title: %s" % movie.getAttribute("title"))
    type = movie.getElementsByTagName('type')[0]
    print ("Type: %s" % type.childNodes[0].data)
    format = movie.getElementsByTagName('format')[0]
    print ("Format: %s" % format.childNodes[0].data)
    rating = movie.getElementsByTagName('rating')[0]
    print ("Rating: %s" % rating.childNodes[0].data)
    description = movie.getElementsByTagName('description')[0]
    print ("Description: %s" % description.childNodes[0].data)
```

执行后会输出：

```
*****Movie*****
Title: AAA
Type: AAAA
Format: DVD
Rating: PG
Description: AAAA
*****Movie*****
Title: BBB
Type: BBB
Format: DVD
Rating: R
Description: BBB
*****Movie*****
Title: CCC
Type: CCC
Format: DVD
Rating: PG
Description: CCC
*****Movie*****
Title: DDD
Type: DDD
Format: VHS
Rating: PG
Description: DDD
```

实例 05-06：使用 DOM 获取 XML 文件中指定元素

下面的实例代码演示了使用 DOM 获取 XML 文件中指定元素的过程。

（1）XML 文件 user.xml 的代码如下。

源码路径：daima\5\05-06\user.xml

```xml
<?xml version="1.0" encoding="UTF-8" ?>
<users>
    <user id="1000001">
        <username>Admin</username>
        <email>admin@live.cn</email>
        <age>23</age>
        <sex>boy</sex>
    </user>
#省略部分代码
</users>
```

（2）实例文件 domuse.py 的功能是解析文件 user.xml 的内容，具体实现代码如下。

源码路径：daima\5\05-06\domuse.py

```python
from xml.dom import minidom

def get_attrvalue(node, attrname):
    return node.getAttribute(attrname) if node else ''

def get_nodevalue(node, index = 0):
    return node.childNodes[index].nodeValue if node else ''

def get_xmlnode(node, name):
    return node.getElementsByTagName(name) if node else []
```

5.2 使用内置模块解析 XML

```python
def get_xml_data(filename = 'user.xml'):
    doc = minidom.parse(filename)
    root = doc.documentElement

    user_nodes = get_xmlnode(root, 'user')
    print ("user_nodes:", user_nodes)

    user_list=[]
    for node in user_nodes:
        user_id = get_attrvalue(node, 'id')
        node_name = get_xmlnode(node, 'username')
        node_email = get_xmlnode(node, 'email')
        node_age = get_xmlnode(node, 'age')
        node_sex = get_xmlnode(node, 'sex')

        user_name =get_nodevalue(node_name[0])
        user_email = get_nodevalue(node_email[0])
        user_age = int(get_nodevalue(node_age[0]))
        user_sex = get_nodevalue(node_sex[0])

        user = {}
        user['id'] , user['username'] , user['email'] , user['age'] , user['sex'] = (
            int(user_id), user_name , user_email , user_age , user_sex
        )
        user_list.append(user)
    return user_list

def test_load_xml():
    user_list = get_xml_data()
    for user in user_list :
        print ('-----------------------------------------------------')
        if user:
            user_str='No.:\t%d\nname:\t%s\nsex:\t%s\nage:\t%s\nEmail:\t%s' % (int(user['id']) , user['username'], user['sex'] ,
                        user['age'] , user['email'])
            print (user_str)

if __name__ == "__main__":
    test_load_xml()
```

执行后会输出：

```
-----------------------------------------------------
No.:    1000001
name: Admin
sex:    boy
age:    23
Email: admin@live.cn
-----------------------------------------------------
No.:    1000002
name: Admin2
sex:    boy
age:    22
Email: admin2@live.cn
-----------------------------------------------------
No.:    1000003
name: Admin3
sex:    boy
age:    27
Email: admin3@live.cn
-----------------------------------------------------
No.:    1000004
name: Admin4
sex:    girl
age:    25
Email: admin4@live.cn
-----------------------------------------------------
No.:    1000005
name: Admin5
sex:    boy
age:    20
Email: admin5@live.cn.
-----------------------------------------------------
Ran 1 test in 0.031s
```

```
OK
-------------------------------------------------
No.:   1000006
name:  Admin6
sex:   girl
age:   23
Email: admin6@live.cn
```

实例 05-07：使用模块 xml.sax.saxutils 创建一个指定元素的 XML 文件

在 Python 程序中，模块 xml.sax.saxutils 中的类和函数在创建直接使用或作为基类的 SAX 应用程序时非常有用。下面的实例文件 xml01.py 演示了使用模块 xml.sax.saxutils 创建一个指定元素的 XML 文件的过程。

源码路径：daima\5\05-07\xml01.xml

```python
COPYRIGHT_TEMPLATE = "Copyright (c) {0} {1}. All rights reserved."

STYLESHEET_TEMPLATE = ('<link rel="stylesheet" type="text/css" '
                      'media="all" href="{0}" />\n')

HTML_TEMPLATE = """<?xml version="1.0"?>
<!DOCTYPE html PUBLIC "-//W3C//DTD XHTML 1.0 Strict//EN" \
"http://www.w3.org/TR/xhtml1/DTD/xhtml1-strict.dtd">
<html xmlns="http://www.w3.org/1999/xhtml" lang="en" xml:lang="en">
<head>
<title>{title}</title>
<!-- {copyright} -->
<meta name="Description" content="{description}" />
<meta name="Keywords" content="{keywords}" />
<meta equiv="content-type" content="text/html; charset=utf-8" />
{stylesheet}\
</head>
<body>

</body>
</html>
"""

class CancelledError(Exception): pass                          #自定义异常


def main():
    information = dict(name=None, year=datetime.date.today().year,
                       filename=None, title=None, description=None,
                       keywords=None, stylesheet=None)
    while True:
        try:
            print("\nMake HTML Skeleton\n")
            populate_information(information)                  #注意两种参数用法的异同
            make_html_skeleton(**information)                  #注意两种参数用法的异同
        except CancelledError:
            print("Cancelled")
        if (get_string("\nCreate another (y/n)?", default="y").lower()
            not in {"y", "yes"}):
            break


def populate_information(information):
    name = get_string("Enter your name (for copyright)", "name",
                      information["name"])
    if not name:
        raise CancelledError()
    year = get_integer("Enter copyright year", "year",
                       information["year"], 2000,
                       datetime.date.today().year + 1, True)
    if year == 0:
        raise CancelledError()                                 #弹出异常
    filename = get_string("Enter filename", "filename")
    if not filename:
        raise CancelledError()
```

```
            if not filename.endswith((".htm", ".html")):
                filename += ".html"
    title = get_string("Enter title", "title")
    if not title:
        raise CancelledError()
    description = get_string("Enter description (optional)",
                             "description")
    keywords = []
    while True:
        keyword = get_string("Enter a keyword (optional)", "keyword")
        if keyword:
            keywords.append(keyword)
        else:
            break
    stylesheet = get_string("Enter the stylesheet filename "
                            "(optional)", "stylesheet")
    if stylesheet and not stylesheet.endswith(".css"):
        stylesheet += ".css"
    information.update(name=name, year=year, filename=filename,
                       title=title, description=description,
                       keywords=keywords, stylesheet=stylesheet)


def make_html_skeleton(year, name, title, description, keywords,
                       stylesheet, filename):
    copyright = COPYRIGHT_TEMPLATE.format(year, xml.sax.saxutils.escape(name))
    #xmlsax.saxutils.escape()函数，接受一个字符串，并返回一个带有HTML字符的字符串（"&"、"<"、">"分别以转义符"&"、
    #"<"、"$gt;"的形式出现）
    title = xml.sax.saxutils.escape(title)
    description = xml.sax.saxutils.escape(description)
    keywords = ",".join([xml.sax.saxutils.escape(k)            #将数据变成字符串，并用逗号隔开
                         for k in keywords]) if keywords else ""
    stylesheet = (STYLESHEET_TEMPLATE.format(stylesheet)
                  if stylesheet else "")
    html = HTML_TEMPLATE.format(**locals())
    fh = None
    try:
        fh = open(filename, "w", encoding="utf8")
        fh.write(html)
    except EnvironmentError as err:
        print("ERROR", err)
    else:
        print("Saved skeleton", filename)
    finally:
        if fh is not None:
            fh.close()


def get_string(message, name="string", default=None,
               minimum_length=0, maximum_length=80):
    message += ": " if default is None else " [{0}]: ".format(default)
    while True:
        try:
            line = input(message)
            if not line:
                if default is not None:
                    return default
                if minimum_length == 0:
                    return ""
                else:
                    raise ValueError("{0} may not be empty".format(
                                     name))
            if not (minimum_length <= len(line) <= maximum_length):
                raise ValueError("{name} must have at least "
                                 "{minimum_length} and at most "
                                 "{maximum_length} characters".format(
                                 **locals()))
            return line
        except ValueError as err:
            print("ERROR", err)


def get_integer(message, name="integer", default=None, minimum=0,
```

```
                        maximum=100, allow_zero=True):
        class RangeError(Exception): pass

        message += ": " if default is None else " [{0}]: ".format(default)
        while True:
            try:
                line = input(message)
                if not line and default is not None:
                    return default
                i = int(line)
                if i == 0:
                    if allow_zero:
                        return i
                    else:
                        raise RangeError("{0} may not be 0".format(name))
                if not (minimum <= i <= maximum):
                    raise RangeError("{name} must be between {minimum} "
                            "and {maximum} inclusive{0}".format(
                            " (or 0)" if allow_zero else "", **locals()))
                return i
            except RangeError as err:
                print("ERROR", err)
            except ValueError as err:
                print("ERROR {0} must be an integer".format(name))

main()
```

执行后可以根据提示输入要创建的文件名和元素，过程如下。

```
Make HTML Skeleton

Enter your name (for copyright): guan
Enter copyright year [2018]: 2018
Enter filename: 123.xml
Enter title: xml
Enter description (optional): aa
Enter a keyword (optional): bb
Enter a keyword (optional): cc
Enter a keyword (optional): dd
Enter a keyword (optional): 12
Enter a keyword (optional): 13
Enter a keyword (optional):
Enter the stylesheet filename (optional): ?
Saved skeleton 123.xml.html

Create another (y/n)? [y]: n
```

实例 05-08：使用模块 xml.parsers.expat 解析 XML 文件

在 Python 程序中，模块 xml.parsers.expat 的功能是动态解析 XML 文件，其中提供了一个单一的扩展类 xmlparser，表示 XML 解析器的当前状态。在创建了 xmlparser 对象后，可以将对象的各种属性设置为处理函数。下面的实例文件 xmlparser.py 演示了使用模块 xml.parsers.expat 解析 XML 文件的过程，具体实现代码如下。

源码路径：daima\5\05-08\xmlparser.py

```
import xml.parsers.expat

def start_element(name, attrs):
    print('Start element:', name, attrs)
def end_element(name):
    print('End element:', name)
def char_data(data):
    print('Character data:', repr(data))

p = xml.parsers.expat.ParserCreate()

p.StartElementHandler = start_element
p.EndElementHandler = end_element
p.CharacterDataHandler = char_data
```

```
p.Parse("""<?xml version="1.0"?>
<parent id="top"><child1 name="paul">Text goes here</child1>
<child2 name="fred">More text</child2>
</parent>""", 1)
```

上述代码分别定义了 3 个函数，即 start_element()、end_element()和 char_data()，分别解析了指定 XML 代码中的开始标记、结束标记和数据元素。

执行后会输出：

```
Start element: parent {'id': 'top'}
Start element: child1 {'name': 'paul'}
Character data: 'Text goes here'
End element: child1
Character data: '\n'
Start element: child2 {'name': 'fred'}
Character data: 'More text'
End element: child2
Character data: '\n'
End element: parent
```

5.3 使用第三方库解析 HTML 和 XML

因为 HTML 和 XML 是互联网应用中较常用的网页标记语言，所以用 Python 处理 HTML 和 XML 页面至关重要。本节将通过具体实例讲解使用第三方库解析 HTML 和 XML 的知识。

实例 05-09：使用库 Beautiful Soup 解析 HTML 代码

Beautiful Soup 是一个可以从 HTML 文件或 XML 文件中提取数据的 Python 库，能够先将 HTML 和 XML 的标签文件解析成树形结构，然后方便地获取指定标签的对应属性。通过使用库 Beautiful Soup 可以大大提高开发效率。可以使用如下两种命令安装库 Beautiful Soup。

```
pip install beautifulsoup4
easy_install beautifulsoup4
```

接下来还需要安装解析器，Beautiful Soup 不但支持 Python 标准库中的 HTML 解析器，而且支持一些第三方的解析器，其中最为常用的是 lxml。根据操作系统不同，可以使用如下命令来安装 lxml。

```
$ apt-get install Python-lxml
$ easy_install lxml
$ pip install lxml
```

下面的实例文件 bs01.py 演示了使用库 Beautiful Soup 解析 HTML 代码的过程。

源码路径：daima\5\05-09\bs01.py

```
from bs4 import BeautifulSoup
html_doc = """
<html><head><title>The Dormouse's story</title></head>
<body>
<p class="title"><b>The Dormouse's story</b></p>

<p class="story">Once upon a time there were three little sisters; and their names were
<a href="http://example.com/elsie" class="sister" id="link1">Elsie</a>,
<a href="http://example.com/lacie" class="sister" id="link2">Lacie</a> and
<a href="http://example.com/tillie" class="sister" id="link3">Tillie</a>;
and they lived at the bottom of a well.</p>

<p class="story">...</p>
"""
soup = BeautifulSoup(html_doc,"lxml")
print(soup)
```

通过上述代码，解析了 html_doc 中的 HTML 代码。执行后会输出解析结果：

```
<html><head><title>The Dormouse's story</title></head>
<body>
<p class="title"><b>The Dormouse's story</b></p>
<p class="story">Once upon a time there were three little sisters; and their names were
<a class="sister" href="http://example.com/elsie" id="link1">Elsie</a>,
```

```
<a class="sister" href="http://example.com/lacie" id="link2">Lacie</a> and
<a class="sister" href="http://example.com/tillie" id="link3">Tillie</a>;
and they lived at the bottom of a well.</p>
<p class="story">...</p>
</body></html>
```

实例 05-10：使用库 Beautiful Soup 解析指定 HTML 标签

在解析 HTML 或 XML 文件时，我们可以使用标签选择器来获得某个具体的标签信息，通过如下代码可以获取不同的标签信息。

```
print(soup.title)
print(type(soup.title))
print(soup.head)
print(soup.p)
```

通过上述这种"soup.标签名"格式即可获得这个标签的内容。需要注意的是，如果文件中有多个这样的标签，则返回的结果是第一个标签的内容。例如，通过 soup.p 可以获取 p 标签，通常文档中有多个 p 标签，在执行 soup.p 后只会返回第一个 p 标签的内容。下面的实例文件 bs02.py 演示了使用库 Beautiful Soup 解析指定 HTML 标签的过程。

源码路径：daima\5\05-10\bs02.py

```
from bs4 import BeautifulSoup

html = '''
<html><head><title>The Dormouse's story</title></head>
<body>
<p class="title"><b>The Dormouse's story</b></p>

<p class="story">Once upon a time there were three little sisters; and their names were
<a href="http://example.com/elsie" class="sister" id="link1">Elsie</a>,
<a href="http://example.com/lacie" class="sister" id="link2">Lacie</a> and
<a href="http://example.com/tillie" class="sister" id="link3">Tillie</a>;
and they lived at the bottom of a well.</p>
<p class="story">...</p>
'''
soup = BeautifulSoup(html,'lxml')
print(soup.title)
print(soup.title.name)
print(soup.title.string)
print(soup.title.parent.name)
print(soup.p)
print(soup.p["class"])
print(soup.a)
print(soup.find_all('a'))
print(soup.find(id='link3'))
```

执行后将输出指定标签的信息：

```
<title>The Dormouse's story</title>
title
The Dormouse's story
head
<p class="title"><b>The Dormouse's story</b></p>
['title']
<a class="sister" href="http://example.com/elsie" id="link1">Elsie</a>
[<a class="sister" href="http://example.com/elsie" id="link1">Elsie</a>, <a class="sister" href="http://example.com/lacie" id="link2">Lacie</a>, <a class="sister" href="http://example.com/tillie" id="link3">Tillie</a>]
<a class="sister" href="http://example.com/tillie" id="link3">Tillie</a>
```

实例 05-11：将 p 标签下的所有子标签存入一个列表中

在库 Beautiful Soup 中，我们可以使用 contents 来处理节点。下面的实例文件 bs03.py 演示了将 p 标签下的所有子标签存入一个列表中的过程。

源码路径：daima\5\05-11\bs03.py

```
html = """
<html>
    <head>
        <title>The Dormouse's story</title>
```

```
        </head>
        <body>
            <p class="story">
                Once upon a time there were three little sisters; and their names were
                <a href="http://example.com/elsie" class="sister" id="link1">
                    <span>Elsie</span>
                </a>
                <a href="http://example.com/lacie" class="sister" id="link2">Lacie</a>
                and
                <a href="http://example.com/tillie" class="sister" id="link3">Tillie</a>
                and they lived at the bottom of a well.
            </p>
            <p class="story">...</p>
"""

from bs4 import BeautifulSoup

soup = BeautifulSoup(html,'lxml')
print(soup.p.contents)
```

执行后会输出：

```
['\n                Once upon a time there were three little sisters; and their names were\n                ', <a class="sister" href="http://example.com/elsie" id="link1">
<span>Elsie</span>
</a>, '\n', <a class="sister" href="http://example.com/lacie" id="link2">Lacie</a>, '\n                and\n                ', <a class="sister" href="http://example.com/tillie" id="link3">Tillie</a>, '\n                and they lived at the bottom of a well.\n            ']
```

也就是说，在列表中会存入图 5-2 所示的框中元素。

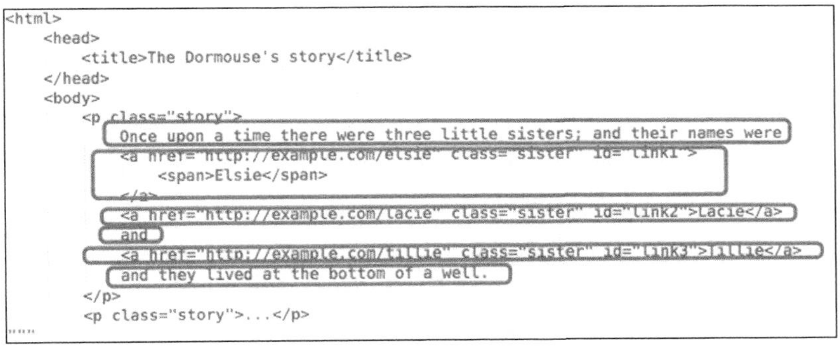

图 5-2　存入框中元素

实例 05-12：获取 p 标签下的所有子节点内容

通过下面的实例文件 bs04.py，使用 children 也可以获取 p 标签下的所有子节点内容，其结果和通过 contents 获取的结果完全一样。但是不同的地方是，soup.p.children 是一个迭代对象，而不是列表，只能通过循环的方式获取所有的信息。

源码路径：daima\5\05-12\bs04.py

```
soup = BeautifulSoup(html,'lxml')
print(soup.p.children)
for i,child in enumerate(soup.p.children):
    print(i,child)
```

实例 05-13：处理标签中的兄弟节点和父节点

通过 soup.a.parent 可以获取父节点的信息，通过 list(enumerate(soup.a.parents))可以获取祖先节点，这个方法返回的结果是一个列表，会将 a 标签的父节点的信息存放到列表中，将父节点的父节点信息也存放到列表中，并且最后还会将整个文件存放到列表中，所有列表的最后一个元素及倒数第二个元素保存的是整个文件的信息。下面的实例文件 bs05.py 演示了处理标签中的兄弟节点和父节点的过程。

源码路径：daima\5\05-13\bs05.py

```
html = """
<html><head><title>The Dormouse's story</title></head>
<body>
<p class="title"><b>The Dormouse's story</b></p>

<p class="story">Once upon a time there were three little sisters; and their names were
<a href="http://example.com/elsie" class="sister" id="link1">Elsie</a>,
<a href="http://example.com/lacie" class="sister" id="link2">Lacie</a> and
<a href="http://example.com/tillie" class="sister" id="link3">Tillie</a>;
and they lived at the bottom of a well.</p>

<p class="story">...</p>
"""
from bs4 import BeautifulSoup
soup = BeautifulSoup(html, 'lxml')
title_tag = soup.title
print(title_tag)
print(title_tag.parent)
#在下面代码中，因为b标签和c标签在同一层:它们是同一个元素的子节点
#所以b和c可以称为兄弟节点。文件以标准格式输出时
#兄弟节点有相同的缩进级别。在代码中也可以使用这种关系
sibling_soup = BeautifulSoup("<a><b>text1</b><c>text2</c></b></a>", 'lxml')
print(sibling_soup.prettify())
#b标签有.next_sibling属性，但是没有.previous_sibling属性，因为b标签在同级节点中是第一个节点
#同理，c标签有.previous_sibling属性，却没有.next_sibling属性
print(sibling_soup.b.next_sibling)
print(sibling_soup.c.previous_sibling)

for sibling in soup.a.next_siblings:
    print(repr(sibling))

for sibling in soup.find(id="link3").previous_siblings:
print(repr(sibling))
```

执行后会输出：

```
<title>The Dormouse's story</title>
<head><title>The Dormouse's story</title></head>
<html>
 <body>
  <a>
   <b>
    text1
   </b>
   <c>
    text2
   </c>
  </a>
 </body>
</html>
<c>text2</c>
```

实例05-14：根据标签名查找文件

通过使用函数 find_all(name,attrs,recursive,text,**kwargs)，可以根据标签名、属性和内容查找文件。下面的实例文件 bs06.py 演示了根据标签名查找文件的过程。

源码路径：daima\5\05-14\bs06.py

```
html='''
####省略部分代码
'''
from bs4 import BeautifulSoup
soup = BeautifulSoup(html, 'lxml')
print(soup.find_all('ul'))
print(type(soup.find_all('ul')[0]))
for ul in soup.find_all('ul'):
    print(ul.find_all('li'))
```

上述代码执行后会返回一个列表形式的结果，并且在最后两行代码中，针对前面的解析结果再次执行 find_all() 操作，从而获取所有的 li 标签信息。

```
[<ul class="list" id="list-1">
```

```
<li class="element">Foo</li>
<li class="element">Bar</li>
<li class="element">Jay</li>
</ul>, <ul class="list list-small" id="list-2">
<li class="element">Foo</li>
<li class="element">Bar</li>
</ul>]
<class 'bs4.element.Tag'>
[<li class="element">Foo</li>, <li class="element">Bar</li>, <li class="element">Jay</li>]
[<li class="element">Foo</li>, <li class="element">Bar</li>]
```

实例 05-15：使用函数 find_all()根据属性查找文件

在使用函数 find_all()时，也可以根据属性 attrs 查找文件。attrs 通过传入字典的方式来查找标签，但是这里有一个特殊情况 class，因为 class 在 Python 中是特殊的字段。如果想要查找 class 相关的信息，则可以做如下更改：attrs={'class_':'element'}或者 soup.find_all('',{"class":"element"})。特殊的标签属性可以不写 attrs，如 id。下面的实例文件 bs07.py 演示了使用函数 find_all()根据属性查找文件的过程。

源码路径：daima\5\05-15\bs07.py

```
html='''
####省略部分代码
'''
from bs4 import BeautifulSoup
soup = BeautifulSoup(html, 'lxml')
print(soup.find_all(attrs={'id': 'list-1'}))
print(soup.find_all(attrs={'name': 'elements'}))
```

执行后会输出：

```
[<ul class="list" id="list-1" name="elements">
<li class="element">Foo</li>
<li class="element">Bar</li>
<li class="element">Jay</li>
</ul>]
[<ul class="list" id="list-1" name="elements">
<li class="element">Foo</li>
<li class="element">Bar</li>
<li class="element">Jay</li>
</ul>]
```

实例 05-16：用函数 find_all()根据 text 查找文件

下面的实例文件 bs08.py 演示了使用函数 find_all()根据 text 查找文件的过程。

源码路径：daima\5\05-16\bs08.py

```
html='''
####省略部分代码
'''
soup = BeautifulSoup(html, 'lxml')
print(soup.find_all(text='Foo'))
```

执行后会输出：

```
['Foo', 'Foo']
```

实例 05-17：使用其他标准选择器

库 Beautiful Soup 还包含如下标准选择器：
- find(name,attrs,recursive,text,**kwargs)：返回匹配结果的第一个元素；
- find_parents()：返回所有祖先节点；
- find_parent()：返回直接父节点；
- find_next_siblings()：返回后面所有兄弟节点；
- find_next_sibling()：返回后面第一个兄弟节点；
- find_previous_siblings()：返回前面所有兄弟节点；

- find_previous_sibling()：返回前面第一个兄弟节点；
- find_all_next()：返回节点后所有符合条件的节点；
- find_next()：返回第一个符合条件的节点；
- find_all_previous()：返回节点后所有符合条件的节点；
- find_previous()：返回第一个符合条件的节点。

下面的实例文件 bs09.py 演示了使用上述标准选择器的过程。

源码路径：daima\5\05-17\bs09.py

```
html = """
#省略部分代码
"""
from bs4 import BeautifulSoup
soup = BeautifulSoup(html, 'lxml')
first_link = soup.a
print(first_link)
print(first_link.find_next_siblings("a"))
first_story_paragraph = soup.find("p", "story")
print(first_story_paragraph.find_next_sibling("p"))
first_link = soup.a
print(first_link)
print(first_link.find_all_next(text=True))
print(first_link.find_next("p"))
first_link = soup.a
print(first_link)
print(first_link.find_all_previous("p"))
print(first_link.find_previous("title"))
```

执行后会输出：

```
<a class="sister" href="http://example.com/elsie" id="link1">Elsie</a>
[<a class="sister" href="http://example.com/lacie" id="link2">Lacie</a>, <a class="sister" href="http://example.com/tillie" id="link3">Tillie</a>]
<p class="story">...</p>
<a class="sister" href="http://example.com/elsie" id="link1">Elsie</a>
['Elsie', ',\n', 'Lacie', ' and\n', 'Tillie', ';\nand they lived at the bottom of a well.', '\n', '...', '\n']
<p class="story">...</p>
<a class="sister" href="http://example.com/elsie" id="link1">Elsie</a>
[<p class="story">Once upon a time there were three little sisters; and their names were
<a class="sister" href="http://example.com/elsie" id="link1">Elsie</a>,
<a class="sister" href="http://example.com/lacie" id="link2">Lacie</a> and
<a class="sister" href="http://example.com/tillie" id="link3">Tillie</a>;
and they lived at the bottom of a well.</p>, <p class="title"><b>The Dormouse's story</b></p>]
<title>The Dormouse's story</title>
```

实例 05-18：使用 select() 直接传入 CSS 选择器

Beautiful Soup 支持大部分的 CSS 选择器，通过在 Tag 或 BeautifulSoup 对象的 select() 函数中传入字符串参数，可以使用 CSS 选择器的语法找到 tag 标签。下面的实例文件 bs10.py 演示了使用 select() 直接传入 CSS 选择器的方式完成标签选择的过程。

源码路径：daima\5\05-18\bs10.py

```
html=''
#省略部分代码
'''
from bs4 import BeautifulSoup
soup = BeautifulSoup(html, 'lxml')
print(soup.select('.panel .panel-heading'))
print(soup.select('ul li'))
print(soup.select('#list-2 .element'))
print(type(soup.select('ul')[0]))
```

执行后会输出：

```
[<div class="panel-heading">
<h4>Hello</h4>
</div>]
[<li class="element">Foo</li>, <li class="element">Bar</li>, <li class="element">Jay</li>, <li class="element">Foo</li>,
 <li class="element">Bar</li>]
```

```
[<li class="element">Foo</li>, <li class="element">Bar</li>]
<class 'bs4.element.Tag'>
```

实例 05-19：使用库 bleach 过滤 HTML 代码

在使用 Python 进行 Web 开发时，必须要考虑防止用户的 XSS（跨站脚本攻击）注入。我们可以自己先写一个白名单，然后通过 Beatiful Soup 等处理 HTML 的库来过滤标签和属性。bleach 便是一个实现上述功能的 Python 库，是一个基于白名单的 HTML 清理和文本链接模块。

我们可以使用如下两种命令安装库 bleach。

```
pip install bleach
easy_install bleach
```

下面的实例文件 ble01.py 演示了使用库 bleach 过滤 HTML 代码的过程。

源码路径：daima\5\05-19\ble01.py

```
import bleach
print(bleach.clean('an <script>evil()</script> example'))
print(bleach.linkify('an http://example.com url'))
```

通过上述代码实现了最基本的 HTML 代码过滤，执行后会输出：

```
an &lt;script&gt;evil()&lt;/script&gt; example
an <a href="http://example.com" rel="nofollow">http://example.com</a> url
```

实例 05-20：使用方法 bleach.clean() 不同参数实现过滤处理

库 bleach 中的重要内置方法 bleach.clean() 是用于对 HTML 片段进行过滤的方法。需要注意的是，该方法过滤的是片段而非整个 HTML 文档，当不传递任何参数时，只用于过滤 HTML 标签，不包括属性、CSS、JSON、XHTML 和 SVG 等其他内容。正因如此，在对一些存在风险的属性的渲染过程中，需要使用模板进行转义。如果你正在清理大量的文本，并传递相同的参数或想要得到更多的可配置选项，则可以考虑使用 bleach.sanitizer.Cleaner 实例。下面的实例文件 ble02.py 演示了使用方法 bleach.clean() 不同参数实现过滤处理的过程。

源码路径：daima\5\05-20\ble02.py

```
import bleach
# tags参数示例
print(bleach.clean(
    u'<b><i>an example</i></b>',
    tags=['b'],
))

# attributes为列表示例
print(bleach.clean(
    u'<p class="foo" style="color: red; font-weight: bold;">blah blah blah</p>',
    tags=['p'],
    attributes=['style'],
    styles=['color'],
))
# attributes为字典示例
attrs = {
    '*': ['class'],
    'a': ['href', 'rel'],
    'img': ['alt'],
}
print(bleach.clean(
    u'<img alt="an example" width=500>',
    tags=['img'],
    attributes=attrs
))

# attributes为函数示例
def allow_h(tag, name, value):
    return name[0] == 'h'
print(bleach.clean(
    u'<a href="http://example.com" title="link">link</a>',
```

```
            tags=['a'],
            attributes=allow_h,
    ))

# styles参数示例
tags = ['p', 'em', 'strong']
attrs = {
    '*': ['style']
}
styles = ['color', 'font-weight']
print(bleach.clean(
    u'<p style="font-weight: heavy;">my html</p>',
    tags=tags,
    attributes=attrs,
    styles=styles
))
# protocols参数示例
print(bleach.clean(
    '<a href="smb://more_text">allowed protocol</a>',
    protocols=['http', 'https', 'smb']
))
print(bleach.clean(
    '<a href="smb://more_text">allowed protocol</a>',
    protocols=bleach.ALLOWED_PROTOCOLS + ['smb']
))

#strip参数示例
print(bleach.clean('<span>is not allowed</span>'))
print(bleach.clean('<b><span>is not allowed</span></b>', tags=['b']))

print(bleach.clean('<span>is not allowed</span>', strip=True))
print(bleach.clean('<b><span>is not allowed</span></b>', tags=['b'], strip=True))

# strip_comments参数示例
html = 'my<!-- commented --> html'
print(bleach.clean(html))
print(bleach.clean(html, strip_comments=False))
```

执行后会输出：

```
<b>&lt;i&gt;an example&lt;/i&gt;</b>
<p style="color: red;">blah blah blah</p>
<img alt="an example">
<a href="http://example.com">link</a>
<p style="font-weight: heavy;">my html</p>
<a href="smb://more_text">allowed protocol</a>
<a href="smb://more_text">allowed protocol</a>
&lt;span&gt;is not allowed&lt;/span&gt;
<b>&lt;span&gt;is not allowed&lt;/span&gt;</b>
is not allowed
<b>is not allowed</b>
my html
my<!-- commented --> html
```

实例 05-21：使用方法 bleach.linkify()添加指定属性

方法 bleach.linkify()会将 HTML 文本中的 URL 形式字符转换为链接，URL 形式字符包括 URL、域名、电子邮箱地址等，但在如下情况下不会进行转换：

❑ 已经以链接格式呈现在文本中；
❑ 该标签属性值包含 URL 格式；
❑ 电子邮箱地址。

下面的实例文件 ble03.py 演示了使用方法 bleach.linkify()添加指定属性的过程。

源码路径：daima\5\05-21\ble03.py

```
from bleach.linkifier import Linker

def set_title(attrs, new=False):
    attrs[(None, u'title')] = u'link in user text'
    return attrs
```

```python
linker = Linker(callbacks=[set_title])
print(linker.linkify('abc http://example.com def'))

#下面的代码将生成的链接设置为内部链接在当前页打开、外部链接在新建页打开
import urllib
from bleach.linkifier import Linker

def set_target(attrs, new=False):
    p = urllib.parse.urlparse(attrs[(None, u'href')])
    if p.netloc not in ['my-domain.com', 'other-domain.com']:
        attrs[(None, u'target')] = u'_blank'
        attrs[(None, u'class')] = u'external'
    else:
        attrs.pop((None, u'target'), None)
    return attrs

linker = Linker(callbacks=[set_target])
print(linker.linkify('abc http://example.com def'))
```

执行后会输出：

```
abc <a href="http://example.com" title="link in user text">http://example.com</a> def
abc <a class="external" href="http://example.com" target="_blank">http://example.com</a> def
```

实例 05-22：使用 callback 参数删除指定属性

通过方法 bleach.linkify() 中的参数 callback，可以实现类似属性白名单的过滤功能，不但可以删除标签中已有属性，而且可以删除那些没有经过链接化的文本内容的标签属性，功能与前面的 clean() 方法类似。下面的实例文件 ble04.py 演示了使用 callback 参数删除指定属性的过程。

源码路径：daima\5\05-22\ble04.py

```python
from bleach.linkifier import Linker

def allowed_attrs(attrs, new=False):
    """Only allow href, target, rel and title."""
    allowed = [
        (None, u'href'),
        (None, u'target'),
        (None, u'rel'),
        (None, u'title'),
        u'_text',
    ]
    return dict((k, v) for k, v in attrs.items() if k in allowed)

linker = Linker(callbacks=[allowed_attrs])
print(linker.linkify('<a style="font-weight: super bold;" href="http://example.com">link</a>'))

#除了删除白名单之外的属性，还可以删除指定属性
def remove_title(attrs, new=False):
    attrs.pop((None, u'title'), None)
    return attrs
linker = Linker(callbacks=[remove_title])
print(linker.linkify('<a href="http://example.com">link</a>'))
print(linker.linkify('<a title="bad title" href="http://example.com">link</a>'))
```

执行后会输出：

```
<a href="http://example.com">link</a>
<a href="http://example.com">link</a>
<a href="http://example.com">link</a>
```

实例 05-23：使用 bleach.linkifier.Linker 处理链接

当使用一套统一的规则进行文本链接化处理时，建议使用 bleach.linkifier.Linker 实例，因为 linkify() 方法的本质就是调用此实例。下面的实例文件 ble05.py 演示了使用 bleach.linkifier.Linke 处理链接的过程。

源码路径：daima\5\05-23\ble05.py

```python
from bleach.linkifier import Linker
```

```
linker = Linker(skip_tags=['pre'])
print(linker.linkify('a b c http://example.com d e f'))
```

执行后会输出：

```
a b c <a href="http://example.com" rel="nofollow">http://example.com</a> d e f
```

实例 05-24：使用 bleach.linkifier.LinkifyFilter 处理链接

方法 bleach.linkify()是通过 bleach.linkifier.LinkifyFilter 实例进行链接化的，跟前面讲解的 bleach.linkifier.Cleaner 一样，此实例也可以当作 html5lib 的 filter 实例使用。例如，可以将此实例传入到 bleach.linkifier.Cleaner 中，使得文本过滤和文本链接化同时进行，一步完成。

下面的实例文件 ble06.py 演示了使用 bleach.linkifier.LinkifyFilter 处理链接的过程。

源码路径：daima\5\05-24\ble06.py

```python
from bleach import Cleaner
from bleach.linkifier import LinkifyFilter
#使用bleach.linkifier.LinkifyFilter的默认配置
cleaner = Cleaner(tags=['pre'])
print(cleaner.clean('<pre>http://example.com</pre>'))

cleaner = Cleaner(tags=['pre'], filters=[LinkifyFilter])
print(cleaner.clean('<pre>http://example.com</pre>'))
#下面演示传参后对比
from functools import partial
from bleach.sanitizer import Cleaner
from bleach.linkifier import LinkifyFilter

cleaner = Cleaner(
    tags=['pre'],
    filters=[partial(LinkifyFilter, skip_tags=['pre'])]
)
print(cleaner.clean('<pre>http://example.com</pre>'))
```

执行后会输出：

```
<pre>http://example.com</pre>
<pre><a href="http://example.com">http://example.com</a></pre>
<pre>http://example.com</pre>
```

实例 05-25：使用库 cssutils 处理 CSS 标记

库 cssutils 是一个 Python 包，用于解析和构建 CSS 代码。库 cssutils 中只有 DOM，没有任何渲染设备。我们可以使用如下两种命令安装库 cssutils。

```
pip install cssutils
easy_install cssutils
```

下面的实例文件 css01.py 演示了使用库 cssutils 处理 CSS 标记的过程。

源码路径：daima\5\05-25\css01.py

```python
import cssutils

css = u'''/* a comment with umlaut &auml; */
    @namespace html "http://www.w3.org/1999/xhtml";
    @variables { BG: #fff }
    html|a { color:red; background: var(BG) }'''
sheet = cssutils.parseString(css)

for rule in sheet:
    if rule.type == rule.STYLE_RULE:
        #遍历属性
        for property in rule.style:
            if property.name == 'color':
                property.value = 'green'
                property.priority = 'IMPORTANT'
                break
        #简易处理
        rule.style['margin'] = '01.0eM' # or: ('1em', 'important')

sheet.encoding = 'ascii'
sheet.namespaces['xhtml'] = 'http://www.w3.org/1999/xhtml'
```

```
sheet.namespaces['atom'] = 'http://www.w3.org/2005/Atom'
sheet.add('atom|title {color: #000000 !important}')
sheet.add('@import "sheets/import.css";')

print(sheet.cssText)
```

执行后会输出：

```
@charset "ascii";
@import "sheets/import.css";
/* a comment with umlaut \E4   */
@namespace xhtml "http://www.w3.org/1999/xhtml";
@namespace atom "http://www.w3.org/2005/Atom";
xhtml|a {
    color: green !important;
    background: #fff;
    margin: 1em
    }
atom|title {
    color: #000 !important
    }
```

实例 05-26：使用 html5lib 解析 HTML 代码

html5lib 是一个兼容标准 HTML 文件、支持片段解析及序列化的库。在本章前面讲解使用库 Beautiful Soup 从 HTML 或 XML 文件中提取数据时，我们使用的解析器是 lxml，而 html5lib 也是 Beautiful Soup 支持的一种解析器。我们可以使用如下两种命令安装库 html5lib。

```
pip install html5lib
easy_install html5lib
```

下面的实例文件 ht501.py 演示了在前面实例文件 ble01.py 中使用 html5lib 解析 HTML 代码的过程。

源码路径：daima\5\05-26\ht501.py

```
from bs4 import BeautifulSoup
html_doc = """
<html><head><title>The Dormouse's story</title></head>
<body>
<p class="title"><b>The Dormouse's story</b></p>

<p class="story">Once upon a time there were three little sisters; and their names were
<a href="http://example.com/elsie" class="sister" id="link1">Elsie</a>,
<a href="http://example.com/lacie" class="sister" id="link2">Lacie</a> and
<a href="http://example.com/tillie" class="sister" id="link3">Tillie</a>;
and they lived at the bottom of a well.</p>

<p class="story">...</p>
"""
soup = BeautifulSoup(html_doc,"html5lib")
print(soup)
```

执行后会输出：

```
<html><head><title>The Dormouse's story</title></head>
<body>
<p class="title"><b>The Dormouse's story</b></p>

<p class="story">Once upon a time there were three little sisters; and their names were
<a class="sister" href="http://example.com/elsie" id="link1">Elsie</a>,
<a class="sister" href="http://example.com/lacie" id="link2">Lacie</a> and
<a class="sister" href="http://example.com/tillie" id="link3">Tillie</a>;
and they lived at the bottom of a well.</p>

<p class="story">...</p>
</body></html>
```

实例 05-27：使用 html5lib 解析 HTML 中的指定标签

我们也可以使用 html5lib 解析 HTML 中的指定标签，下面的实例文件 ht502.py 演示了这一功能。

源码路径：daima\5\05-27\ht502.py

```
html = """
<html><head><title>The Dormouse's story</title></head>
<body>
<p class="title"><b>The Dormouse's story</b></p>

<p class="story">Once upon a time there were three little sisters; and their names were
<a href="http://example.com/elsie" class="sister" id="link1">Elsie</a>,
<a href="http://example.com/lacie" class="sister" id="link2">Lacie</a> and
<a href="http://example.com/tillie" class="sister" id="link3">Tillie</a>;
and they lived at the bottom of a well.</p>

<p class="story">...</p>
"""

from bs4 import BeautifulSoup

#添加一个解析器
soup = BeautifulSoup(html,'html5lib')
print(soup.title)
print(soup.title.name)
print(soup.title.text)
print(soup.body)

#从文件中找到所有a标签的内容
for link in soup.find_all('a'):
    print(link.get('href'))


#从文件中找到所有文字内容
print(soup.get_text())
```

执行后会输出：

```
<title>The Dormouse's story</title>
title
The Dormouse's story
<body>
<p class="title"><b>The Dormouse's story</b></p>

<p class="story">Once upon a time there were three little sisters; and their names were
<a class="sister" href="http://example.com/elsie" id="link1">Elsie</a>,
<a class="sister" href="http://example.com/lacie" id="link2">Lacie</a> and
<a class="sister" href="http://example.com/tillie" id="link3">Tillie</a>;
and they lived at the bottom of a well.</p>

<p class="story">...</p>
</body>
http://example.com/elsie
http://example.com/lacie
http://example.com/tillie
The Dormouse's story

The Dormouse's story

Once upon a time there were three little sisters; and their names were
Elsie,
Lacie and
Tillie;
and they lived at the bottom of a well.

...
```

实例 05-28：使用库 MarkupSafe 构建安全 HTML

库 MarkupSafe 为 XML/HTML/XHTML 标记提供了安全字符串功能，我们可以使用如下两种命令安装库 MarkupSafe。

```
pip install MarkupSafe
easy_install MarkupSafe
```

下面的实例文件 mark01.py 演示了使用库 MarkupSafe 构建安全 HTML 的过程。

源码路径：daima\5\05-28\mark01.py

```python
from markupsafe import Markup, escape
#实现支持HTML字符串的Unicode子类
print(escape("<script>alert(document.cookie);</script>"))
tmpl = Markup("<em>%s</em>")
print(tmpl % "Peter > Lustig")

#可以通过重写__html__()自定义等效HTML标记
class Foo(object):
    def __html__(self):
        return '<strong>Nice</strong>'

print(escape(Foo()))
print(Markup(Foo()))
```

执行后会输出：

```
&lt;script&gt;alert(document.cookie);&lt;/script&gt;
<em>Peter &gt; Lustig</em>
<strong>Nice</strong>
<strong>Nice</strong>
```

实例 05-29：使用库 MarkupSafe 实现格式化

下面的实例文件 mark02.py 演示了使用库 MarkupSafe 实现格式化的过程。

源码路径：daima\5\05-29\mark02.py

```python
from markupsafe import Markup, escape

class User(object):

    def __init__(self, id, username):
        self.id = id
        self.username = username

    def __html_format__(self, format_spec):
        if format_spec == 'link':
            return Markup('<a href="/user/{0}">{1}</a>').format(
                self.id,
                self.__html__(),
            )
        elif format_spec:
            raise ValueError('Invalid format spec')
        return self.__html__()

    def __html__(self):
        return Markup('<span class=user>{0}</span>').format(self.username)

user = User(1, 'foo')
print(Markup('<p>User: {0:link}').format(user))
```

执行后会输出：

```
<p>User: <a href="/user/1"><span class=user>foo</span></a>
```

实例 05-30：使用库 pyquery 实现字符串初始化

pyquery 是一个解析 HTML 的库，是 Python 对 jQuery 的封装。如果读者拥有前端开发经验，那么应该接触过 jQuery。pyquery 是 Python 严格仿照 jQuery 的实现，其语法与 jQuery 几乎完全相同。我们可以使用如下两种命令安装库 pyquery。

```
pip install pyquery
easy_install pyquery
```

下面的实例文件 pyq01.py 演示了使用库 pyquery 实现字符串初始化的过程。

源码路径：daima\5\05-30\pyq01.py

```html
html = '''
<div>
    <ul>
        <li class="item-0">first item</li>
        <li class="item-1"><a href="link2.html">second item</a></li>
```

```html
            <li class="item-0 active"><a href="link3.html"><span class="bold">third item</span></a></li>
            <li class="item-1 active"><a href="link4.html">fourth item</a></li>
            <li class="item-0"><a href="link5.html">fifth item</a></li>
     </div>
'''
```
```python
from pyquery import PyQuery as pq
doc = pq(html)
print(doc)
print(type(doc))
print(doc('li'))
```

在上述代码中，由于 pyquery 写起来比较麻烦，所以我们导入的时候添加了别名 pq。

```python
from pyquery import PyQuery as pq
```

上述代码中的 doc 其实就是一个 pyquery 对象，我们可以通过 doc 进行元素的选择，其实它就是一个 CSS 选择器，所以 CSS 选择器的规则都可以用，直接使用 doc(标签名)就可以获取该标签的所有内容。如果想要获取 class，则使用 doc('.class_name')，如果想获取 id，则使用 doc('#id_name')。

执行后会输出：

```html
<div>
    <ul>
        <li class="item-0">first item</li>
        <li class="item-1"><a href="link2.html">second item</a></li>
        <li class="item-0 active"><a href="link3.html"><span class="bold">third item</span></a></li>
        <li class="item-1 active"><a href="link4.html">fourth item</a></li>
        <li class="item-0"><a href="link5.html">fifth item</a></li>
    </ul>
</div>
<class 'pyquery.pyquery.PyQuery'>
<li class="item-0">first item</li>
        <li class="item-1"><a href="link2.html">second item</a></li>
        <li class="item-0 active"><a href="link3.html"><span class="bold">third item</span></a></li>
        <li class="item-1 active"><a href="link4.html">fourth item</a></li>
        <li class="item-0"><a href="link5.html">fifth item</a></li>
```

实例 05-31：使用 pyquery 解析 HTML 内容

下面的实例文件 pyq02.py 演示了使用 pyquery 解析 HTML 内容的过程。

源码路径：daima\5\05-31\pyq02.py

```python
from pyquery import PyQuery as pyq

html = '''
<html>
    <title>这是标题</title>
<body>
    <p id="hi">Hello</p>
    <ul>
        <li>list1</li>
        <li>list2</li>
    </ul>
</body>
</html>
'''
jq = pyq(html)
print(jq('title'))                  #获取title标签的源码
print(jq('title').text())           #获取title标签的内容
print(jq('#hi').text())             #获取id为hi的标签的内容

li = jq('li')                       #处理多个元素
for i in li:
    print(pyq(i).text())
```

执行后会输出：

```
<title>这是标题</title>
这是标题
Hello
list1
list2
```

实例 05-32：使用库 pyquery 解析本地 HTML 文件和网络页面

下面的实例文件 pyq03.py 演示了使用库 pyquery 解析本地 HTML 文件和网络页面的过程。

源码路径：daima\5\05-32\pyq03.py

```python
from pyquery import PyQuery as pq
doc = pq(filename='123.html')
print(doc)
print(doc('head'))

doc1 = pq(url="http://www.baidu.com",encoding='utf-8')
print(doc1)
print(doc1('head'))
```

在上述代码中，我们在 pq() 中可以传入 URL 参数，也可以传入文件参数，当然这里的文件通常是一个本地 HTML 文件，如 pq(filename='index.html')。上述代码执行后会分别解析本地文件 123.html 和百度主页：

```
<html><head><title>The Dormouse's story</title></head>
<body>
<p class="title"><b>The Dormouse's story</b></p>

<p class="story">Once upon a time there were three little sisters; and their names were
<a href="http://example.com/elsie" class="sister" id="link1">Elsie</a>,
<a href="http://example.com/lacie" class="sister" id="link2">Lacie</a> and
<a href="http://example.com/tillie" class="sister" id="link3">Tillie</a>;
and they lived at the bottom of a well.</p>

<p class="story">...</p>

</body>
</html>
<head><title>The Dormouse's story</title></head>

<html> <head><meta http-equiv="content-type" content="text/html;charset=utf-8"/><meta http-equiv="X-UA-Compatible" content="IE=Edge"/><meta content="always" name="referrer"/><link rel="stylesheet" type="text/css" href="http://s1.bdstatic.com/r/www/cache/bdorz/baidu.min.css"/><title>百度一下，你就知道</title></head> <body link="#0000cc"> <div id="wrapper"> <div id="head"> <div class="head_wrapper"> <div class="s_form"> <div class="s_form_wrapper"> <div id="lg"> <img hidefocus="true" src="//www.baidu.com/img/bd_logo1.png" width="270" height="129"/> </div> <form id="form" name="f" action="//www.baidu.com/s" class="fm"> <input type="hidden" name="bdorz_come" value="1"/> <input type="hidden" name="ie" value="utf-8"/> <input type="hidden" name="f" value="8"/> <input type="hidden" name="rsv_bp" value="1"/> <input type="hidden" name="rsv_idx" value="1"/> <input type="hidden" name="tn" value="baidu"/><span class="bg s_ipt_wr"><input id="kw" name="wd" class="s_ipt" value="" maxlength="255" autocomplete="off" autofocus=""/></span><span class="bg s_btn_wr"><input type="submit" id="su" value="百度一下" class="bg s_btn"/></span> </form> </div> </div> <div id="u1"> <a href="http://news.baidu.com" name="tj_trnews" class="mnav">新闻</a> <a href="http://www.hao123.com" name="tj_trhao123" class="mnav">hao123</a> <a href="http://map.baidu.com" name="tj_trmap" class="mnav">地图</a> <a href="http://v.baidu.com" name="tj_trvideo" class="mnav">视频</a> <a href="http://tieba.baidu.com" name="tj_trtieba" class="mnav">贴吧</a> <noscript> <a href="http://www.baidu.com/bdorz/login.gif?login&amp;tpl=mn&amp;u=http%3A%2F%2Fwww.baidu.com%2f%3fbdorz_come%3d1" name="tj_login" class="lb">登录</a> </noscript> <script>document.write('&lt;a href="http://www.baidu.com/bdorz/login.gif?login&amp;tpl=mn&amp;u='+ encodeURIComponent(window.location.href+ (window.location.search === "" ? "?" : "&amp;")+ "bdorz_come=1")+ '" name="tj_login" class="lb"&gt;登录&lt;/a&gt;');</script> <a href="//www.baidu.com/more/" name="tj_briicon" class="bri" style="display: block;">更多产品</a> </div> </div> </div> <div id="ftCon"> <div id="ftConw"> <p id="lh"> <a href="http://home.baidu.com">关于百度</a> <a href="http://ir.baidu.com">About Baidu</a> </p> <p id="cp">©2017 Baidu <a href="http://www.baidu.com/duty/">使用百度前必读</a> <a href="http://jianyi.baidu.com/" class="cp-feedback">意见反馈</a> 京ICP证030173号 <img src="//www.baidu.com/img/gs.gif"/> </p> </div> </div> </div> </body> </html>
<head><meta http-equiv="content-type" content="text/html;charset=utf-8"/><meta http-equiv="X-UA-Compatible" content="IE=Edge"/><meta content="always" name="referrer"/><link rel="stylesheet" type="text/css" href="http://s1.bdstatic.com/r/www/cache/bdorz/baidu.min.css"/><title>百度一下，你就知道</title></head>
```

实例 05-33：使用库 pyquery 实现基于 CSS 选择器查找

下面的实例文件 pyq04.py 演示了使用库 pyquery 实现基于 CSS 选择器查找的过程。

源码路径：daima\5\05-33\pyq04.py

```
html = '''
<div id="container">
    <ul class="list">
         <li class="item-0">first item</li>
         <li class="item-1"><a href="link2.html">second item</a></li>
         <li class="item-0 active"><a href="link3.html"><span class="bold">third item</span></a></li>
```

```
            <li class="item-1 active"><a href="link4.html">fourth item</a></li>
            <li class="item-0"><a href="link5.html">fifth item</a></li>
        </ul>
    </div>
'''
from pyquery import PyQuery as pq
doc = pq(html)
print(doc('#container .list li'))
```

在上述代码中需要注意 doc('#container .list li'),这三者并不是必须要挨着,只要是层级关系即可。

执行后会输出:

```
<li class="item-0">first item</li>
            <li class="item-1"><a href="link2.html">second item</a></li>
            <li class="item-0 active"><a href="link3.html"><span class="bold">third item</span></a></li>
            <li class="item-1 active"><a href="link4.html">fourth item</a></li>
            <li class="item-0"><a href="link5.html">fifth item</a></li>
```

实例 05-34:使用库 pyquery 查找子节点

在使用库 pyquery 时,可以通过已经查找的节点来查找其后的子节点或父节点,而不用从头开始查找。下面的实例文件 pyq05.py 演示了使用库 pyquery 查找子节点的过程。

源码路径:daima\5\05-34\pyq05.py

```
html = '''
<div id="container">
    <ul class="list">
        <li class="item-0">first item</li>
        <li class="item-1"><a href="link2.html">second item</a></li>
        <li class="item-0 active"><a href="link3.html"><span class="bold">third item</span></a></li>
        <li class="item-1 active"><a href="link4.html">fourth item</a></li>
        <li class="item-0"><a href="link5.html">fifth item</a></li>
    </ul>
</div>
'''
from pyquery import PyQuery as pq
doc = pq(html)
items = doc('.list')
print(type(items))
print(items)
lis = items.find('li')
print(type(lis))
print(lis)
```

执行后会输出:

```
<class 'pyquery.pyquery.PyQuery'>
<ul class="list">
         <li class="item-0">first item</li>
         <li class="item-1"><a href="link2.html">second item</a></li>
         <li class="item-0 active"><a href="link3.html"><span class="bold">third item</span></a></li>
         <li class="item-1 active"><a href="link4.html">fourth item</a></li>
         <li class="item-0"><a href="link5.html">fifth item</a></li>
     </ul>

<class 'pyquery.pyquery.PyQuery'>
<li class="item-0">first item</li>
         <li class="item-1"><a href="link2.html">second item</a></li>
         <li class="item-0 active"><a href="link3.html"><span class="bold">third item</span></a></li>
         <li class="item-1 active"><a href="link4.html">fourth item</a></li>
         <li class="item-0"><a href="link5.html">fifth item</a></li>
```

从上述输出结果可以看出,通过 pyquery 找到的其实还是一个 pyquery 对象,可以继续查找,上述代码中的 items.find('li') 则表示查找 ul 中的所有的 li 节点。当然这里通过 children 可以实现同样的效果,并且通过 children() 函数得到的结果也是一个 pyquery 对象。例如:

```
li = items.children()
print(type(li))
print(li)
```

同时，在children()中也可以用CSS选择器，例如：

```
li2 = items.children('.active')    print(li2)
```

实例05-35：使用库pyquery查找父节点

在使用库pyquery时，通过parent()可以找到父节点的内容。下面的实例文件pyq06.py演示了使用库pyquery查找父节点的过程。

源码路径：daima\5\05-35\pyq06.py

```
html = '''
<div id="container">
    <ul class="list">
         <li class="item-0">first item</li>
         <li class="item-1"><a href="link2.html">second item</a></li>
         <li class="item-0 active"><a href="link3.html"><span class="bold">third item</span></a></li>
         <li class="item-1 active"><a href="link4.html">fourth item</a></li>
         <li class="item-0"><a href="link5.html">fifth item</a></li>
    </ul>
</div>
'''
from pyquery import PyQuery as pq
doc = pq(html)
items = doc('.list')
container = items.parent()
print(type(container))
print(container)
```

执行后会输出：

```
<class 'pyquery.pyquery.PyQuery'>
<div id="container">
    <ul class="list">
         <li class="item-0">first item</li>
         <li class="item-1"><a href="link2.html">second item</a></li>
         <li class="item-0 active"><a href="link3.html"><span class="bold">third item</span></a></li>
         <li class="item-1 active"><a href="link4.html">fourth item</a></li>
         <li class="item-0"><a href="link5.html">fifth item</a></li>
    </ul>
</div>
```

从输出结果可以看出，结果返回了两部分内容，一部分是父节点的信息，另一部分是父节点的父节点的信息，即祖先节点的信息。同样道理，我们通过parents()查找时可以以添加CSS选择器的方式来进行内容的筛选。

实例05-36：使用库pyquery获取兄弟节点信息

在使用库pyquery时，通过siblings()可以获取所有的兄弟节点，当然这里不包括自己。同理，在siblings()中可以通过CSS选择器进行筛选。下面的实例文件pyq07.py演示了使用库pyquery获取兄弟节点信息的过程。

源码路径：daima\5\05-36\pyq07.py

```
html = '''
<div class="wrap">
    <div id="container">
        <ul class="list">
             <li class="item-0">first item</li>
             <li class="item-1"><a href="link2.html">second item</a></li>
             <li class="item-0 active"><a href="link3.html"><span class="bold">third item</span></a></li>
             <li class="item-1 active"><a href="link4.html">fourth item</a></li>
             <li class="item-0"><a href="link5.html">fifth item</a></li>
        </ul>
    </div>
</div>
'''
from pyquery import PyQuery as pq
doc = pq(html)
li = doc('.list .item-0.active')
print(li.siblings())
```

第 5 章 结构化标记处理实战

在上述代码中，doc('.list .item-0.active')中的 tem-0 和 active 是紧挨着的，表示两者是并的关系，这样满足条件的就剩下一个了：内容为 thired item 的节点。

执行后会输出：

```
<li class="item-1"><a href="link2.html">second item</a></li>
       <li class="item-0">first item</li>
       <li class="item-1 active"><a href="link4.html">fourth item</a></li>
       <li class="item-0"><a href="link5.html">fifth item</a></li>
```

第 6 章

应用程序开发实战

网络协议是为计算机网络进行数据交换而建立的规则、标准或约定的集合,现实中常用的网络协议有 HTTP、FTP 和 SMTP 等。另外,随着互联网技术的发展和普及,各种各样的互联网数据被展示在人们的面前,如邮件、JSON 数据等。本章将通过具体实例的实现过程,详细讲解开发互联网应用程序的知识,为读者学习本书后面的知识打下基础。

6.1 使用 webbrowser 实现浏览器操作

在 Python 中，内置模块 webbrowser 提供了一个可以向用户展示 Web 文件的高级接口。在绝大多数情况下，只需使用本模块中的 open()函数即可。

实例 06-01：分别调用 IE 和谷歌浏览器打开百度网主页

下面的实例文件 liulan.py 演示了使用模块 webbrowser 分别调用 IE 和谷歌浏览器打开百度网主页的过程。

源码路径：daima\6\06-01\liulan.py

```
import webbrowser
webbrowser.open('www.baidu.com')
#例如作者的谷歌浏览器安装位置是：C:\Program Files (x86)\Google\Chrome\Application\chrome.exe
chromePath = r'C:\Program Files (x86)\Google\Chrome\Application\chrome.exe'
webbrowser.register('chrome', None, webbrowser.BackgroundBrowser(chromePath))
#这里的chrome可以用其他任意名字，如chrome111，这里将想打开的浏览器保存到chrome
webbrowser.get('chrome').open('www.baidu.com',new=1,autoraise=True)
```

实例 06-02：调用默认浏览器每隔 5s 打开一次指定网页

下面的实例文件 liulan11.py 演示了使用模块 webbrowser 调用默认浏览器每隔 5s 打开一次指定网页的过程。

源码路径：daima\6\06-02\liulan11.py

```
import webbrowser
import time
import os

counter=1
while(1):
    counter=counter+1
    time.sleep(5)
    webbrowser.open("http://www.toppr.net",0,False)
    if counter==10:
        counter=
        os.system('pkill firefox')
```

在上述代码中，webbrowser.open()函数有 3 个参数，第 1 个参数是要打开的 URL，第 2 个参数为 0 代表刷新（具体情况要看浏览器配置，可能不能刷新），1 代表在新的浏览器中打开，2 代表在新的标签页中打开。执行上述代码后会调用内置浏览器，并按照 5s 每次的频率打开网页 http://www.toppr.net。

6.2 使用 urllib 包

在计算机网络模型中，Socket 编程属于底层网络协议开发的内容。虽然编写网络程序需要从底层开始构建，但是自行处理相关协议是一件比较麻烦的事情。其实对于大多数程序员来说，最常见的网络编程开发是针对应用协议进行的。在 Python 程序中，使用内置的包 urllib 和 http 可以完成 HTTP 层程序的开发工作。

实例 06-03：在百度搜索关键词中得到第一页链接

在 Python 程序中，urllib.request 模块定义了通过身份验证、重定向、Cookies 等方法打开 URL 的方法和类。在 urllib.request 模块中，方法 urlopen()的功能是打开一个 URL，将返回一个 HTTPResponse 对象，可以像操作文件一样使用 read()、readline()和 close()等对 URL 进行操作。下面的实例文件 url.py 演示了使用 urlopen()方法在百度搜索关键词中得到第一页链接的过程。

源码路径：daima\6\06-03\url.py

```
from urllib.request import urlopen         #导入Python的内置模块
from urllib.parse import urlencode         #导入Python的内置模块
import re                                  #导入Python的内置模块
##wd = input('输入一个要搜索的关键字：')
wd= 'www.toppr.net'                        #初始化变量wd
wd = urlencode({'wd':wd})                  #对URL进行编码
url = 'http://www.baidu.com/s?' + wd       #初始化url变量
page = urlopen(url).read()                 #打开变量url的网页并读取内容
#定义变量content，对网页进行编码处理，并实现特殊字符处理
content = (page.decode('utf-8')).replace("\n","").replace("\t","")
title = re.findall(r'<h3 class="t".*?h3>', content)            #正则表达式处理
title = [item[item.find('href=')+6:item.find('target=')] for item in title]   #正则表达式处理
title = [item.replace(' ',"").replace("'","") for item in title]              #正则表达式处理
for item in title:                         #遍历title
    print(item)                            #显示遍历值
```

在上述实例代码中，使用方法 urlencode()对搜索的关键字"www.toppr.net"进行 URL 编码，在拼接到百度的网址后，使用 urlopen()方法发出访问请求并取得结果，最后通过将结果进行解码、正则搜索与字符串处理后输出。如果将程序中的"##"标注的注释去除而把其后一句注释掉，就可以在执行时自主输入搜索的关键词。执行效果如图 6-1 所示。

```
http://www.baidu.com/link?url=hm6N8CdYPCSxsCsreajusLxba8mRVPAgc1D_WBhkYb7
http://www.baidu.com/link?url=N1f7T18n1Q0pke8pH8CIzg0V_wjqTKRtQ2NXLs-wUzyLHM0UknbUf1sJT3DLE2G0m6JW5G1RoBx-GbF6epS7sa
http://www.baidu.com/link?url=cbZcgLHZSTFBp6tFwWGwTuVq6xE3FcjM_d-cIH5qRNrkkXaETLwKKj9n9Rhlvvi8
http://www.baidu.com/link?url=0AHbz_vI3wIC_ocpmRc3jzcjIeu3gDeImuXcGfKu1zKGtaZ50-KR-HfGchsHSyGY
http://www.baidu.com/link?url=h591VC_3X6t7hm6eptcTS0dxFe5c4Z7XznvLzpqkJ1Z6a01WptFh4IS37h6LhzIC
```

图 6-1　执行效果

注意：urllib.response 模块是 urllib 使用的响应类，定义了和 urllib.request 模块类似的接口、方法和类，包括 read()和 readline()。限于篇幅，这里不再进行讲解。

实例 06-04：使用 urllib 实现 HTTP 身份验证

在 Python 程序中，urllib.parse 模块提供了一些用于处理 URL 字符串的功能。假设有一个需要身份验证（登录名和密码）的 Web 站点，通过验证的最简单方法是在 URL 中使用登录信息进行访问，如 http://username:passwd@www.python.org。但是这种方法的缺点是它不具有可编程性。通过使用 urllib 可以很好地解决这个问题，假设合法的登录信息如下。

```
LOGIN = 'admin'
PASSWD = "admin"
URL = 'http://localhost'
REALM = 'Secure AAA'
```

此时便可以通过下面的实例文件 pa.py，使用 urllib 实现 HTTP 身份验证。

源码路径：daima\6\06-04\pa.py

```
import urllib.request, urllib.error, urllib.parse

①LOGIN = 'admin'
  PASSWD = "admin"
  URL = 'http://localhost'
②REALM = 'Secure AAA'

③def handler_version(url):
    hdlr = urllib.request.HTTPBasicAuthHandler()
    hdlr.add_password(REALM,
        urllib.parse.urlparse(url)[1], LOGIN, PASSWD)
    opener = urllib.request.build_opener(hdlr)
    urllib.request.install_opener(opener)
```

```
④      return url
⑤def request_version(url):
       import base64
       req = urllib.request.Request(url)
       b64str = base64.b64encode(
           bytes('%s:%s' % (LOGIN, PASSWD), 'utf-8'))[:-1]
       req.add_header("Authorization", "Basic %s" % b64str)
⑥      return req
⑦for funcType in ('handler', 'request'):
       print('*** Using %s:' % funcType.upper())
       url = eval('%s_version' % funcType)(URL)
       f = urllib.request.urlopen(url)
       print(str(f.readline(), 'utf-8'))
⑧f.close()
```

①～②中实现普通的初始化功能，设置合法的登录验证信息。

③～④中定义函数 handler_version()，添加验证信息后建立一个 URL 开启器，安装该开启器的目的是方便所有已打开的 URL 都能用到这些验证信息。

⑤～⑥中定义函数 request_version()创建一个 Request 对象，并在 HTTP 请求中添加简单的 Base64 编码的验证头信息。在 for 循环中调用 urlopen()时，该请求用于替换其中的 URL 字符串。

⑦～⑧中分别打开了给定的 URL，通过验证后会显示服务器返回的 HTML 页面的第 1 行（转储了其他行）。如果验证信息无效，则会返回一个 HTTP 错误（并且不会有 HTML 页面）。

6.3 使用内置模块 http

在 Python 程序中，http 包实现了对 HTTP 的封装。http 包主要包含如下模块。

- http.client：底层的 HTTP 客户端，可以为 urllib.request 模块所用。
- http.server：提供了基于 socketserver 模块的基本 HTTP 服务器类。
- http.cookies：Cookies 的管理工具。
- http.cookiejar：提供了 Cookies 的持久化支持。

实例 06-05：访问指定的网站

http.client 模块定义了实现 HTTP 和 HTTPS 客户端的类。通常来说，不能直接使用 http.client 模块，需要通过模块 urllib.request 调用该模块来处理使用 HTTP 和 HTTPS 的 URL。下面的实例文件 fang.py 演示了使用 http.client.HTTPConnection 对象访问指定网站的过程。

源码路径：daima\6\06-05\fang.py

```
from http.client import HTTPConnection #导入内置模块
#基于HTTP访问客户端
mc = HTTPConnection('www.baidu.com:80')
mc.request('GET','/')                  #设置GET请求方法
res = mc.getresponse()                 #获取访问的网页
print(res.status,res.reason)           #输出响应的状态
print(res.read().decode('utf-8'))      #显示获取的内容
```

上述实例代码只实现了一个基本的访问实例，首先实例化 http.client.HTTPConnection，指定请求方法为 GET，然后使用 getresponse()方法获取访问的网页，并输出响应的状态。执行效果如图 6-2 所示。

6.3 使用内置模块 http

```
200 OK
<!DOCTYPE html><!--STATUS OK-->
<html>
<head>
    <meta http-equiv="content-type" content="text/html;charset=utf-8">
    <meta http-equiv="X-UA-Compatible" content="IE=Edge">
    <link rel="dns-prefetch" href="//s1.bdstatic.com"/>
    <link rel="dns-prefetch" href="//t1.baidu.com"/>
    <link rel="dns-prefetch" href="//t2.baidu.com"/>
    <link rel="dns-prefetch" href="//t3.baidu.com"/>
    <link rel="dns-prefetch" href="//t10.baidu.com"/>
    <link rel="dns-prefetch" href="//t11.baidu.com"/>
    <link rel="dns-prefetch" href="//t12.baidu.com"/>
    <link rel="dns-prefetch" href="//b1.bdstatic.com"/>
    <title>百度一下,你就知道</title>
    <link href="http://s1.bdstatic.com/r/www/cache/static/home/css/index.css" rel="stylesheet" type="text/css" />
    <!--[if lte IE 8]><style index="index">#content{height:480px\9}#m{top:260px\9}</style><![endif]-->
    <!--[if IE 8]><style index="index">#ul a.mnav,#ul a.mnav:visited{font-family:simsun}</style><![endif]-->
    <script>var hashMatch = document.location.href.match(/#+(.*wd=[^&].+)/);if (hashMatch && hashMatch[0] && hashMatch[1])
    <script>function h(obj){obj.style.behavior='url(#default#homepage)';var a = obj.setHomePage('//www.baidu.com/');}</scri
```

图 6-2 执行效果

实例 06-06：使用 http.client 模块中 GET 方式获取数据

下面的实例文件 httpmo.py 演示了使用 http.client 模块中 GET 方式获取数据的过程。

源码路径：daima\6\06-06\httpmo.py

```python
import http.client
conn = http.client.HTTPSConnection("www.python.org")
conn.request("GET", "/")
r1 = conn.getresponse()
print(r1.status, r1.reason)

data1 = r1.read()    #返回全部内容
#以块的方式读取数据
conn.request("GET", "/")
r1 = conn.getresponse()
while not r1.closed:
    print(r1.read(200))   # 200字节

#无效请求的示例
conn.request("GET", "/parrot.spam")
r2 = conn.getresponse()
print(r2.status, r2.reason)
data2 = r2.read()
conn.close()
```

实例 06-07：综合使用模块 http 和 urllib

下面的实例文件 wangluo.py 演示了综合使用模块 http 和 urllib 的过程。

源码路径：daima\6\06-07\wangluo.py

```python
import http.client
import urllib.parser
#初始化一个HTTPS链接
conn = http.client.HTTPSConnection("www.python.org")
#指定请求的方法和请求的链接地址
conn.request("GET","/doc/")
#得到返回的HTTP响应
r1 = conn.getresponse()
# HTTP状态码
print(r1.status,r1.reason)
# HTTP头部
print(r1.getheaders())
# body部分
print(r1.read())
#如果连接没有关闭，输出前200字节
if not r1.closed:
    print(r1.read(200))
#关闭连接后才能重新请求
```

```
    conn.close()
    #请求一个不存在的文件或地址
    conn.request("GET","/parrot.spam")
    r2 = conn.getresponse()
    print(r2.status,r2.reason)
    conn.close()
    #使用EAD求，但是不会返回任何数据
    conn = http.client.HTTPSConnection("www.python.org")
    conn.request("HEAD","/")
    res = conn.getresponse()
    print(res.status,res.reason)
    data = res.read()
    print(len(data))
    conn.close()
    #使用POST请求，提交的数据放在body部分
    params = urllib.parse.urlencode({'@number':12524,'@type':'issue','@action':'show'})
    #POST请求数据，要带上Content-type字段，以告知消息主体以何种方式编码
    headers = {"Content-type":"application/x-www-form-urlencoded","Accept":"text/plain"}
    conn = http.client.HTTPConnection("bugs.python.org")
    conn.request("POST","/",params,headers)
    response = conn.getresponse()
    #访问被重定向
    print(response.status,response.reason)
    print(response.read().decode("utf-8"))
    conn.close()
```

实例06-08：发送 HTTP GET 请求到远端服务器

在现实应用中，有时需要以客户端的形式通过 HTTP 访问多种服务，例如，下载数据或同一个基于 REST 的 API 进行交互。对于简单的任务来说，使用 urllib.request 模块就可以实现。要发送 HTTP GET 请求到远端服务器上，只需通过下面的实例文件 fang1.py 即可实现。

源码路径：daima\6\06-08\fang1.py

```
from urllib import request, parse

#要访问的URL
url = 'http://httpbin.org/get'

parms = {
    'name1' : 'value1',
    'name2' : 'value2'
}

#对查询字符串进行编码
querystring = parse.urlencode(parms)

#发出GET请求并读取响应
u = request.urlopen(url+'?' + querystring)
resp = u.read()

import json
from pprint import pprint

json_resp = json.loads(resp.decode('utf-8'))
pprint(json_resp)
```

执行后会输出：

```
{'args': {'name1': 'value1', 'name2': 'value2'},
 'headers': {'Accept-Encoding': 'identity',
             'Connection': 'close',
             'Host': 'httpbin.org',
             'User-Agent': 'Python-urllib/3.6'},
 'origin': '27.211.158.101',
 'url': 'http://httpbin.org/get?name1=value1&name2=value2'}
```

实例06-09：使用 POST 方法在请求主体中发送查询参数

如果需要使用 POST 方法在请求主体（request body）中发送查询参数，则可以将参数编码后作为可选参数提供给 urlopen()方法。下面的实例文件 fang2.py 即可实现该功能。

源码路径：daima\6\06-09\fang2.py

```python
from urllib import request, parse

#访问URL
url = 'http://httpbin.org/post'

# Dictionary of query parameters (if any)
parms = {
    'name1' : 'value1',
    'name2' : 'value2'
}

#对查询字符串进行编码
querystring = parse.urlencode(parms)

#发出POST请求并读取回复信息
u = request.urlopen(url, querystring.encode('ascii'))
resp = u.read()

import json
from pprint import pprint

json_resp = json.loads(resp.decode('utf-8'))
pprint(json_resp)
```

实例 06-10：在发出的请求中提供自定义的 HTTP 头

如果需要在发出的请求中提供自定义的 HTTP 头，如修改 user-agent 字段，此时可以创建一个包含字段值的字典来实现，并创建一个 Request 实例然后将其传给 urlopen()。下面的实例文件 fang3.py 即可实现该功能。

源码路径：daima\6\06-10\fang3.py

```python
import requests

#要访问的URL
url = 'http://httpbin.org/post'

# Dictionary of query parameters (if any)
parms = {
    'name1' : 'value1',
    'name2' : 'value2'
}

#浏览器代理
headers = {
    'User-agent' : 'none/ofyourbusiness',
    'Spam' : 'Eggs'
}

resp = requests.post(url, data=parms, headers=headers)

#请求返回的已解码文本
text = resp.text

from pprint import pprint
pprint(resp.json)
```

在上述代码中，因为需要交互的服务比较复杂，所以用到了 requests 库。

执行后会输出：

```
<bound method Response.json of <Response [200]>>
```

6.4 FTP 传输、SMTP 服务器和 XML-RPC 服务器

作为一门功能强大的开发语言，Python 通过内置模块可以实现 FTP 传输、SMTP 服务器和 XML-RPC 服务器功能。

实例 06-11：创建一个 FTP 文件传输客户端

当使用 Python 编写 FTP 文件传输客户端程序时，需要将相应的库 ftplib 导入程序中，然后实例化一个 ftplib.FTP 类对象，所有的 FTP 操作（如登录、传输文件和注销等）都要使用这个对象。下面的实例文件 ftp.py 演示了使用 ftplib 库创建一个 FTP 文件传输客户端的过程。

源码路径：daima\6\06-11\fb.py

```python
from ftplib import FTP                              #导入FTP
bufsize = 1024                                      #设置缓冲区的大小
def Get(filename):                                  #定义下载文件函数Get()
    command = 'RETR ' + filename                    #初始化变量command
    #下载FTP文件
    ftp.retrbinary(command, open(filename,'wb').write, bufsize)
    print('下载成功')                               #下载成功提示
def Put(filename):                                  #定义上传文件函数Put()
    command = 'STOR ' + filename                    #初始化变量command
    filehandler = open(filename,'rb')               #打开指定文件
    ftp.storbinary(command,filehandler,bufsize)     #实现文件上传操作
    filehandler.close()                             #关闭连接
    print('上传成功')                               #显示提示
def PWD():                                          #定义获取当前目录函数PWD()
    print(ftp.pwd())                                #返回当前所在位置
def Size(filename):                                 #定义获取文件大小函数Size()
    print(ftp.size(filename))                       #显示文件大小
def Help():                                         #定义帮助函数Help()
    print('''                                       #开始显示帮助
                Simple Python FTP
    cd        进入文件夹
    delete    删除文件
    dir       获取当前文件列表
    get       下载文件
    help      帮助
    mkdir     创建文件夹
    put       上传文件
    pwd       获取当前目录
    rename    重命名文件
    rmdir     删除文件夹
    size      获取文件大小
    ''')
server = input('请输入FTP服务器地址:')               #信息输入
ftp = FTP(server)                                   #获取服务器地址
username = input('请输入用户名:')                   #输入用户名
password = input('请输入密码:')                     #输入密码
ftp.login(username,password)                        #使用用户名和密码登录FTP服务器
print(ftp.getwelcome())                             #显示欢迎信息
#定义一个字典actions，用于保存操作命令
actions  = {'dir':ftp.dir, 'pwd': PWD, 'cd':ftp.cwd, 'get':Get,
            'put':Put, 'help':Help, 'rmdir': ftp.rmd,
            'mkdir': ftp.mkd, 'delete':ftp.delete,
            'size':Size, 'rename':ftp.rename}
while True:                                         #执行循环操作
    print('pyftp>')                                 #显示提示符
    cmds = input()                                  #获取用户的输入
    cmd = str.split(cmds)                           #使用空格分隔用户输入的内容
    try:                                            #异常处理
        if len(cmd) == 1:                           #验证输入命令中是否有参数
            if str.lower(cmd[0]) == 'quit':         #如果输入命令是quit则退出循环
                break
            else:
                actions[str.lower(cmd[0])]()        #调用与输出命令对应的函数
        elif len(cmd) == 2:                         #处理只有一个参数的命令
            actions[str.lower(cmd[0])](cmd[1])      #调用与输入命令对应的函数
        elif len(cmd) == 3:                         #处理有两个参数的命令
            actions[str.lower(cmd[0])](cmd[1],cmd[2]) #调用与输入命令对应的函数
        else:                                       #如果是其他情况
            print('输入错误')                       #显示错误提示
```

```
        except:
            print('命令出错')
    ftp.quit()                                          #退出客户端
```

执行上述实例代码后，会要求输入 FTP 服务器的地址、用户名和密码。如果输入上述信息正确则完成 FTP 服务器登录，并显示一个“pyftp>”提示符，等待用户输入命令。如果输入“dir”和“pwd”这两个命令，则会调用执行对应命令的函数。在测试、执行本实例代码时，需要有一个 FTP 服务器及登录该服务器的用户名和密码。如果读者没有互联网中的 FTP 服务器，可以尝试在本地计算机中通过 IIS 配置一个 FTP 服务器，然后进行测试。执行效果如图 6-3 所示。

```
请输入FTP服务器地址:118.184.21.197
请输入用户名:to
请输入密码:24cp
220 Microsoft FTP Service
pyftp>
dir
11-07-16  11:39AM       <DIR>          backup_c21323
11-07-16  11:38AM               476469 backup_c21323.rar
05-26-16  09:29AM       <DIR>          data
11-07-16  10:24AM             21819039 study.rar
11-07-16  12:56PM              3853663 study.sql
05-29-16  08:09AM       <DIR>          W3SVC81
05-26-16  09:33AM             26458130 web.rar
05-26-16  09:29AM       <DIR>          wwwlog
11-07-16  11:14AM       <DIR>          wwwroot
pyftp>
```

图 6-3 执行效果

实例 06-12：使用模块 smtpd 创建一个 SMTP 服务器

下面的实例文件 custom_server.py 先演示了使用模块 smtpd 创建一个 SMTP 服务器的过程。

源码路径：daima\6\06-12\custom_server.py

```
import smtpd
import asyncore

class CustomSMTPServer(smtpd.SMTPServer):

    def process_message(self, peer, mailfrom, rcpttos, data):
        print('Receiving message from:', peer)
        print('Message addressed from:', mailfrom)
        print('Message addressed to   :', rcpttos)
        print('Message length         :', len(data))

server = CustomSMTPServer(('127.0.0.1', 1025), None)
print('Server now running... ')
asyncore.loop()
```

然后通过下面的实例文件 senddata.py，从 SMTP 服务器发送指定的电子邮件信息。

源码路径：daima\6\06-12\senddata.py

```
import smtplib
import email.utils
from email.mime.text import MIMEText

#创建一个发送的信息
msg = MIMEText('This is the body of the message')
msg['to'] = email.utils.formataddr(('Recipient', 'casey.oneill@gmail.com'))
msg['From'] = email.utils.formataddr(('Author', 'caseyoneill78@hotmail.com'))
msg['Subject'] = 'Simple test message'

server = smtplib.SMTP('127.0.0.1', 1025)
server.set_debuglevel(True) # show communication with server
try:
    server.sendmail('caseyoneill78@hotmail.com', ['casey.oneill@gmail.com'], msg.as_string())
finally:
    server.quit()
```

实例 06-13：使用模块 xmlrpc.server 实现 XML-RPC 客户端和服务器相互通信

在 Python 中，模块 xmlrpc.server 为使用 Python 编写的 XML-RPC 服务器提供了一个基本的服务器框架。XML-RPC 服务器可以使用 SimpleXMLRPCServer 独立执行，也可以使用 CGIXMLRPCRequestHandler 嵌入到 CGI 环境中。在下面的实例中，演示了使用模块 xmlrpc.server 实现 XML-RPC 客户端和服务器相互通信的过程。其中，服务器文件 custom_server.py 的具体实现代码如下。

源码路径：daima\6\06-13\custom_server.py

```python
from xmlrpc.server import SimpleXMLRPCServer
from xmlrpc.server import SimpleXMLRPCRequestHandler

#限制特定路径
class RequestHandler(SimpleXMLRPCRequestHandler):
    rpc_paths = ('/RPC2',)

#创建server
server = SimpleXMLRPCServer(("localhost", 8000),
                            requestHandler=RequestHandler)
server.register_introspection_functions()

#注册函数pow()，将使用这个值
server.register_function(pow)

#在不同名称下注册函数
def adder_function(x,y):
    return x + y
server.register_function(adder_function, 'add')

class MyFuncs:
    def mul(self, x, y):
        return x * y

server.register_instance(MyFuncs())

#执行服务器的主循环
server.serve_forever()
```

客户端文件 CLI.py 的具体实现代码如下。

源码路径：daima\6\06-13\CLI.py

```python
import xmlrpc.client

s = xmlrpc.client.ServerProxy('http://localhost:8000')
print(s.pow(2,3))    # 返回2**3 = 8
print(s.add(2,3))    # 返回5
print(s.mul(5,2))    # 返回5*2 = 10

#输出可用的函数列表
print(s.system.listMethods())
```

在执行调试时先执行服务器文件，然后执行客户端文件，输出结果如下。

```
8
5
10
['add', 'mul', 'pow', 'system.listMethods', 'system.methodHelp', 'system.methodSignature']
```

6.5 开发电子邮件系统

从互联网诞生那一刻起，人与人之间日常交互又多了一种新的渠道。从此以后，交流变得更加迅速快捷，更具有实时性。后来，很多网络通信产品出现在市场上，如 QQ、MSN 和电子邮件系统，其中电子邮件更是深受人们的追捧。使用 Python 可以开发出功能强大的电子邮件系统，本节将通过具体实例来讲解使用 Python 开发电子邮件系统的过程。

实例 06-14：获取指定电子邮箱中最新两封电子邮件的主题和发件人

在计算机应用中，使用 POP3 协议可以登录电子邮箱服务器收取电子邮件。在 Python 程序中，内置模块 poplib 提供了对 POP3 协议的支持。现在市面中大多数电子邮箱客户端都提供了 POP3 收取电子邮件的方式，例如，Outlook 等电子邮箱客户端就是如此。开发者可以使用 Python 中的 poplib 模块开发出一个支持 POP3 协议的客户端程序。要想使用 Python 获取某个电子邮箱中电子邮件主题和发件人的信息，首先应该知道自己所使用的电子邮箱的 POP3 服务器的地址

6.5 开发电子邮件系统

和端口。

一般来说，电子邮箱服务器的地址格式如下。

pop.主机名.域名

端口的默认值是 110，例如，126 邮箱的 POP3 服务器地址为 pop.126.com，端口为默认值110。在下面的实例代码中，演示了使用 poplib 库获取指定电子邮箱中的最新两封电子邮件的主题和发件人的方法。实例文件 pop.py 的具体实现代码如下。

源码路径：daima\6\06-14\pop.py

```python
from poplib import POP3              #导入内置电子邮件处理模块
import re,email,email.header         #导入内置文件处理模块
from p_email import mypass           #导入内置模块
def jie(msg_src,names):              #定义解码电子邮件内容函数jie()
    msg = email.message_from_bytes(msg_src)
    result = {}                      #变量初始化
    for name in names:               #遍历name
        content = msg.get(name)      #获取name
        info = email.header.decode_header(content) #定义变量info
        if info[0][1]:
            if info[0][1].find('unknown-') == -1: #如果是已知编码
                result[name] = info[0][0].decode(info[0][1])
            else:                    #如果是未知编码
                try:                 #异常处理
                    result[name] = info[0][0].decode('gbk')
                except:
                    result[name] = info[0][0].decode('utf-8')
        else:
            result[name] = info[0][0]  #获取解码结果
    return result                    #返回解码结果
if __name__ == "__main__":
    pp = POP3("pop.sina.com")        #实例化电子邮箱服务器类
    pp.user('guanxijing820111@sina.com')  #传入电子邮箱地址
    pp.pass_(mypass)                 #密码设置
    total,totalnum = pp.stat()       #获取电子邮箱的状态
    print(total,totalnum)            #显示统计信息
    for i in range(total-2,total):   #遍历获取最新的两封电子邮件
        hinfo,msgs,octet = pp.top(i+1,0) #返回字节类型的内容
        b=b""
        for msg in msgs:             #遍历msg
            b += msg+b'\n'
        items = jie(b,['subject','from']) #调用函数jie()返回电子邮件主题
        print(items['subject'],'\nFrom:',items['from']) #调用函数jie()返回发件人的信息
        print()                      #输出空行
    pp.close()                       #关闭连接
```

在上述实例代码中，函数 jie() 的功能是使用 email 库来解码电子邮件头，用 POP3 对象的方法连接 POP3 服务器并获取电子邮箱中的电子邮件总数。在程序中获取最新的两封电子邮件的电子邮件头，然后传递给函数 jie() 进行分析，并返回电子邮件的主题和发件人的信息。执行效果如图 6-4 所示。

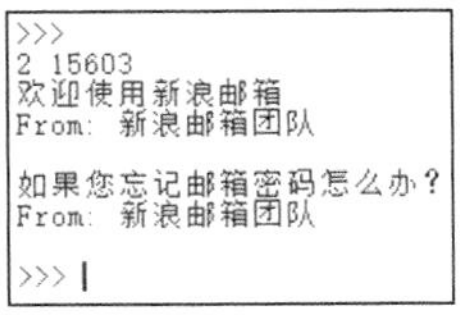

```
>>>
2 15603
欢迎使用新浪邮箱
From: 新浪邮箱团队

如果您忘记邮箱密码怎么办？
From: 新浪邮箱团队
>>>
```

图 6-4 执行效果

实例 06-15：向指定电子邮箱发送电子邮件

当使用 Python 发送电子邮件时，需要找到所使用电子邮件的 SMTP 服务器的地址和端口。例如，新浪邮箱的 SMTP 服务器的地址为 smtp.sina.com，端口为默认值 25。下面的实例文件 sm.py 演示了向指定电子邮箱发送电子邮件的过程。为了防止电子邮件被当作垃圾电子邮件丢弃，我们设置在登录认证后再发送电子邮件。实例文件 sm.py 的具体实现代码如下。

源码路径：daima\6\06-15\sm.py

```python
import smtplib,email                 #导入内置模块
from p_email import mypass           #导入内置模块
#使用email库构建一封电子邮件
chst = email.charset.Charset(input_charset='utf-8')
header = ("From: %s\nTo: %s\nSubject: %s\n\n" #电子邮件主题
```

177

第6章 应用程序开发实战

```
            % ("guanxijing820111@sina.com",     #电子邮箱地址
              "好人",                             #收件人
              chst.header_encode("Python smtplib 测试！"))) #电子邮件头
body = "你好！"                                   #电子邮件内容
email_con = header.encode('utf-8') + body.encode('utf-8') #构建电子邮件完整内容，中文编码处理
smtp = smtplib.SMTP("smtp.sina.com")              #电子邮箱服务器
smtp.login("guanxijing820111@sina.com",mypass) #以用户名和密码登录电子邮箱
#开始发送电子邮件
smtp.sendmail("guanxijing820111@sina.com","371972484@qq.com",email_con)
smtp.quit()                                       #退出系统
```

在上述实例代码中，使用新浪的 SMTP 服务器电子邮箱 guanxijing820111@sina.com 发送电子邮件，收件人的电子邮箱地址是 371972484@qq.com。首先使用 email.charset.Charset 对象对电子邮件头进行编码，然后创建 SMTP 对象，并通过验证的方式给 371972484@qq.com 发送一封测试电子邮件。因为电子邮件的主体内容含有中文字符，所以使用 encode()函数进行编码。执行效果如图 6-5 所示。

图 6-5　执行效果

实例 06-16：发送带附件功能的电子邮件

在 Python 程序中，内置标准库 email 的功能是管理电子邮件，具体来说，不是实现向 SMTP 服务器、NNTP 服务器或其他服务器发送任何电子邮件的功能，而是实现诸如 smtplib 和 nntplib 之类的库的功能。库 email 的主要功能是解析邮件服务器中的邮件信息，并根据用户的需求生成邮件内容。通过使用 email，可以向消息中添加子对象，从消息中删除子对象，完全重新排列内容。下面的实例文件 youjian.py 演示了使用库 email 和 smtplib 发送带附件功能电子邮件的过程。

源码路径：daima\6\06-16\youjian.py

```
import smtplib
from email.mime.multipart import MIMEMultipart
from email.mime.text import MIMEText
from email.mime.image import MIMEImage

sender = '***'
receiver = '***'
subject = 'python email test'
smtpserver = 'smtp.163.com'
username = '***'
password = '***'

msgRoot = MIMEMultipart('related')
msgRoot['Subject'] = 'test message'

# 构造附件
att = MIMEText(open('h:\\python\\1.jpg', 'rb').read(), 'base64', 'utf-8')
att["Content-Type"] = 'application/octet-stream'
att["Content-Disposition"] = 'attachment; filename="1.jpg"'
msgRoot.attach(att)

smtp = smtplib.SMTP()
smtp.connect('smtp.163.com')
smtp.login(username, password)
smtp.sendmail(sender, receiver, msgRoot.as_string())
smtp.quit()
```

实例 06-17：使用库 envelopes 向指定电子邮箱发送电子邮件

库 envelopes 是 Python 处理电子邮件的一个第三方库，是对 Python 内置库 email 和 smtplib 的封装。对于 Linux 操作系统，直接使用如下命令即可安装 envelopes。

```
pip install envelopes
```

对于 Windows 操作系统，需要先下载 envelopes 的源码文件压缩包，解压之后使用如下命令即可安装 envelopes。

```
python setup.py install
```

下面的实例文件 evelopes01.py 演示了使用库 envelopes 向指定电子邮箱发送电子邮件的过程。

源码路径：daima\6\06-17\envelopes01.py

```python
from envelopes import Envelope, GMailSMTP

envelope = Envelope(   # 实例化Envelope
    from_addr=(u'from@example.com', u'From Example'),
    #必选参数，发件人信息。前面是发送电子邮箱，后面是发送人；只有发送电子邮箱也可以
    to_addr=(u'to@example.com', u'To Example'),
    #必选参数，发送多人可以直接使用(u'user1@example.com', u'user2@example.com')
    subject=u'Envelopes demo',        #必选参数，电子邮件标题
    html_body=u'<h1>活着之上</h1>'#可选参数，带HTML的电子邮件正文
    text_body=u"I'm a helicopter!",   #可选参数，文本格式的电子邮件正文
    cc_addr=u'boss1@example.com',     #可选参数，抄送人，也可以是列表
    bcc_addr=u'boss2@example.com',    #可选参数，隐藏抄送人，也可以是列表
    headers=u'',                      #可选参数，电子邮件头内容，字典
    charset=u'',                      #可选参数，电子邮件字符集
)
envelope.add_attachment('/Users/bilbo/Pictures/helicopter.jpg') #增加附件，注意文件是完整路径，也可以加入多个附件

envelope.send('smtp.163.com', login='from@example.com',
              password='password', tls=True)  #发送电子邮件，通过SMTP服务器，登录电子邮箱

#或者使用共享Gmail连接发送电子邮件
gmail = GMailSMTP('from@example.com', 'password')
gmail.send(envelope)
```

实例 06-18：使用库 envelopes 构建 Flask Web 电子邮件发送程序

下面的实例文件 envelopes02.py 演示了使用库 envelopes 构建 Flask Web 电子邮件发送程序的过程。

源码路径：daima\6\06-18\envelopes02.py

```python
from envelopes import Envelope, SMTP
import envelopes.connstack
from flask import Flask, jsonify
import os

app = Flask(__name__)
app.config['DEBUG'] = True
conn = SMTP('127.0.0.1', 1025)

@app.before_request
def app_before_request():
    envelopes.connstack.push_connection(conn)

@app.after_request
def app_after_request(response):
    envelopes.connstack.pop_connection()
    return response

@app.route('/mail', methods=['POST'])
def post_mail():
    envelope = Envelope(
        from_addr='%s@localhost' % os.getlogin(),
        to_addr='%s@localhost' % os.getlogin(),
        subject='Envelopes in Flask demo',
```

```
                text_body="I'm a helicopter!"
            )
            smtp = envelopes.connstack.get_current_connection()
            smtp.send(envelope)
            return jsonify(dict(status='ok'))

    if __name__ == '__main__':
        app.run()
```

实例 06-19：创建一个带有 HTTP REST 接口的 SMTP 服务器

Inbox 是一个用 Python 实现的简单 SMTP 服务器，它是异步的。单个实例处理电子邮件的速度达到 1000 封/秒以上。我们可以使用如下命令安装 Inbox。

```
pip install Inbox
```

下面的实例文件 Inbox01.py 使用库 Inbox 创建一个带有 HTTP REST 接口的 SMTP 服务器，用于访问电子邮件信息。

源码路径：daima\6\06-19\Inbox01.py

```
@inbox.collate
def handle(to, sender, body):
    parser = email.parser.Parser()
    mail = parser.parsestr(body)
    message = {}
    message['to'] = to
    message['sender'] = sender
    message['subject'] = mail['subject']
    message['received'] = time.strftime("%Y-%m-%d %H:%M:%S")
    message['content'] = ""
    message['attachments'] = []

    for part in mail.walk():
        if part.get_content_maintype() == "multipart":
            continue

        if not part.get_filename():
            if part.get_content_maintype() == "text":
                message['content'] += part.get_payload(decode=False)
        else:
            attachment = {}
            attachment['filename'] = part.get_filename()
            attachment['type'] = part.get_content_type()
            payload = part.get_payload(decode=True)
            attachment['payload-id'] = storage.store_attachment(payload, part.get_content_type())
            message['attachments'].append(attachment)

    storage.store_mail(message)

server = HttpServer('0.0.0.0', 8123, storage)

handle = open('123.txt', 'wb')
sys.stderr = handle

inbox.serve(address='0.0.0.0', port=4467)
```

6.6 解析 JSON 数据

JSON 是指 JavaScript Object Notation，是一种轻量级的数据交换格式。JSON 是基于 ECMAScript 的一个子集。

实例 06-20：将 Python 字典转换为 JSON 对象

下面的实例文件 js.py 演示了将 Python 字典转换为 JSON 对象的过程。

6.6 解析 JSON 数据

源码路径：daima\6\06-20\js.py

```
import json
#将字典转换为JSON对象
data = {
    'no' : 1,
    'name' : 'laoguan',
    'url' : 'http://www.toppr.net'
}
JSON_str = JSON.dumps(data)
print ("Python 原始数据：", repr(data))
print ("JSON 对象：", JSON_str)
```

执行效果如图 6-6 所示。通过执行效果可以看出，简单类型通过编码后跟其原始的 repr() 输出结果非常相似。

```
Python 原始数据： {'url': 'http://www.toppr.net', 'no': 1, 'name': 'laoguan'}
JSON 对象： {"url": "http://www.toppr.net", "no": 1, "name": "laoguan"}
```

图 6-6　执行效果

实例 06-21：将 JSON 编码的字符串转换为 Python 数据结构

基于上面的实例 06-20，下面的实例文件 fan.py 演示了将 JSON 编码的字符串转换为 Python 数据结构的过程。

源码路径：daima\6\06-21\fan.py

```
import JSON
#将字典转换为JSON对象
data1 = {
    'no' : 1,
    'name' : 'laoguan',
    'url' : 'http://www.toppr.net'
}
JSON_str = JSON.dumps(data1)
print ("Python 原始数据：", repr(data1))
print ("JSON 对象：", json_str)
# 将 JSON 对象转换为字典
data2 = JSON.loads(json_str)
print ("data2['name']: ", data2['name'])
print ("data2['url']: ", data2['url'])
```

执行效果如图 6-7 所示。

```
Python 原始数据： {'name': 'laoguan', 'url': 'http://www.toppr.net', 'no': 1}
JSON 对象： {"name": "laoguan", "url": "http://www.toppr.net", "no": 1}
data2['name']:  laoguan
data2['url']:  http://www.toppr.net
```

图 6-7　执行效果

实例 06-22：编写自定义类解析 JSON 数据

在 Python 程序中，如果要处理的是 JSON 文件而不是字符串，那么可以使用函数 JSON.dump()和函数 JSON.load()来编码和解码 JSON 数据。例如：

```
#写入JSON数据
with open('data.JSON', 'w') as f:
    JSON.dump(data, f)
#读取数据
with open('data.JSON', 'r') as f:
    data = JSON.load(f)
```

下面的实例文件 JSONparser.py 演示了编写自定义类解析 JSON 数据的过程。

源码路径：daima\6\06-22\JSONparser.py

```
def txt2str(file='JSONdata2.txt'):
    '''
    打开指定的JSON文件
    '''
    fp=open(file,encoding='UTF-8')
    allLines = fp.readlines()
```

```python
            fp.close()
            str=""
            for eachLine in allLines:
                    #eachLine=ConvertCN(eachLine)
                    #转换成字符串
                    for i in range(0,len(eachLine)):
                            #if eachLine[i]!=' ' and eachLine[i]!= '   ' and eachLine[i]!='\n':
                            #删除空格和换行符，但是双引号中的空格不能删除
                            str+=eachLine[i]
            return str

class JSONparser:
        def __init__(self, str=None):
                self._str = str
                self._index=0

        def _skipBlank(self):
                '''
                跳过空格符、换行符或制表符：\n\t\r
                '''
                while self._index<len(self._str) and self._str[self._index] in ' \n\t\r':
                        self._index=self._index+1
        def parse(self):
                '''
                进行解析的主要函数
                '''
                self._skipBlank()
                if self._str[self._index]=='{':
                        self._index+=1
                        return self._parse_object()
                elif self._str[self._index] == '[':
                        self._index+=1
                        return self._parse_array()
                else:
                        print("JSON format error!")
        def _parse_string(self):
                '''
                找出两个双引号中的字符串
                '''
                begin = end =self._index
                #找到字符串的范围
                while self._str[end]!='"':
                        if self._str[end]=='\\': #重点！出现\，表明其后面是配合\的转义符号，如\"、\t、\r，主要针对\"的情况
                                end+=1
                                if self._str[end] not in '"\\bfnrtu':
                                        print
                        end+=1
                self._index = end+1
                return self._str[begin:end]

        def _parse_number(self):
                '''
                数字没有双引号
                '''
                begin = end = self._index
                end_str=' \n\t\r,}]' #数字结束的字符串
                while self._str[end] not in end_str:
                        end += 1
                number = self._str[begin:end]

                #进行转换
                if '.' in number or 'e' in number or 'E' in number :
                        res = float(number)
                else:
                        res = int(number)
                self._index = end
                return res

        def _parse_value(self):
                '''
```

```python
        解析值，包括字符串、数字
        """
        c = self._str[self._index]

        #解析对象
        if c == '{':
            self._index+=1
            self._skipBlank()
            return self._parse_object()
        #解析数组
        elif c == '[':
            self._index+=1
            self._skipBlank()
            return self._parse_array()
        #解析字符串
        elif c == '"':
            self._index += 1
            self._skipBlank()
            return self._parse_string()
        #解析null
        elif c=='n' and self._str[self._index:self._index+4] == 'null':
            self._index+=4
            return None
        #解析布尔值true
        elif c=='t' and self._str[self._index:self._index+4] == 'true':
            self._index+=4
            return True
        #解析布尔值false
        elif c=='f' and self._str[self._index:self._index+5] == 'false':
            self._index+=5
            return False
        #剩下的情况为数字
          else:
            return self._parse_number()

    def _parse_array(self):
        """
        解析数组
        """
        arr=[]
        self._skipBlank()
        #空数组
        if self._str[self._index]==']':
            self._index +=1
            return arr
        while True:
            val = self._parse_value()      #获取数组中的值，可能是字符串、数组等
            arr.append(val)                #添加到数组中
            self._skipBlank()              #跳过空白
            if self._str[self._index] == ',':
                self._index += 1
                self._skipBlank()
            elif self._str[self._index] ==']':
                self._index += 1
                return arr
            else:
                print("array parse error!")
                return None

    def _parse_object(self):
        """
        解析对象
        """
        obj={}
        self._skipBlank()
        #空对象
        if self._str[self._index]=='}':
            self._index +=1
            return obj
        #elif self._str[self._index] !='"':
            #报错

        self._index+=1 #跳过当前的双引号
```

```python
            while True:
                key = self._parse_string() #获取key值
                self._skipBlank()
                
                self._index = self._index+1#跳过冒号
                self._skipBlank()
                
                #self._index = self._index+1#跳过双引号
                #self._skipBlank()
                #获取value值，目前假设只有value的字符串和数字
                obj[key]= self._parse_value()
                self._skipBlank()
                #print key,":",obj[key]
                #对象结束了，执行break语句
                if self._str[self._index]=='}':
                    self._index +=1
                    break
                elif self._str[self._index]==',':
                    self._index +=1
                    self._skipBlank()
            self._index +=1#跳过下一个对象的第一个双引号
        return obj#返回对象

    def display(self):
        displayStr=""
        self._skipBlank()
        while self._index<len(self._str):
            displayStr=displayStr+self._str[self._index]
            self._index=self._index+1
            self._skipBlank()
        print(displayStr)


def _to_str(pv):
    '''把Python变量转换成字符串'''
    _str=''
    if type(pv) == type({}):
        #处理对象
        _str+='{'
        _noNull = False
        for key in pv.keys():
            if type(key) == type(''):
                _noNull = True #对象非空
                _str+='"'+key+'":'+_to_str(pv[key])+','
        if _noNull:
            _str = _str[:-1] #把最后的逗号去掉
        _str +='}'
        
    elif type(pv) == type([]):
        #处理数组
        _str+='['
        if len(pv) >0: #数组不为空，方便后续格式合并
            _str += _to_str(pv[0])
            for i in range(1,len(pv)):
                _str+=','+_to_str(pv[i])#因为已经合并了第一个，所以可以加逗号
        _str+=']'
        
    elif type(pv) == type(''):
        #字符串
        _str = '"'+pv+'"'
    elif pv == True:
        _str+='true'
    elif pv == False:
        _str+='false'
    elif pv == None:
        _str+='null'
    else:
        _str = str(pv)
    return _str
```

上述代码中提及的 JSON 主体内容主要指两大类：对象和数组。因为一个 JSON 格式的字符串不是一个对象就是一个数组，所以类 JSONparser 中有_parse_object()和_parse_array()两个函

数。首先通过 parse()函数直接判断开始的符号为花括号"{"还是方括号"[",进而决定调用_parse_object()还是_parse_array()。在标准 JSON 格式中，key 是字符串，使用双引号标识，其中_parse_string()函数专门用来解析 key。而 JSON 中的 value 则相对复杂一些，类型可以是对象、数组、字符串、数字、布尔值和 null。

在上述代码中，如果遇到的字符为花括号"{"，则调用_parse_object()函数；如果遇到的字符为方括号"["，则调用_parse_array()函数，并且把 value 解析统一封装到函数_parse_value()中。

本实例执行后会输出：

```
test
['JSON Test Pattern pass1', {'object with 1 member': ['array with 1 element']}, {}, [], -42, True, False, None, {'integer': 1234567890, 'real': -9876.54321, 'e': 1.23456789e-13, 'E': 1.23456789e+34, '': -inf, 'zero': 0, 'one': 1, 'space': ' ', 'singlequote': "\\'", 'singlequote2': "''", 'quote': '\\"', 'backslash': '\\\\', 'controls': '\\b\\f\\n\\r\\t', 'slash': '/ & \\/', 'alpha': 'abcdefghijklmnopqrstuvwyz', 'ALPHA': 'ABCDEFGHIJKLMNOPQRSTUVWYZ', 'digit': '0123456789', 'special': "`1~!@#$%^&*()_+-=\{:[,]\}|;.</>?", 'hex': '\\u0123\\u4567\\u89AB\\uCDEF\\uabcd\\uef4A', 'true': True, 'false': False, 'null': None, 'array': [], 'object': {}, 'address': '50 St. James Street', 'url': 'http://www.JSON.org/', 'comment': '// /* <!-- --', '# -- --> */': '', 's p a c e d': [1, 2, 3, 4, 5, 6, 7], 'compact': [1, 2, 3, 4, 5, 6, 7], 'jsontext': '{\\"object with 1 member\\":[\\"array with 1 element\\"]}', '\\/\\\\\\"\\uCAFE\\uBABE\\uAB98\\uFCDE\\ubcda\\uef4A\\b\\f\\n\\r\\t`1~!@#$%^&*()_+-=[]{}|;:\',./<>?': 'A key can be any string'}, 0.5, 98.6, 99.44, 1066, 'rosebud']
[{"JSON Test Pattern pass1",{"object with 1 member":["array with 1 element"]},{},[],-42,true,false,null,{"integer":1234567890,"real":-9876.54321,"e":1.23456789e-13,"E":1.23456789e+34,"":-inf,"zero":false,"one":true,"space":"","singlequote":"'","singlequote2":"''","quote":"\\"","backslash":"\\\\","controls":"\\b\\f\\n\\r\\t","slash":"/ & \\/","alpha":"abcdefghijklmnopqrstuvwyz","ALPHA":"ABCDEFGHIJKLMNOPQRSTUVWYZ","digit":"0123456789","special":"`1~!@#$%^&*()_+-={:[,]}|;.</>?","hex":"\\u0123\\u4567\\u89AB\\uCDEF\\uabcd\\uef4A","true":true,"false":false,"null":null,"array":[],"object":{},"address":"50 St. James Street","url":"http://www.JSON.org/","comment":"// /* <!-- --","# -- --> */":"","s p a c e d":[true,2,3,4,5,6,7],"compact":[true,2,3,4,5,6,7],"jsontext":"{\\"object with 1 member\\":[\\"array with 1 element\\"]}","\\/\\\\\\"\\uCAFE\\uBABE\\uAB98\\uFCDE\\ubcda\\uef4A\\b\\f\\n\\r\\t`1~!@#$%^&*()_+-=[] {}|;:',./<>?":"A key can be any string"},0.5,98.6,99.44,1066,"rosebud"]

Json format error!
None
null
```

6.7 实现数据编码和解码

在 Python 中，我们可以使用模块 base64 实现数据的编码和解码功能。模块 base64 为在 RFC 3548 中指定的编码提供编码和解码功能，该编码定义了 Base16、Base32 和 Base64 算法，以及事实上的标准 ASCII85 和 Base85 编码。

实例 06-23：实现数据"编码/解码"操作

下面的实例文件 base.py 演示了使用模块 base64 实现数据"编码/解码"操作的过程。

源码路径：daima\6\06-23\base.py

```python
import base64

a = 'this is a test'
b = base64.b64encode(a.encode(encoding='utf-8'))
print(b)
print('---------------')
print(base64.b64decode(b))

x = 'http://www.baidu.com'
y = base64.urlsafe_b64encode(x.encode(encoding='utf-8'))
print(y)
print('*****************')
print(base64.urlsafe_b64decode(y))
```

执行后会输出：

```
b'dGhpcyBpcyBhIHRlc3Q='
---------------
b'this is a test'
b'aHR0cDovL3d3dy5iYWlkdS5jb20='
*****************
b'http://www.baidu.com'
```

实例 06-24：实现 bytes 类型和 base64 类型的相互转换

下面的实例文件 base02.py 演示了使用模块 base64 实现 bytes 类型和 base64 类型相互转换的过程。

源码路径：daima\6\06-24\base02.py

```python
"""
base64实现
"""

import base64
import string

# base64字符集
base64_charset = string.ascii_uppercase + string.ascii_lowercase + string.digits + '+/'


def encode(origin_bytes):
    """
    将bytes类型编码为base64
    :参数origin_bytes:需要编码的bytes
    :返回:base64字符串
    """

    #将每一位bytes转换为二进制字符串
    base64_bytes = ['{:0>8}'.format(str(bin(b)).replace('0b', '')) for b in origin_bytes]

    resp = ''
    nums = len(base64_bytes) // 3
    remain = len(base64_bytes) % 3

    integral_part = base64_bytes[0:3 * nums]
    while integral_part:
        #取3字节，每6位，转换为4个整数
        tmp_unit = ''.join(integral_part[0:3])
        tmp_unit = [int(tmp_unit[x: x + 6], 2) for x in [0, 6, 12, 18]]
        #取对应base64字符
        resp += ''.join([base64_charset[i] for i in tmp_unit])
        integral_part = integral_part[3:]

    if remain:
        #补齐3字节，每个字节补充 0000 0000
        remain_part = ''.join(base64_bytes[3 * nums:]) + (3 - remain) * '0' * 8
        #取3字节，每6位，转换为4个整数
        #剩余1字节可构造2个base64字符，补充==；剩余2字节可构造3个base64字符，补充=
        tmp_unit = [int(remain_part[x: x + 6], 2) for x in [0, 6, 12, 18]][:remain + 1]
        resp += ''.join([base64_charset[i] for i in tmp_unit]) + (3 - remain) * '='

    return resp


def decode(base64_str):
    """
    解码base64字符串
    :参数base64_str:base64字符串
    :返回:解码后的bytearray对象；若传入参数不是合法base64字符串，返回空bytearray对象
    """
    if not valid_base64_str(base64_str):
        return bytearray()

    #对每一个base64字符取索引，并转换为二进制字符串
    base64_bytes = ['{:0>6}'.format(str(bin(base64_charset.index(s))).replace('0b', '')) for s in base64_str if
                    s != '=']
    resp = bytearray()
    nums = len(base64_bytes) // 4
    remain = len(base64_bytes) % 4
    integral_part = base64_bytes[0:4 * nums]

    while integral_part:
        #取4个6位base64字符，作为3字节
```

```
                tmp_unit = ''.join(integral_part[0:4])
                tmp_unit = [int(tmp_unit[x: x + 8], 2) for x in [0, 8, 16]]
                for i in tmp_unit:
                    resp.append(i)
                integral_part = integral_part[4:]

        if remain:
            remain_part = ''.join(base64_bytes[nums * 4:])
            tmp_unit = [int(remain_part[i * 8:(i + 1) * 8], 2) for i in range(remain - 1)]
            for i in tmp_unit:
                resp.append(i)

        return resp


    def valid_base64_str(b_str):
        """
        验证是否为合法base64字符串
        :参数b_str: 待验证的base64字符串
        :返回:是否合法
        """
        if len(b_str) % 4:
            return False

        for m in b_str:
            if m not in base64_charset:
                return False
        return True


    if __name__ == '__main__':
        s = '我的目标是星辰大海. One piece, all Blue'.encode()
        local_base64 = encode(s)
        print('使用本地base64加密: ', local_base64)
        b_base64 = base64.b64encode(s)
        print('使用base64加密: ', b_base64.decode())

        print('使用本地base64解密: ', decode(local_base64).decode())
        print('使用base64解密: ', base64.b64decode(b_base64).decode())
```

执行后会输出：

```
使用本地base64加密: 5oiR55qE55uu5qCH5piv5pif6L6w5aSn5rW3LiBPbmUgcGllY2UsIGFsbCBCbHVl
使用base64加密:   5oiR55qE55uu5qCH5piv5pif6L6w5aSn5rW3LiBPbmUgcGllY2UsIGFsbCBCbHVl
使用本地base64解密: 我的目标是星辰大海. One piece, all Blue
使用base64解密:   我的目标是星辰大海. One piece, all Blue
```

实例 06-25：生成由某地址可表示的全部 IP 地址的范围

在 Python 程序中，模块 ipaddress 提供了创建、操作 IPv4 和 IPv6 地址和网络的功能。

模块 ipaddress 中的函数和类可以直接处理与 IP 地址相关的各种任务，包括检查两个主机是否在同一子网上，遍历特定子网中的所有主机，检查字符串是否表示有效 IP 地址或网络定义等。

假如有一个类似于 192.168.1.1/27 这样的 CIDR（Classless InterDomain Routing）网络地址，要想生成由该地址可表示的全部 IP 地址的范围（如 192.168.1.1/27,192.168.1.2,…,192.168.1.95）。下面的实例文件 ipaddress01.py 演示了使用模块 ipaddress 实现上述功能的过程。

源码路径：daima\6\06-25\ipaddress01.py

```
① import ipaddress
   net = ipaddress.ip_network('192.168.1.1/27')
   print(net)
   for a in net:
       print(a)

   net6 = ipaddress.ip_network('19:2168:1:90ab:cd:ef01:92:168/125')
   print(net6)
   for a in net6:
② print(a)

③ print(net.num_addresses)
   print(net[0])
```

```
    print(net[1])
    print(net[-1])
④ print(net[-2])

⑤ a = ipaddress.ip_address('192.168.1.1')
    print(a in net)
    b = ipaddress.ip_address('192.168.1.123')
⑥ print(b in net)

⑦ inet = ipaddress.ip_interface('192.168.1.73/27')
    print(inet.network)
⑧ print(inet.ip)
```

①～②中将"192.168.1.89"CIDR 网络地址生成了由该地址可表示的全部 IP 地址的范围。

③～④中使用 ip_network 对象实现像数组那样的索引操作。

⑤～⑥中检查成员的归属。

⑦～⑧中在 IP 地址加上一个网络号后，可以用来指定一个 IP 接口。

6.8 实现身份验证

实例 06-26：获取指定字符串的数据指纹

在 Python 程序中，模块 hashlib 为不同的安全哈希（secure hash algorithm）和信息摘要算法（message digest algorithm）实现了一个公共的、通用的接口。也可以说，模块 hashlib 是一个统一的入口，因为 hashlib 模块不仅仅整合了 md5 模块和 sha 模块的功能，而且提供了对更多算法的函数实现，如 MD5、SHA1、SHA224、SHA256、SHA384 和 SHA512。

下面的实例文件 jiami01.py 演示了以 MD5 算法为例，使用 hashlib 模块获取字符串"Hello, World"的摘要信息（也叫数据指纹）的过程。

源码路径：daima\6\06-26\jiami01.py

```
import hashlib

pwd = "Hello, World"
checkcode = hashlib.md5(pwd.encode("utf-8")).hexdigest()
print(checkcode)

print(type(pwd), len(pwd), pwd)
```

实例 06-27：利用 hmac 模块实现简单且高效的身份验证

在 Python 程序中，HMAC 算法也是一种单项加密算法，并且它是基于各种哈希算法的，只是它可以在运算过程中使用一个密钥来增强安全性。hmac 模块实现了 HAMC 算法，提供了相应的函数和方法，且与 hashlib 提供的基本一致。

假如我们希望有一种简单的方式：可以对在分布式系统中连接到各个服务器上的客户端进行身份验证，但是又不想使用像 ssl 那样的复杂组件，此时可以利用 hmac 模块实现一个握手连接来达到简单且高效的身份验证目的。下面的实例文件 jiami02.py 演示了使用 hmac 模块实现上述功能的过程。

源码路径：daima\6\06-27\jiami02.py

```
import hmac
import os

def client_authenticate(connection, secret_key):
    '''
    向远程服务验证客户端
    这里的连接表示网络连接
    secret_key表示只有客户机/服务器端知道的密钥
    '''
```

```
        message = connection.recv(32)
        hash = hmac.new(secret_key, message)
        digest = hash.digest()
        connection.send(digest)

def server_authenticate(connection, secret_key):
        message = os.urandom(32)
        connection.send(message)
        hash = hmac.new(secret_key, message)
        digest = hash.digest()
        response = connection.recv(len(digest))
        return hmac.compare_digest(digest,response)
```

通过上述代码，在发起连接时，服务器将一段由随机字节组成的消息（在本例中是由 os.urandom()返回的）发送给客户端。客户端和服务器通过 hmac 模块及双方事先都知道的密钥计算出随机数据的加密 hash。客户端将它计算出的摘要值（digest）发送给服务器，而服务器对摘要值进行比较，以此来决定接受还是拒绝这个连接。

实例 06-28：Socket 服务器和客户端的加密认证

对摘要值进行比较时需要使用 hmac.compare_digest()函数，这个函数的实现可避免遭受基于时序分析的攻击（timing-analysis attack），因此应该用它来对摘要值进行比较而不能用普通的比较操作符（==）。要使用这些函数，我们可以将它们合并到已有的有关网络或消息处理的代码中。例如，如果用到了 Socket，则可以用下面的服务器文件 server.py 进行测试。

源码路径：daima\6\06-28\server.py

```
from socket import socket, AF_INET, SOCK_STREAM
from jiami02 import server_authenticate

secret_key = b'peekaboo'

def echo_handler(client_sock):
    if not server_authenticate(client_sock, secret_key):
        client_sock.close()
        return
    while True:
        msg = client_sock.recv(8192)
        if not msg:
            break
        client_sock.sendall(msg)

def echo_server(address):
    s = socket(AF_INET, SOCK_STREAM)
    s.bind(address)
    s.listen(5)
    while True:
        c,a = s.accept()
        echo_handler(c)

print('Echo server running on port 18000')

echo_server(('', 18000))
```

而在客户端，则可以用下面的服务器文件 client.py 进行测试。

源码路径：daima\6\06-28\client.py

```
from socket import socket, AF_INET, SOCK_STREAM
from jiami02 import client_authenticate

secret_key = b'peekaboo'

s = socket(AF_INET, SOCK_STREAM)
s.connect(('localhost', 18000))
client_authenticate(s, secret_key)
s.send(b'Hello World')
resp = s.recv(1024)
print('Got:', resp)
```

内部消息系统及进程间通信中常常会用 HMAC 来验证身份。例如，如果正在编写的系统需要实现跨集群的多进程间通信，就可以使用这种方法来确保只有获得许可的进程才能互相通信。事实上，在 multiprocessing 库中，当同子进程建立通信时，其内部也是采用基于 HMAC 的身份验证方式。

6.9　使用第三方库处理 HTTP

本节将通过具体实例的实现过程，详细讲解使用常用的第三方库处理 HTTP 的知识。

实例 06-29：使用库 aiohttp 实现异步处理

Python 的标准库 asyncio 可以实现单线程并发 I/O 操作，实现了 TCP、UDP、SSL 等协议，而 aiohttp 则是基于 asyncio 实现的 HTTP 库，实现了异步 HTTP 网络处理。

可以使用如下两种命令安装库 aiohttp。

```
pip install aiohttp
easy_install aiohttp
```

下面的实例演示了使用库 aiohttp 实现异步处理的过程。

（1）客户端文件 aiocli01.py 的功能是从网站 http://python.org 中检索信息，具体实现代码如下。

源码路径：daima\6\06-29\aiocli01.py

```python
import aiohttp
import asyncio
import async_timeout

async def fetch(session, url):
    async with async_timeout.timeout(10):
        async with session.get(url) as response:
            return await response.text()

async def main():
    async with aiohttp.ClientSession() as session:
        html = await fetch(session, 'http://python.org')
        print(html)

loop = asyncio.get_event_loop()
loop.run_until_complete(main())
```

（2）服务器文件 aioser01.py 的功能是获取客户端访问者的信息，并输出欢迎信息，具体实现代码如下。

源码路径：daima\6\06-29\aioser01.py

```python
from aiohttp import web

async def handle(request):
    name = request.match_info.get('name', "Anonymous")
    text = "Hello, " + name
    return web.Response(text=text)

app = web.Application()
app.add_routes([web.get('/', handle),
                web.get('/{name}', handle)])

web.run_app(app)
```

库 aiohttp 的最大用处是在服务器上，下面的实例文件 aio02.py 演示了使用库 aiohttp 创建一个指定地址和端口的服务器的过程。

源码路径：daima\6\06-29\aio02.py

```python
import asyncio
```

```
from aiohttp import web

async def index(request):
    await asyncio.sleep(0.5)
    return web.Response(body=b'<h1>Index</h1>')

async def hello(request):
    await asyncio.sleep(0.5)
    text = '<h1>hello, %s!</h1>' % request.match_info['name']
    return web.Response(body=text.encode('utf-8'))

async def init(loop):
    app = web.Application(loop=loop)
    app.router.add_route('GET', '/', index)
    app.router.add_route('GET', '/hello/{name}', hello)
    srv = await loop.create_server(app.make_handler(), '127.0.0.1', 8000)
    print('Server started at http://127.0.0.1:8000...')
    return srv

loop = asyncio.get_event_loop()
loop.run_until_complete(init(loop))
loop.run_forever()
```

执行后会启动创建的服务器：

```
Server started at http://127.0.0.1:8000...
```

实例 06-30：使用库 aiohttp 爬取指定 CSDN 博客中技术文章地址

下面的实例文件 aio03.py 的作用是使用库 aiohttp 爬取指定博客中技术文章地址。

源码路径：daima\6\06-30\aio03.py

```
import urllib.request as request
from bs4 import BeautifulSoup as bs
import asyncio
import aiohttp

@asyncio.coroutine
async def getPage(url,res_list):
    print(url)
    headers = {'User-Agent':'Mozilla/4.0 (compatible; MSIE 5.5; Windows NT)'}
    # conn = aiohttp.ProxyConnector(proxy="http://127.0.0.1:8087")
    async with aiohttp.ClientSession() as session:
        async with session.get(url,headers=headers) as resp:
            assert resp.status==200
            res_list.append(await resp.text())


class parseListPage():
    def __init__(self,page_str):
        self.page_str = page_str
    def __enter__(self):
        page_str = self.page_str
        page = bs(page_str,'lxml')
        #  获取文章地址
        articles = page.find_all('div',attrs={'class':'article_title'})
        art_urls = []
        for a in articles:
            x = a.find('a')['href']
            art_urls.append('http://blog.csdn.net'+x)
        return art_urls
    def __exit__(self, exc_type, exc_val, exc_tb):
        pass


page_num = 100
page_url_base = 'http://blog.csdn.net/u0114751***/article/list/'
page_urls = [page_url_base + str(i+1) for i in range(page_num)]
loop = asyncio.get_event_loop()
ret_list = []
tasks = [getPage(host,ret_list) for host in page_urls]
loop.run_until_complete(asyncio.wait(tasks))
```

```
articles_url = []
for ret in ret_list:
    with parseListPage(ret) as tmp:
        articles_url += tmp
ret_list = []

tasks = [getPage(url, ret_list) for url in articles_url]
loop.run_until_complete(asyncio.wait(tasks))
loop.close()
```

执行后会输出：

```
http://blog.csdn.net/u0114751***/article/list/29
http://blog.csdn.net/u0114751***/article/list/30
http://blog.csdn.net/u0114751***/article/list/73
http://blog.csdn.net/u0114751***/article/list/2
#省略爬取结果
```

实例 06-31：使用库 requests 返回指定 URL 请求

库 requests 是用 Python 基于 urllib 编写的，采用 Apache License,Version 2.0 开源协议的 HTTP 库。requests 比 urllib 更加方便，可以节约开发者大量的时间。可以使用如下两种命令安装库 requests。

```
pip install requests
easy_install requests
```

下面的实例文件 Requests01.py 演示了使用库 requests 返回指定 URL 请求的过程。

源码路径：daima\6\06-31\Requests01.py

```
import requests

r = requests.get(url='http://www.toppr.net')              #最基本的GET请求
print(r.status_code)                                       #获取返回状态
r = requests.get(url='http://www.toppr.net', params={'wd': 'python'})  #带参数的GET请求
print(r.url)
print(r.text)                                              #输出解码后的返回数据
```

上述代码创建了一个名为 r 的 response 对象，我们可以从这个对象中获取所有想要的信息。

执行后会输出：

```
200
http://www.toppr.net/?wd=python
<!DOCTYPE html PUBLIC "-//W3C//DTD XHTML 1.0 Transitional//EN" "http://www.w3.org/TR/xhtml1/DTD/xhtml1-transitional.dtd">
<html xmlns="http://www.w3.org/1999/xhtml">
<head>
<meta http-equiv="X-UA-Compatible" content="IE=edge">
<meta http-equiv="Content-Type" content="text/html; charset=gbk" />
<title>门户 - Powered by Discuz!</title>

<meta name="keywords" content="门户" />
<meta name="description" content="门户 " />
<meta name="generator" content="Discuz! X3.2" />
#省略后面的结果
```

上述实例只是演示了 get()接口的用法，其实其他接口的用法也十分简单，例如：

```
requests.get('https://github.com/timeline.json')          #GET请求
requests.post("http://httpbin.org/post")                  #POST请求
requests.put("http://httpbin.org/put")                    #PUT请求
requests.delete("http://httpbin.org/delete")              #DELETE请求
requests.head("http://httpbin.org/get")                   #HEAD请求
requests.options("http://httpbin.org/get")                #OPTIONS请求
```

实例 06-32：提交的数据是向指定地址传送的 data 里面的数据

如果想查询 http://httpbin.org/get 页面的具体参数，则需要在 URL 里面加上这个参数。假如我们想看有没有 Host=httpbin.org 这条参数，URL 应该是 http://httpbin.org/get?Host=httpbin.org。下面的实例文件 Requests02.py 演示了提交的数据是向指定地址传送的 data 里面的数据。

源码路径：daima\6\06-32\Requests02.py

```python
import requests
url = 'http://httpbin.org/get'
data = {
    'name': 'zhangsan',
    'age': '25'
}
response = requests.get(url, params=data)
print(response.url)
print(response.text)
```

执行后会输出：

```
http://httpbin.org/get?name=zhangsan&age=25
{
  "args": {
    "age": "25",
    "name": "zhangsan"
  },
  "headers": {
    "Accept": "*/*",
    "Accept-Encoding": "gzip, deflate",
    "Connection": "close",
    "Host": "httpbin.org",
    "User-Agent": "python-requests/2.6.4"
  },
  "origin": "39.71.61.153",
  "url": "http://httpbin.org/get?name=zhangsan&age=25"
}
```

实例 06-33：使用 GET 和 POST 方式处理 JSON 数据

库 requests 也能够处理 JSON 数据，其内置方法 response.JSON() 等同于 JSON.loads(response.text) 方法。下面的实例文件 Requests03.py 演示了分别使用 GET 和 POST 方式处理 JSON 数据的过程。

源码路径：daima\6\06-33\Requests03.py

```python
import requests
import JSON
r = requests.post('https://api.github.com/some/endpoint', data=json.dumps({'some': 'data'}))
print(r.JSON())
response = requests.get("http://httpbin.org/get")
print(type(response.text))
print(response.JSON())
print(JSON.loads(response.text))
```

执行后会输出：

```
{'message': 'Not Found', 'documentation_url': 'https://developer.github.com/v3'}
<class 'str'>
{'args': {}, 'headers': {'Accept': '*/*', 'Accept-Encoding': 'gzip, deflate', 'Connection': 'close', 'Host': 'httpbin.org', 'User-Agent': 'python-requests/2.6.4'}, 'origin': '39.71.61.153', 'url': 'http://httpbin.org/get'}
{'args': {}, 'headers': {'Accept': '*/*', 'Accept-Encoding': 'gzip, deflate', 'Connection': 'close', 'Host': 'httpbin.org', 'User-Agent': 'python-requests/2.6.4'}, 'origin': '39.71.61.153', 'url': 'http://httpbin.org/get'}
```

实例 06-34：添加 headers 获取知乎页面信息

在使用库 requests 获取网页信息时，很多验证网站需要用户提供 headers（头部信息）。例如，想访问知乎页面，我们可以尝试不加 headers 信息进行访问，下面的实例文件 Requests04.py 演示了此过程。

源码路径：daima\6\06-34\Requests04.py

```python
import requests
url = 'https://www.zhihu.com/'
response = requests.get(url)
response.encoding = "utf-8"
print(response.text)
```

上述代码执行后会提示发生内部服务器错误，也就说连知乎登录页面的 HTML 代码都无法下载。

```
<html><body><h1>500 Server Error</h1>
An internal server error occured.
</body></html>
```

要想成功访问知乎页面,就必须添加对应的 headers 信息。我们可以通过浏览器工具查看知乎页面的 headers 信息,如图 6-8 所示。

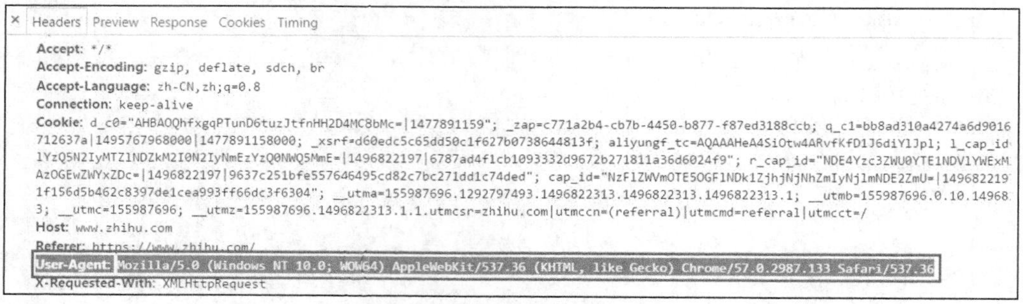

图 6-8 查看知乎页面的 headers 信息

下面的实例文件 Requests05.py 演示了添加 headers 获取知乎页面信息的过程。

源码路径:daima\6\06-34\Requests05.py

```
import requests
url = 'https://www.zhihu.com/'
headers = {
        'User-Agent': 'Mozilla/5.0 (Windows NT 10.0; WOW64) AppleWebKit/537.36 (KHTML, like Gecko)
                Chrome/57.0.2987.133 Safari/537.36'
}
response = requests.get(url, headers=headers)
print(response.text)
```

上述代码执行后会成功获取知乎页面的信息:

```
<!doctype html>
<html lang="zh" data-hairline="true" data-theme="light"><head><meta charSet="utf-8"/><title data-react-helmet="true">知乎
- 发现更大的世界</title><meta name="viewport" content="width=device-width,initial-scale=1,maximum-scale=1"/><meta name=
"renderer" content="webkit"/><meta name="force-rendering" content="webkit"/><meta http-equiv="X-UA-Compatible" content="IE=
edge,chrome=1"/><meta name="google-site-verification" content="FTeR0c8arOPKh8c5DYh_9uu98_zJbaWw53J-Sch9MTg"/><link
rel="shortcut icon"
            #省略后面的结果
```

实例 06-35:使用自定义的编码格式进行解码

在使用 requests 的内置方法后,会返回一个 response 对象,里面存储了服务器响应的内容,如上述实例中的 response.text。下面的实例文件 Requests06.py 访问 r.text 时,会使用其响应的文本编码进行解码,并允许修改其文本编码让 r.text 使用自定义的编码格式进行解码。

源码路径:daima\6\06-35\Requests06.py

```
import requests
r = requests.get('http://www.toppr.net')
print(r.text, '\n{}\n'.format('*'*79), r.encoding)
r.encoding = 'GBK'
print(r.text, '\n{}\n'.format('*'*79), r.encoding)
```

实例 06-36:访问远程页面信息

在使用库 requests 时,可以通过代理或证书来访问并获取指定网站的信息。下面的实例文件 Requests07.py 演示了使用手工设置的证书来访问远程页面信息的过程。

源码路径:daima\6\06-36\Requests07.py

```
import requests
response = requests.get('https://www.12306.cn', cert=('/path/server.crt', '/path/key'))
print(response.status_code)
```

下面的实例文件 Requests08.py 演示了用设置的普通代理来访问远程页面信息的过程。

源码路径：daima\6\06-36\Requests08.py

```
import requests
proxies = {
    "http": "http://127.0.0.1:9743",
    "https": "https://127.0.0.1:9743",
}
response = requests.get("https://www.taobao.com", proxies=proxies)
print(response.status_code)
```

实例 06-37：使用库 grequests 同时处理一组请求

库 grequests 是 requests 和 gevent 的联合体，用于异步处理 HTTP 请求。使用库 requests 可以请求一个网站上的数据，正常来说，请求会一条一条地执行。下面的实例文件 gr01.py 演示了使用库 grequests 同时处理一组请求的过程。

源码路径：daima\6\06-37\gr01.py

```
import grequests
urls = [
    'http://httpbin.org',
]
#创建一组未发送请求
rs = (grequests.get(u) for u in urls)
#同时发送这组请求
print(grequests.map(rs))
```

执行后会输出：

[<Response [200]>]

实例 06-38：使用库 grequests 提升访问请求性能

因为 Python 存在全局锁，所以并不推荐使用多线程这样的伪并行的方式。虽然 coroutines（协程）也是伪并行的方式，即线性请求，当其中一个请求遇到 I/O 等待的时候切换到另一个 coroutines 继续执行，等到 I/O 返回之后再接收信息，避免长时间的等待和阻塞。但是库 grequests 可以在 Python 中使用 gevent 基于 coroutines 解决这个问题。下面的实例文件 gr02.py 演示了使用库 grequests 提升访问请求性能的过程。

源码路径：daima\6\06-38\gr02.py

```
import grequests
import requests
import cProfile

urls = [
    'http://www.xiachufang.com/downloads/baidu_pip/2016030101.json',
    'http://www.xiachufang.com/downloads/baidu_pip/2016030102.json',
    'http://www.xiachufang.com/downloads/baidu_pip/2016030103.json',
    'http://www.xiachufang.com/downloads/baidu_pip/2016030104.json',
    'http://www.xiachufang.com/downloads/baidu_pip/2016030105.json',
    'http://www.xiachufang.com/downloads/baidu_pip/2016030106.json',
    'http://www.xiachufang.com/downloads/baidu_pip/2016030107.json',
    'http://www.xiachufang.com/downloads/baidu_pip/2016030108.json',
    'http://www.xiachufang.com/downloads/baidu_pip/2016030109.json',
    'http://www.xiachufang.com/downloads/baidu_pip/2016030110.json',
    'http://www.xiachufang.com/downloads/baidu_pip/2016030111.json',
    'http://www.xiachufang.com/downloads/baidu_pip/2016030116.json',
    'http://www.xiachufang.com/downloads/baidu_pip/2016030113.json',
    'http://www.xiachufang.com/downloads/baidu_pip/2016030114.json',
    'http://www.xiachufang.com/downloads/baidu_pip/2016030115.json',
    'http://www.xiachufang.com/downloads/baidu_pip/2016030116.json',
    'http://www.xiachufang.com/downloads/baidu_pip/2016030117.json',
    'http://www.xiachufang.com/downloads/baidu_pip/2016030118.json',
    'http://www.xiachufang.com/downloads/baidu_pip/2016030119.json',
    'http://www.xiachufang.com/downloads/baidu_pip/2016030120.json',
    'http://www.xiachufang.com/downloads/baidu_pip/2016030121.json',
```

```
        'http://www.xiachufang.com/downloads/baidu_pip/2016030122.json',
        'http://www.xiachufang.com/downloads/baidu_pip/2016030123.json',
        'http://www.xiachufang.com/downloads/baidu_pip/2016030200.json',
]
def haha(urls):
    rs = (grequests.get(u) for u in urls)
    return grequests.map(rs)

cProfile.run("haha(urls)")

def hehe(urls):
    hehe = [requests.get(i) for i in urls]
    return hehe

cProfile.run("haha(urls)")
```

通过上述代码，异步访问了 24 个 URL 信息，在作者的计算机中进行了 10 次计算，平均花费 395ms：

```
93430 function calls (91789 primitive calls) in 0.395 seconds
```

如果在上述代码中使用函数 hehe()，那么在使用同步、普通的 requests 库请求的情况下，同样在作者计算机中测试 10 次，平均花费 629ms。如果中途遇到单个连接不稳或者超时的情况，则甚至会花费 1s。

```
93391 function calls (91750 primitive calls) in 0.629 seconds
```

在 I/O 越密集、等待 I/O 时间长的请求量越大的情况下，grequests 的性能提升效果越明显。使用并发或者协程可以至少提升 5 倍以上性能，这时我们更应该使用异步或并行操作来缩短 I/O 的等待时间。

实例 06-39：使用库 httplib2 获取网页数据

httplib2 是一个第三方的开源库，比 Python 内置库 http.client 更完整地实现了 HTTP，同时相比 urllib.request 提供了更好的抽象。一旦拥有了 HTTP 对象，将非常简单地获取网页数据，只需将需要的数据的地址作为参数来调用 request()方法即可，这会对该 URL 执行一个 HTTP GET 请求。下面的实例文件 http201.py 演示了使用库 httplib2 获取网页数据的过程。

源码路径：daima\6\06-39\http201.py

```
import httplib2
#获取HTTP对象
h =httplib2.Http()
#发出同步请求，并获取内容
resp, content = h.request("http://www.baidu.com/")
print(resp)
print(content)
```

实例 06-40：使用库 httplib2 处理网页缓存数据

缓存是很多机制都必有的功能，和 Python 内置库相比，库 httplib2 的最大优势是可以处理缓存。下面的实例文件 http202.py 演示了使用库 httplib2 处理网页缓存数据的过程。

源码路径：daima\6\06-40\http202.py

```
import httplib2
#获取HTTP对象
h =httplib2.Http('.cache')
#发出同步请求，并获取内容
resp, content = h.request("http://www.baidu.com")
print(resp)
print("......"*3)
httplib2.debuglevel = 1
h1 = httplib2.Http('.cache')
resp,content = h1.request('http://www.baidu.com')

print(resp)
print('debug',resp.fromcache)
```

上述代码执行后会输出网页的源码，并获取带有缓存的 HTTP 对象 h1，缓存被存储在当前环境的.cache 目录下。

```
{'date': 'Fri, 20 Apr 2018 08:06:48 GMT', 'content-type': 'text/html; charset=utf-8', 'transfer-encoding': 'chunked', 'connection':
'Keep-Alive', 'vary': 'Accept-Encoding', 'set-cookie': 'BAIDUID=AAEB2299BC40EA4AFC6AEA020A6FC971:FG=1; expires=Thu,
31-Dec-37 23:55:55 GMT; max-age=2147483647; path=/; domain=.baidu.com, BIDUPSID=AAEB2299BC40EA4AFC6AEA020
A6FC971; expires=Thu, 31-Dec-37 23:55:55 GMT; max-age=2147483647; path=/; domain=.baidu.com, PSTM=1524211608;
expires=Thu, 31-Dec-37 23:55:55 GMT; max-age=2147483647; path=/; domain=.baidu.com, BDSVRTM=0; path=/, BD_HOME=0;
path=/, H_PS_PSSID=26193_1469_21111_20928; path=/; domain=.baidu.com', 'p3p': 'CP=" OTI DSP COR IVA OUR IND COM "',
'cache-control': 'private', 'cxy_all': 'baidu+28c185a0926430a3884a32911dee55ed', 'expires': 'Fri, 20 Apr 2018 08:06:31 GMT',
'x-powered-by': 'HPHP', 'server': 'BWS/1.1', 'x-ua-compatible': 'IE=Edge,chrome=1', 'bdpagetype': '1', 'bdqid': '0xab4be86f00008f3b',
'status': '200', 'content-length': '115109', '-content-encoding': 'gzip', 'content-location': 'http://www.baidu.com'}
................
connect: (www.baidu.com, 80) ************
send: b'GET / HTTP/1.1\r\nHost: www.baidu.com\r\nuser-agent: Python-httplib2/0.11.3 (gzip)\r\naccept-encoding: gzip,
deflate\r\n\r\n'
reply: 'HTTP/1.1 200 OK\r\n'
header: Date header: Content-Type header: Transfer-Encoding header: Connection header: Vary header: Set-Cookie header:
Set-Cookie header: Set-Cookie header: Set-Cookie header: Set-Cookie header: Set-Cookie header: P3P header: Cache-Control header:
Cxy_all header: Expires header: X-Powered-By header: Server header: X-UA-Compatible header: BDPAGETYPE header: BDQID
header: Content-Encoding {'date': 'Fri, 20 Apr 2018 08:06:49 GMT', 'content-type': 'text/html; charset=utf-8', 'transfer-encoding':
'chunked', 'connection': 'Keep-Alive', 'vary': 'Accept-Encoding', 'set-cookie': 'BAIDUID=166923F37969C306C0ABB71432D96615:
FG=1; expires=Thu, 31-Dec-37 23:55:55 GMT; max-age=2147483647; path=/; domain=.baidu.com, BIDUPSID=166923F37969C306
C0ABB71432D96615;    expires=Thu,    31-Dec-37    23:55:55    GMT;    max-age=2147483647;    path=/;    domain=.baidu.com,
PSTM=1524211609; expires=Thu, 31-Dec-37 23:55:55 GMT; max-age=2147483647; path=/; domain=.baidu.com, BDSVRTM=0;
path=/, BD_HOME=0; path=/, H_PS_PSSID=26254_1466_13289_21095_26105; path=/; domain=.baidu.com', 'p3p': 'CP=" OTI DSP
COR IVA OUR IND COM "', 'cache-control': 'private', 'cxy_all': 'baidu+6b76b5a82a8a96ac81598ed88b6af038', 'expires': 'Fri, 20 Apr
2018 08:06:31 GMT', 'x-powered-by': 'HPHP', 'server': 'BWS/1.1', 'x-ua-compatible': 'IE=Edge,chrome=1', 'bdpagetype': '1', 'bdqid':
'0xbf58353700009f7f', 'status': '200', 'content-length': '114516', '-content-encoding': 'gzip', 'content-location': 'http://www.baidu.com'}
debug True
```

此时 debug 的值是 True，说明数据是从本地的.cache 读取的，没经过原网站。

```
resp,content = h1.request('http://www.topper.net/sitemap.xml',headers={'cache-control':'no-cache' })
```

库 httplib2 允许用户添加任意的 HTTP 头部到发出的请求中，为了跳过所有缓存（不仅是用户本地的磁盘缓存，也包括任何处于用户和远程服务器之间代理服务器的缓存），只需在 headers 字典中加入 no-cache 头即可。

实例 06-41：使用 POST 发送构造数据

下面的实例文件 http203.py 演示了使用 POST 发送构造数据的过程。

源码路径：daima\6\06-41\http203.py

```python
from urllib.parse import urlencode

import httplib2

httplib2.debuglevel = 1

h = httplib2.Http('.cache')

data = {'status': 'Test update from Python 3'}

h.add_credentials('diveintomark', 'MY_SECRET_PASSWORD', 'identi.ca')

resp, content = h.request('https://www.baidu.com',
    'POST',
    urlencode(data),
    headers={'Content-Type': 'application/x-www-form-urlencoded'})
```

执行后会输出：

```
connect: (www.baidu.com, 443)
send: b'POST / HTTP/1.1\r\nHost: www.baidu.com\r\nContent-Length: 32\r\ncontent-type: application/x-www-form-urlencoded\
r\nuser-agent: Python-httplib2/0.11.3 (gzip)\r\naccept-encoding: gzip, deflate\r\n\r\n'
send: b'status=Test+update+from+Python+3'
reply: 'HTTP/1.1 302 Found\r\n'
header: Bdpagetype header: Connection header: Content-Length header: Content-Type header: Date header: Location header:
Server header: Set-Cookie header: X-Ua-Compatible
```

在上述代码中，add_credentials()方法的第 3 个参数是该证书有效的域名。我们应该总是指定这个参数，如果省略了这个参数，并且之后重用这个 httplib2.Http 对象访问另一个需要认证的网站，则可能会导致 httplib2 将一个网站的用户名、密码泄露给其他网站。httplib2 返回的数据总是字节串，不是字符串。为了将其转化为字符串，需要用合适的字符编码进行解码。例如：

```
print(content.decode('utf-8'))
```

实例 06-42：使用库 urllib3 中的 request()方法创建请求

库 urllib3 是一个具有线程安全连接池、支持文件上传的清晰友好的 HTTP 库。可以使用如下两种命令安装库 urllib3。

```
pip install urllib3
easy_install urllib3
```

下面的实例文件 urllib301.py 演示了使用库 urllib3 中的 request()方法创建请求的过程。

源码路径：daima\6\06-42\urllib301.py

```python
import urllib3
import requests
#忽略警告： InsecureRequestWarning: Unverified HTTPS request is being made. Adding certificate verification is strongly
#advised.
requests.packages.urllib3.disable_warnings()
#通过一个PoolManager实例来生成请求，由该实例对象处理与线程池的连接及线程安全的所有细节
http = urllib3.PoolManager()
#通过request()方法创建一个请求
r = http.request('GET', 'http://www.baidu.com/')
print(r.status) # 200
#获得HTML源码，UTF-8解码
print(r.data.decode())
```

通过上述代码获取了百度主页的信息：

```
200
<!DOCTYPE html><!--STATUS OK-->
<html>
<head>
        <meta http-equiv="content-type" content="text/html;charset=utf-8">
        <meta http-equiv="X-UA-Compatible" content="IE=Edge">
        <link rel="dns-prefetch" href="//s1.bdstatic.com"/>
        <link rel="dns-prefetch" href="//t1.baidu.com"/>
        <link rel="dns-prefetch" href="//t2.baidu.com"/>
        <link rel="dns-prefetch" href="//t3.baidu.com"/>
        <link rel="dns-prefetch" href="//t10.baidu.com"/>
#省略后面的结果
```

实例 06-43：在 request()方法中添加 head 头创建请求

下面的实例文件 urllib302.py 演示了在 request()方法中添加 head 头创建请求的过程。

源码路径：daima\6\06-43\urllib302.py

```python
import urllib3
header = {
        'User-Agent': 'Mozilla/5.0 (Windows NT 6.1; Win64; x64) AppleWebKit/537.36 (KHTML, like Gecko) Chrome/63.0.3239.108 Safari/537.36'
    }
http = urllib3.PoolManager()
r = http.request('GET',
                 'https://www.baidu.com/s?',
                 fields={'wd': 'hello'},
                 headers=header)
print(r.status) # 200
print(r.data.decode())
```

实例 06-44：使用库 urllib3 中的 post()方法创建请求

下面的实例文件 urllib303.py 演示了使用库 urllib3 中的 post()方法创建请求的过程。

源码路径：daima\6\06-44\urllib303.py

```python
import urllib3
http = urllib3.PoolManager()
#还可以通过request()方法向请求中添加一些其他信息
header = {
        'User-Agent': 'Mozilla/5.0 (Windows NT 6.1; Win64; x64) AppleWebKit/537.36 (KHTML, like Gecko)
```

```
                        Chrome/63.0.3239.108 Safari/537.36'
}
r = http.request('POST',
                        'http://httpbin.org/post',
                        fields={'hello':'world'},
                        headers=header)
print(r.data.decode())

#对于POST和PUT请求，需要手动对传入数据进行编码，然后加在URL之后
encode_arg = urllib.parse.urlencode({'arg': '我的'})
print(encode_arg.encode())
r = http.request('POST',
                        'http://httpbin.org/post?'+encode_arg,
                        headers=header)
#解码
print(r.data.decode('unicode_escape'))
```

执行后会输出：

```
{
   "args": {},
   "data": "",
   "files": {},
   "form": {
     "hello": "world"
   },
   "headers": {
     "Accept-Encoding": "identity",
     "Connection": "close",
     "Content-Length": "129",
     "Content-Type": "multipart/form-data; boundary=b33b20053e6444ee947a6b7b3f4572b2",
     "Host": "httpbin.org",
     "User-Agent": "Mozilla/5.0 (Windows NT 6.1; Win64; x64) AppleWebKit/537.36 (KHTML, like Gecko)
        Chrome/63.0.3239.108 Safari/537.36"
   },
   "JSON": null,
   "origin": "39.71.61.153",
   "url": "http://httpbin.org/post"
}

b'arg=%E6%88%91%E7%9A%84'
{
   "args": {
     "arg": "我的"
   },
   "data": "",
   "files": {},
   "form": {},
   "headers": {
     "Accept-Encoding": "identity",
     "Connection": "close",
     "Content-Length": "0",
     "Host": "httpbin.org",
     "User-Agent": "Mozilla/5.0 (Windows NT 6.1; Win64; x64) AppleWebKit/537.36 (KHTML, like Gecko)
        Chrome/63.0.3239.108 Safari/537.36"
   },
   "JSON": null,
   "origin": "39.71.61.153",
   "url": "http://httpbin.org/post?arg=我的"
}
```

实例 06-45：使用库 urllib3 发送 JSON 数据

下面的实例文件 urllib304.py 演示了使用库 urllib3 发送 JSON 数据的过程。

源码路径：daima\6\06-45\urllib304.py

```
import urllib3
http = urllib3.PoolManager()
data={'attribute':'value'}
encode_data= JSON.dumps(data).encode()

r = http.request('POST',
                        'http://httpbin.org/post',
```

```
                body=encode_data,
                headers={'Content-Type':'application/JSON'}
            )
print(r.data.decode('unicode_escape'))
```

执行后会输出：

```
{
  "args": {},
  "data": "{\"attribute\": \"value\"}",
  "files": {},
  "form": {},
  "headers": {
    "Accept-Encoding": "identity",
    "Connection": "close",
    "Content-Length": "22",
    "Content-Type": "application/JSON",
    "Host": "httpbin.org"
  },
  "JSON": {
    "attribute": "value"
  },
  "origin": "39.71.61.153",
  "url": "http://httpbin.org/post"
}
```

6.10 使用第三方库处理 URL

本节将通过具体实例的实现过程，详细讲解使用 Python 第三方库处理 URL 的过程。

实例 06-46：使用库 furl 优雅地处理 URL 分页

库 furl 是一个让处理 URL 更简单的小型 Python 库，可以更加优雅地操作 URL。可以使用如下命令安装 furl。

```
pip install furl
```

下面的实例文件 url01.py 演示了使用库 furl 优雅地处理 URL 分页的过程。

源码路径：daima\6\06-46\url01.py

```
from furl import furl
f = furl('http://www.google.com/?page=1')
f.scheme, f.host, f.port, f.path, f.query, f.fragment
print(f.url)
f.args
f.args['page']
f.args['page'] = 2
print(f.url)
```

执行后会输出：

```
http://www.google.com/?page=1
http://www.google.com/?page=2
```

实例 06-47：使用库 furl 处理 URL 参数

通过使用库 furl，我们可以灵活地处理 URL 参数。例如，下面的实例文件 url02.py 演示了使用库 furl 处理 URL 参数的过程。

源码路径：daima\6\06-47\url02.py

```
from furl import furl
f= furl('http://www.baidu.com/?bid=12331')
#输出参数
print(f.args)
#增加参数
f.args['haha']='123'
print(f.args)
#修改参数
f.args['haha']='124'
print(f.args)
#删除参数
```

```
del f.args['haha']
print(f.args)
```
执行后会输出：
```
{'bid': '12331'}
{'bid': '12331', 'haha': '123'}
{'bid': '12331', 'haha': '124'}
{'bid': '12331'}
```

实例 06-48：使用内联方法处理 URL 参数

在使用库 furl 时，可以使用内联方法来处理 URL 参数。下面的实例文件 url03.py 演示了使用内联方法处理 URL 参数的过程。

源码路径：daima\6\06-48\url03.py

```python
from furl import furl
f= furl('http://www.baidu.com/?bid=12331')
#增加参数
print(furl('http://www.baidu.com/?bid=12331').add({'haha':'123'}).url)
#设置参数（只保留设置的参数）
print(furl('http://www.baidu.com/?bid=12331').set({'haha':'123'}).url)
#移除参数
print(furl('http://www.baidu.com/?bid=12331').remove(['bid']).url)
```

执行后会输出：
```
http://www.baidu.com/?bid=12331&haha=123
http://www.baidu.com/?haha=123
http://www.baidu.com/
```

实例 06-49：使用库 purl 处理 3 种构造类型 URL

库 purl 包含一个简单的、不可变的 URL 类，提供了简洁的 API 来询问和处理 URL。下面的实例文件 purl01.py 演示了使用库 purl 处理 3 种构造类型 URL 的过程。

源码路径：daima\6\06-49\purl01.py

```python
from purl import URL
#字符串构造器
from_str = URL('https://www.baidu.com/search?q=testing')
print(from_str)
#关键字构造器
from_kwargs = URL(scheme='https', host='www.baidu.com', path='/search', query='q=testing')
print(from_kwargs)
#联合使用
from_combo = URL('https://www.baidu.com').path('search').query_param('q', 'testing')
print(from_combo)
```

执行后会输出：
```
https://www.baidu.com/search?q=testing
https://www.baidu.com/search?q=testing
https://www.baidu.com/search?q=testing
```

实例 06-50：使用库 purl 返回各个 URL 对象

在库 purl 中，URL 对象是不可变的，所有的修改器方法都返回一个新的实例。下面的实例文件 purl02.py 演示了使用库 purl 返回各个 URL 对象的过程。

源码路径：daima\6\06-50\purl02.py

```python
from purl import URL
u = URL('https://www.baidu.com/search?q=testing')
print(u.scheme())
print(u.host())
print(u.domain())
print(u.username())
u.password()
print(u.netloc())
u.port()
print(u.path())
print(u.query())
print(u.fragment())
```

```
print(u.path_segment(0))
print(u.path_segments())
print(u.query_param('q'))
print(u.query_param('q', as_list=True))
print(u.query_param('lang', default='GB'))
print(u.query_params)
print(u.has_query_param('q'))
print(u.has_query_params(('q', 'r')))
print(u.subdomains())
print(u.subdomain(0))
```

执行后会输出：

```
https
www.baidu.com
www.baidu.com
None
www.baidu.com
/search
q=testing

search
('search',)
testing
['testing']
GB
{'q': ['testing']}
True
False
['www', 'baidu', 'com']
www
```

实例 06-51：使用库 purl 修改 URL 参数值

在库 purl 中，每个访问器方法被重载为一个类似于 jQuery API 的修改器方法。下面的实例文件 purl03.py 演示了使用库 purl 修改 URL 参数值的过程。

源码路径：daima\6\06-51\purl03.py

```
from purl import URL
u = URL.from_string('https://baidu.com/codeinthehole')
#输出
print(u.path_segment(0))
#修改器（创建一个新实例）
new_url = u.path_segment(0, 'tangentlabs')
print(new_url)
print(new_url is u)
print(new_url.path_segment(0))
```

执行后会输出：

```
codeinthehole
https://baidu.com/tangentlabs
False
tangentlabs
```

实例 06-52：在当前路径末尾添加字段

除了用重载方法外，可以使用方法 add_path_segment()在当前路径的末尾添加一个字段。下面的实例文件 purl04.py 演示了使用方法 add_path_segment()在当前路径末尾添加字段的过程。

源码路径：daima\6\06-52\purl04.py

```
from purl import URL
u = URL().scheme('http').domain('www.example.com').path('/some/path').query_param('q', 'search term')
print(u)
new_url = u.add_path_segment('here')
print( new_url.as_string())
```

执行后会输出：

```
http://www.example.com/some/path?q=search+term
http://www.example.com/some/path/here?q=search+term
```

6.10 使用第三方库处理 URL

实例 06-53：使用库 webargs 处理 URL 参数

库 webargs 是一个解析 HTTP 请求参数的 Python 框架，内置流行的 Web 框架，包括 Flask、Django、Bottle、Tornado、Pyramid、webapp2、Falcon 和 aiohttp。可以使用如下命令安装 webargs。

```
pip install webargs
```

下面的实例文件 webargs01.py 演示了在 Flask 程序中使用库 webargs 处理 URL 参数的过程。

源码路径：daima\6\06-53\webargs01.py

```python
from flask import Flask
from webargs import fields
from webargs.flaskparser import use_args

app = Flask(__name__)

hello_args = {
    'name': fields.Str(required=True)
}

@app.route('/')
@use_args(hello_args)
def index(args):
    return 'Hello ' + args['name']

if __name__ == '__main__':
    app.run()
```

在浏览器中输入 http://127.0.0.1:5000/?name='World'后会显示执行效果，如图 6-9 所示。

图 6-9　执行效果

实例 06-54：在 aiohttp 程序中使用库 webargs

下面的实例文件 webargs02.py 演示了在 aiohttp 程序中使用库 webargs 的过程。

源码路径：daima\6\06-54\webargs02.py

```python
hello_args = {
    'name': fields.Str(missing='Friend')
}
@asyncio.coroutine
@use_args(hello_args)
def index(request, args):
    return json_response({'message': 'Welcome, {}!'.format(args['name'])})

add_args = {
    'x': fields.Float(required=True),
    'y': fields.Float(required=True),
}
@asyncio.coroutine
@use_kwargs(add_args)
def add(request, x, y):
    return json_response({'result': x + y})

dateadd_args = {
    'value': fields.Date(required=False),
    'addend': fields.Int(required=True, validate=validate.Range(min=1)),
    'unit': fields.Str(missing='days', validate=validate.OneOf(['minutes', 'days']))
}
@asyncio.coroutine
@use_kwargs(dateadd_args)
def dateadd(request, value, addend, unit):
    value = value or dt.datetime.utcnow()
```

```python
        if unit == 'minutes':
            delta = dt.timedelta(minutes=addend)
        else:
            delta = dt.timedelta(days=addend)
        result = value + delta
        return json_response({'result': result.isoformat()})


def create_app():
    app = web.Application()
    app.router.add_route('GET', '/', index)
    app.router.add_route('POST', '/add', add)
    app.router.add_route('POST', '/dateadd', dateadd)
    return app

def run(app, port=5001):
    loop = asyncio.get_event_loop()
    handler = app.make_handler()
    f = loop.create_server(handler, '0.0.0.0', port)
    srv = loop.run_until_complete(f)
    print('serving on', srv.sockets[0].getsockname())
    try:
        loop.run_forever()
    except KeyboardInterrupt:
        pass
    finally:
        loop.run_until_complete(handler.finish_connections(1.0))
        srv.close()
        loop.run_until_complete(srv.wait_closed())
        loop.run_until_complete(app.finish())
    loop.close()
```

实例 06-55：在 Tornado 程序中使用库 webargs

下面的实例文件 webargs03.py 演示了在 Tornado 程序中使用库 webargs 的过程。

源码路径：daima\6\06-55\webargs03.py

```python
class HelloHandler(tornado.web.RequestHandler):

    hello_args = {
        'name': fields.Str()
    }

    def post(self, id):
        reqargs = parser.parse(self.hello_args, self.request)
        response = {
            'message': 'Hello {}'.format(reqargs['name'])
        }
        self.write(response)

application = tornado.web.Application([
    (r"/hello/([0-9]+)", HelloHandler),
], debug=True)

if __name__ == "__main__":
    application.listen(8888)
    tornado.ioloop.IOLoop.instance().start()
```

第 7 章

数据持久化操作实战

数据持久化是指永久地将数据存储在磁盘上。数据持久化技术是实现动态软件项目的必需手段，在软件项目中通过数据持久化可以存储海量的数据。因为软件显示的内容是从数据库中读取的，所以开发者可以通过修改数据库内容而实现动态交互功能。在 Python 软件项目中，数据库在其实现过程中起了一个中间媒介的作用。本章将通过具体实例的实现过程，向读者介绍实现 Python 数据持久化操作的知识。

7.1 操作 SQLite3 数据库

从 Python 3.x 开始，标准库已经内置了 sqlite3 模块，可以支持 SQLite3 数据库的访问和相关的数据库操作。在需要操作 SQLite3 数据库数据时，只需在程序中导入 sqlite3 模块即可。

实例 07-01：使用方法 cursor.execute()执行指定 SQL 语句

使用 sqlite3 模块，可以满足开发者在 Python 程序中使用 SQLite 数据库的需求。例如，下面的实例文件 e.py 演示了使用方法 cursor.execute()执行指定 SQL 语句的过程。

源码路径：daima\7\07-01\e.py

```python
import sqlite3

con = sqlite3.connect(":memory:")
cur = con.cursor()
cur.execute("create table people (name_last, age)")

who = "Yeltsin"
age = 72

cur.execute("insert into people values (?, ?)", (who, age))

cur.execute("select * from people where name_last=:who and age=:age", {"who": who, "age": age})

print(cur.fetchone())
```

执行后会输出：

```
('Yeltsin', 72)
```

实例 07-02：使用方法 cursor.executemany()执行指定的 SQL 命令

在内置模块 sqlite3 中，方法 cursor.executemany(sql, seq_of_parameters)的功能是对 seq_of_parameters 中的所有参数或映射执行一个 SQL 命令。例如，下面的实例文件 f.py 演示了使用方法 cursor.executemany()执行指定 SQL 命令的过程。

源码路径：daima\7\07-02\f.py

```python
import sqlite3

class IterChars:
    def __init__(self):
        self.count = ord('a')

    def __iter__(self):
        return self

    def __next__(self):
        if self.count > ord('z'):
            raise StopIteration
        self.count += 1
        return (chr(self.count - 1),)

con = sqlite3.connect(":memory:")
cur = con.cursor()
cur.execute("create table characters(c)")

theIter = IterChars()
cur.executemany("insert into characters(c) values (?)", theIter)

cur.execute("select c from characters")
print(cur.fetchall())
```

执行后会输出：

```
[('a',), ('b',), ('c',), ('d',), ('e',), ('f',), ('g',), ('h',), ('i',), ('j',), ('k',), ('l',), ('m',), ('n',), ('o',), ('p',), ('q',), ('r',), ('s',), ('t',), ('u',), ('v',), ('w',), ('x',), ('y',), ('z',)]
```

实例 07-03：同时执行多个 SQL 语句

内置模块 sqlite3 中，方法 cursor.executescript()的功能是一旦接收到脚本就会执行多个 SQL 语句。首先执行 COMMIT（提交）语句，然后执行作为参数传入的 SQL 脚本。所有的 SQL 语句应该用分号";"分隔。例如，下面的实例文件 g.py 演示了使用方法 cursor.executescript()同时执行多个 SQL 语句的过程。

源码路径：daima\7\07-03\g.py

```python
import sqlite3

con = sqlite3.connect(":memory:")
cur = con.cursor()
cur.executescript("""
    create table person(
        firstname,
        lastname,
        age
    );

    create table book(
        title,
        author,
        published
    );

    insert into book(title, author, published)
    values (
        'Dirk Gently''s Holistic Detective Agency',
        'Douglas Adams',
        1987
    );
""")
```

实例 07-04：使用方法 create_function()执行指定函数

在内置模块 sqlite3 中，方法 connection.create_function(name, num_params, func)的功能是创建一个自定义的函数，随后可以在 SQL 语句中以函数名 name 来调用它。参数 num_params 表示此方法接受的参数数量，如果 num_params 为-1，则函数可以接受任意数量的参数。参数 func 是一个可以被调用的 SQL 函数。例如，下面的实例文件 b.py 演示了使用方法 create_function()执行指定函数的过程。

源码路径：daima\7\07-04\b.py

```python
import sqlite3
import hashlib

def md5sum(t):
    return hashlib.md5(t).hexdigest()

con = sqlite3.connect(":memory:")
con.create_function("md5", 1, md5sum)
cur = con.cursor()
cur.execute("select md5(?)", (b"foo",))
print(cur.fetchone()[0])
```

执行后会输出：

```
acbd18db4cc2f85cedef654fccc4a4d8
```

实例 07-05：创建用户定义的聚合函数

在内置模块 sqlite3 中，方法 connection.create_aggregate(name, num_params, aggregate_class)的功能是创建一个用户定义的聚合函数。聚合类必须实现 step()方法，参数 num_params 表示此方法可以接受参数的数量，如果 num_params 为-1，则函数可以接受任意数量的参数。参数也可以是 finalize()方法，表示可以返回 SQLite 支持的任何类型，如 bytes、str、int、float 和 None。例如，下面的实例文件 c.py 演示了使用方法 create_aggregate()创建用户定义的聚合函数的过程。

源码路径：daima\7\07-05\c.py

```python
import sqlite3

class MySum:
    def __init__(self):
        self.count = 0

    def step(self, value):
        self.count += value

    def finalize(self):
        return self.count

con = sqlite3.connect(":memory:")
con.create_aggregate("mysum", 1, MySum)
cur = con.cursor()
cur.execute("create table test(i)")
cur.execute("insert into test(i) values (1)")
cur.execute("insert into test(i) values (2)")
cur.execute("select mysum(i) from test")
print(cur.fetchone()[0])
```

执行后会输出：

3

实例 07-06：用自定义排序规则以"错误方式"进行排序

在内置模块 sqlite3 中，方法 connection.create_collation(name, callable)的功能是用指定的 name 和 callable 创建一个排序规则，以传递两个字符串参数给可调用对象。如果第一个字符串比第二个字符串小则返回-1，如果两者相等则返回 0，如果第一个字符串比第二个字符串大则返回 1。需要注意的是，可调用对象将会以 Python 字节串的方式得到它的参数，一般为 UTF-8 编码。例如，下面的实例文件 d.py 演示了使用方法 create_collation()用自定义排序规则以"错误方式"进行排序的过程。

源码路径：daima\7\07-06\d.py

```python
import sqlite3

def collate_reverse(string1, string2):
    if string1 == string2:
        return 0
    elif string1 < string2:
        return 1
    else:
        return -1

con = sqlite3.connect(":memory:")
con.create_collation("reverse", collate_reverse)

cur = con.cursor()
cur.execute("create table test(x)")
cur.executemany("insert into test(x) values (?)", [("a",), ("b",)])
cur.execute("select x from test order by x collate reverse")
for row in cur:
    print(row)
con.close()
```

执行后会输出：

('b',)
('a',)

实例07-07：生成一个 SQLite Shell

在内置模块 sqlite3 中，方法 complete_statement(sql)的功能是，如果字符串 sql 包含一个或多个以分号结束的完整的 SQL 语句则返回 True。该方法不会验证 SQL 的语法正确性，只是检查没有未关闭的字符串常量或以分号的语句结束。例如，下面的实例文件 a.py 演示了使用方法 complete_statement(sql)生成一个 SQLite Shell 的过程。

源码路径：daima\7\07-07\a.py

```python
import sqlite3

con = sqlite3.connect(":memory:")
con.isolation_level = None
cur = con.cursor()

buffer = ""

print("Enter your SQL commands to execute in sqlite3.")
print("Enter a blank line to exit.")

while True:
    line = input()
    if line == "":
        break
    buffer += line
    if sqlite3.complete_statement(buffer):
        try:
            buffer = buffer.strip()
            cur.execute(buffer)

            if buffer.lstrip().upper().startswith("SELECT"):
                print(cur.fetchall())
        except sqlite3.Error as e:
            print("An error occurred:", e.args[0])
        buffer = ""

con.close()
```

执行后会输出：

```
Enter your SQL commands to execute in sqlite3.
Enter a blank line to exit.
```

实例07-08：返回数据库中的列名称列表

在内置模块 sqlite3 中，方法 Row.keys()的功能是返回列名称的列表。在查询后，该方法返回的是 Cursor.description 中每个元组的第一个成员。例如，下面的实例文件 h.py 演示了使用 Row 对象和 keys()方法返回数据库中列名称列表的过程。

源码路径：daima\7\07-08\h.py

```python
import sqlite3
conn = sqlite3.connect(":memory:")
c = conn.cursor()
c.execute('''create table stocks
(date text, trans text, symbol text,
 qty real, price real)''')
c.execute("""insert into stocks
          values ('2018-01-05','BUY','RHAT',100,35.14)""")
conn.commit()
c.close()

conn.row_factory = sqlite3.Row
c = conn.cursor()
print(c.execute('select * from stocks'))
r = c.fetchone()
print(type(r))
print(tuple(r))
print(len(r))
print(r[2])
```

```
print(r.keys())
print(r['qty'])
for member in r:
    print(member)
```

执行后会输出：

```
<sqlite3.Cursor object at 0x0000021B3B291F10>
<class 'sqlite3.Row'>
('2018-01-05', 'BUY', 'RHAT', 100.0, 35.14)
5
RHAT
['date', 'trans', 'symbol', 'qty', 'price']
100.0
2018-01-05
BUY
RHAT
100.0
35.14
```

实例 07-09：操作 SQLite3 数据库

下面的实例文件 sqlite.py 演示了使用 sqlite3 模块操作 SQLite3 数据库的过程。

源码路径：daima\7\07-09\sqlite.py

```python
import sqlite3                                          #导入内置模块
import random                                           #导入内置模块
#初始化变量src，设置用于随机生成字符串的所有字符
src = 'abcdefghijklmnopqrstuvwxyz'
def get_str(x,y):                                       #生成字符串函数get_str()
    str_sum = random.randint(x,y)                       #生成x和y之间的随机整数
    astr = ''                                           #为变量astr赋值
    for i in range(str_sum):                            #遍历随机数
        astr += random.choice(src)                      #累计求和生成的随机数
    return astr                                         #返回和
def output():                                           #函数output()用于输出数据库的表中的信息
    cur.execute('select * from biao')                   #查询表biao中的所有信息
    for sid,name,ps in cur:                             #查询表中的3个字段sid、name和ps
        print(sid,' ',name,' ',ps)                      #输出3个字段的查询结果

def output_all():                                       #函数output_all()用于输出数据库的表中的所有信息
    cur.execute('select * from biao')                   #查询表biao中的所有信息
    for item in cur.fetchall():                         #获取查询到的所有信息
        print(item)                                     #输出获取到的信息

def get_data_list(n):                                   #函数get_data_list()用于生成查询列表
    res = []                                            #列表初始化
    for i in range(n):                                  #遍历列表
        res.append((get_str(2,4),get_str(8,12)))        #生成列表
    return res                                          #返回生成的列表
if __name__ == '__main__':
    print("建立连接...")                                #输出提示信息
    con = sqlite3.connect(':memory:')                   #开始建立和数据库的连接
    print("建立游标...")
    cur = con.cursor()                                  #获取游标
    print('创建一张表biao...')                          #输出提示信息
    #在数据库中创建表biao，设置表中的各个字段
    cur.execute('create table biao(id integer primary key autoincrement not null,name text,passwd text)')
    print('插入一条记录...')                            #输出提示信息
    #插入1条记录
    cur.execute('insert into biao (name,passwd)values(?,?)',(get_str(2,4),get_str(8,12),))
    print('显示所有记录...')                            #输出提示信息
    output()                                            #输出数据库中的记录
    print('批量插入多条记录...')                        #输出提示信息
    #插入多条记录
    cur.executemany('insert into biao (name,passwd)values(?,?)',get_data_list(3))
    print("显示所有记录...")                            #输出提示信息
    output_all()                                        #输出数据库中的记录
    print('更新一条记录...')                            #输出提示信息
    #修改表biao中的一条记录
    cur.execute('update biao set name=? where id=?',('aaa',1))
    print('显示所有记录...')                            #输出提示信息
    output()                                            #输出数据库中的记录
    print('删除一条记录...')                            #输出提示信息
```

```
#删除表biao中的一条记录
cur.execute('delete from    biao where id=?',(3,))
print('显示所有记录: ')                          #输出提示信息
output()                                         #输出数据库中的记录
```

在上述实例代码中,首先定义了两个能够生成随机字符串的函数,生成的随机字符串作为数据库中存储的数据。然后定义 output()和 output_all()方法,功能是分别通过遍历游标、调用游标的方式来获取数据库表中的所有记录并输出。然后在主程序中,依次通过建立连接,获取连接的游标,通过游标的 execute()和 executemany()等方法来执行 SQL 语句,以实现插入一条记录、插入多条记录、更新记录和删除记录的功能。最后依次关闭游标和数据库连接。

执行后会输出:

```
建立连接...
建立游标...
创建一张表biao...
插入一条记录...
显示所有记录...
1    bld    zbynubfxt
批量插入多条记录...
显示所有记录...
(1, 'bld', 'zbynubfxt')
(2, 'owd', 'lqpperrey')
(3, 'vc', 'fqrbarwsotra')
(4, 'yqk', 'oyzarvrv')
更新一条记录...
显示所有记录...
1    aaa    zbynubfxt
2    owd    lqpperrey
3    vc     fqrbarwsotra
4    yqk    oyzarvrv
删除一条记录...
显示所有记录:
1    aaa    zbynubfxt
2    owd    lqpperrey
4    yqk    oyzarvrv
```

实例 07-10:将自定义类 Point 适配 SQLite3 数据库

因为 SQLite 只默认支持有限的类,要使用其他 Python 类型与 SQLite 进行交互,就必须适应它们为 sqlite3 模块支持的 SQLite 类型之一:None、int、float、str 或 bytes。开发者可以编写一个自定义类,假设编写了如下类。

```
class Point:
    def __init__(self, x, y):
        self.x, self.y = x, y
```

想要在某个 SQLite 列中存储类 Point,首先选择一个支持的类型,这个类型可以被用来表示 Point。假定使用 str,并用分号来分隔坐标。需要给类加一个__conform__(self, protocl)方法,该方法必须返回转换后的值。参数 protocol 为 PrepareProtocol 类型。例如,下面的实例文件 i.py 演示了将自定义类 Point 适配 SQLite3 数据库的过程。

源码路径: daima\7\07-10\i.py

```python
import sqlite3

class Point:
    def __init__(self, x, y):
        self.x, self.y = x, y

def adapt_point(point):
    return "%f;%f" % (point.x, point.y)

sqlite3.register_adapter(Point, adapt_point)

con = sqlite3.connect(":memory:")
cur = con.cursor()

p = Point(4.0, -3.2)
```

```
cur.execute("select ?", (p,))
print(cur.fetchone()[0])
```
执行后会输出：
4.000000;-3.200000

实例 07-11：使用函数 register_adapter()注册适配器函数

有一种可能性是创建一个函数，用来将类型转成字符串，然后使用函数 register_adapter()来注册该函数。例如，下面的实例文件 j.py 演示了使用函数 register_adapter()注册适配器函数的过程。

源码路径：daima\7\07-11\j.py

```python
import sqlite3

class Point:
    def __init__(self, x, y):
        self.x, self.y = x, y

def adapt_point(point):
    return "%f;%f" % (point.x, point.y)

sqlite3.register_adapter(Point, adapt_point)

con = sqlite3.connect(":memory:")
cur = con.cursor()

p = Point(4.0, -3.2)
cur.execute("select ?", (p,))
print(cur.fetchone()[0])
```

执行后会输出：
4.000000;-3.200000

实例 07-12：将 datetime.datetime 对象保存为 UNIX 时间戳

sqlite3 模块为 Python 内置的 datetime.date 和 datetime.datetime 类型设置了两个默认适配器。假设想要将 datetime.datetime 对象不以 ISO 形式存储，而是保存为 UNIX 时间戳，则可以通过下面的实例文件 k.py 实现。

源码路径：daima\7\07-12\k.py

```python
import sqlite3
import datetime
import time

def adapt_datetime(ts):
    return time.mktime(ts.timetuple())

sqlite3.register_adapter(datetime.datetime, adapt_datetime)

con = sqlite3.connect(":memory:")
cur = con.cursor()

now = datetime.datetime.now()
cur.execute("select ?", (now,))
print(cur.fetchone()[0])
```

执行后会输出：
1513239416.0

实例 07-13：将自定义 Python 类型转换成 SQLite 类型

在 Python 程序中，可以通过编写适配器将自定义 Python 类型转换成 SQLite 类型。再次以前面的 Point 类进行举例，假设在 SQLite 中以字符串的形式存储以分号分隔的 x、y。我们可以先定义如下的转换器函数 convert_point()，用于接收字符串参数，并从中构造一个 Point 对象。

7.1 操作 SQLite3 数据库

无论将数据类型转换成 SQLite 的哪种数据类型，转换器函数总是使用 bytes 对象调用。

```python
def convert_point(s):
    x, y = map(float, s.split(b";"))
    return Point(x, y)
```

接下来需要让 sqlite3 模块知道从数据库中实际选择的是一个 Point，可以通过如下两种方法实现这个功能。

- ❏ 隐式地通过声明的类型。
- ❏ 显式地通过列名。

下面的实例文件 l.py 演示了上述两种方法的实现过程。

源码路径：daima\7\07-13\l.py

```python
import sqlite3

class Point:
    def __init__(self, x, y):
        self.x, self.y = x, y

    def __repr__(self):
        return "(%f;%f)" % (self.x, self.y)

def adapt_point(point):
    return ("%f;%f" % (point.x, point.y)).encode('ascii')

def convert_point(s):
    x, y = list(map(float, s.split(b";")))
    return Point(x, y)

sqlite3.register_adapter(Point, adapt_point)

#注册转换器
sqlite3.register_converter("point", convert_point)

p = Point(4.0, -3.2)

con = sqlite3.connect(":memory:", detect_types=sqlite3.PARSE_DECLTYPES)
cur = con.cursor()
cur.execute("create table test(p point)")

cur.execute("insert into test(p) values (?)", (p,))
cur.execute("select p from test")
print("with declared types:", cur.fetchone()[0])
cur.close()
con.close()

con = sqlite3.connect(":memory:", detect_types=sqlite3.PARSE_COLNAMES)
cur = con.cursor()
cur.execute("create table test(p)")

cur.execute("insert into test(p) values (?)", (p,))
cur.execute('select p as "p [point]" from test')
print("with column names:", cur.fetchone()[0])
cur.close()
con.close()
```

执行后会输出：

```
with declared types: (4.000000;-3.200000)
with column names: (4.000000;-3.200000)
```

实例 07-14：使用默认适配器和转换器

Python 的 datetime 模块中有对 date 和 datetime 类型的默认的适配器，将 ISO 日期/ISO 时间戳发送给 SQLite。默认的转换器以 "date" 为名注册给 datetime.date，以 "timestamp" 为名注册给 datetime.datetime。这样在大多数情况下，在 Python 中使用 date/timestamp 时不需要额外的动作，适配器的格式兼容于 SQLite 的 date()/time()函数。

例如，下面的实例文件 m.py 演示了使用默认适配器和转换器的过程。

源码路径：daima\7\07-14\m.py

```python
import sqlite3
import datetime

con = sqlite3.connect(":memory:", detect_types=sqlite3.PARSE_DECLTYPES|sqlite3.PARSE_COLNAMES)
cur = con.cursor()
cur.execute("create table test(d date, ts timestamp)")

today = datetime.date.today()
now = datetime.datetime.now()

cur.execute("insert into test(d, ts) values (?, ?)", (today, now))
cur.execute("select d, ts from test")
row = cur.fetchone()
print(today, "=>", row[0], type(row[0]))
print(now, "=>", row[1], type(row[1]))

cur.execute('select current_date as "d [date]", current_timestamp as "ts [timestamp]"')
row = cur.fetchone()
print("current_date", row[0], type(row[0]))
print("current_timestamp", row[1], type(row[1]))
```

执行后会输出：

```
2017-12-14 => 2017-12-14 <class 'datetime.date'>
2017-12-14 16:37:55.125396 => 2017-12-14 16:37:55.125396 <class 'datetime.datetime'>
current_date 2017-12-14 <class 'datetime.date'>
current_timestamp 2017-12-14 08:37:55 <class 'datetime.datetime'>
```

> 注意：如果存储在 SQLite 中的时间戳的小数部分大于 6 个数字，则它的值将由时间戳转换器截断至微秒的精度。

实例 07-15：使用 isolation_level 开启智能 commit

通过 connect() 函数调用的 isolation_level 参数或连接的 isolation_level 属性，可以控制 sqlite3 隐式地执行哪种 BEGIN 语句（或完全不执行）。如果需要自动提交模式，则需要将 isolation_level 设置为 None；在其他情况下，保留其默认值，这将产生一个简单的 BEGIN 语句；或者将其设置成 SQLite 支持的隔离级别：DEFERRED、IMMEDIATE、或者 EXCLUSIVE。

当 isolation_level 为 None 时，开启自动 commit；当 isolation_level 为非 None 时，通过设置 BEGIN 的类型，开启智能 commit。例如，下面的实例文件 n.py 演示了使用 isolation_level 开启智能 commit 的过程。

源码路径：daima\7\07-15\n.py

```python
import sqlite3

con = sqlite3.connect(":memory:",isolation_level=None)
cur = con.cursor()
cur.execute("create table people (num, age)")

num = 1
age = 2 * num

while num <= 1000000:
    cur.execute("insert into people values (?, ?)", (num, age))
    num += 1
    age = 2 * num

cur.execute("select count(*) from people")

print(cur.fetchone())
```

执行后会输出：

```
(1000000,)
real    0m10.693s
user    0m10.569s
sys     0m0.099s
```

实例 07-16：手动开始 commit（提交执行）操作

智能 commit 的优点是速度快，在单进程情况下执行良好。其缺点是当多个进程并发操作数据库时，会出现这边写完另一边读不出来的问题。为了改正智能 commit 的缺点，在批量操作前可以手动开始事务，之后手动开始 commit 操作。下面的实例文件 o.py 解决了这一问题。

源码路径：daima\7\07-16\o.py

```
import sqlite3

con = sqlite3.connect(":memory:")
cur = con.cursor()
cur.execute("create table people (num, age)")

num = 1
age = 2 * num

while num <= 1000000:
    cur.execute("insert into people values (?, ?)", (num, age))
    con.commit()    # 关键在这里
    num += 1
    age = 2 * num

cur.execute("select count(*) from people")

print(cur.fetchone())
```

执行后会输出：
```
(1000000,)
real    0m20.797s
user    0m20.611s
sys     0m0.156s
```

实例 07-17：使用模块 apsw 创建并操作 SQLite 数据库数据

在开发 Python 程序的过程中，可以使用第三方库 apsw 实现和 SQLite 数据库的连接和操作功能。模块 apsw 是一个第三方库，实现了对 Python SQLite 的封装。开发者可以通过如下命令安装 apsw。

```
pip install apsw
```

例如，下面的实例文件 apsw01.py 演示了使用模块 apsw 创建并操作 SQLite 数据库数据的过程。

源码路径：daima\7\07-17\apsw01.py

```
import apsw
con=apsw.Connection(":memory:")
cur=con.cursor()
for row in cur.execute("create table foo(x,y,z);insert into foo values (?,?,?);"
                       "insert into foo values(?,?,?);select * from foo;drop table foo;"
                       "create table bar(x,y);insert into bar values(?,?);"
                       "insert into bar values(?,?);select * from bar;",
                       (1,2,3,4,5,6,7,8,9,10)):
    print(row)
```

上述代码执行后会输出添加到 SQLite 数据库中的数据：
```
(1, 2, 3)
(4, 5, 6)
(7, 8)
(9, 10)
```

实例 07-18：同时批处理上千条数据

下面的实例文件 apsw02.py 演示了使用模块 apsw 向 SQLite 数据库中同时批处理上千条数据的过程。

源码路径：daima\7\07-18\apsw02.py

```python
import threading, apsw
import queue
import sys
class TestThr(threading.Thread):
    def __init__(self):
        threading.Thread.__init__(self)
        self.IQ = queue.Queue()
        self.OQ = queue.Queue()

    def run(self):
        try:
            print("*THREAD: Thread started")
            while self.IQ.empty(): pass
            self.IQ.get()
            print("*THREAD: <<< Prepare database")
            con = apsw.Connection('test.db')
            c = con.cursor()
            try:
                c.execute('create table a(a integer)')
                c.execute('end')
            except:
                pass
            c.execute('begin')
            c.execute('delete from a')
            c.execute('end')
            print("*THREAD: >>> Prepare database")

            self.OQ.put(1)
            while self.IQ.empty(): pass
            self.IQ.get()
            print("*THREAD: <<< Fillup 1000 values")
            c.execute('begin')
            print("*THREAD: Trans. started")
            for i in range(1000):
                c.execute('insert into a values(%d)' % i)
            print("*THREAD: >>> Fillup 1000 values")
            self.OQ.put(1)
            while self.IQ.empty(): pass
            self.IQ.get()
            c.execute('end')
            print("*THREAD: Trans. finished")
            self.OQ.put(1)
            while self.IQ.empty(): pass
            self.IQ.get()
            print("*THREAD: <<< Fillup 1000 values")
            c.execute('begin')
            print("Trans. started")
            for i in range(1000, 2000):
                c.execute('insert into a values(%d)' % i)
            print("*THREAD: >>> Fillup 1000 values")
            c.execute('end')
            print("*THREAD: Trans. finished")
            self.OQ.put(1)
            while self.IQ.empty(): pass
            self.IQ.get()
            print("*THREAD: Thread end")

            self.OQ.put(1)
        except:
            print(sys.exc_info())
            sys.exit()

con = apsw.Connection('test.db')
c = con.cursor()

t = TestThr()
t.IQ.put(1)
t.start()
while t.OQ.empty(): pass
t.OQ.get()

def ReadLastRec():
```

```
        rec = None
        for rec in c.execute('select * from a'): pass
        print("- MAIN: Read last record", rec)

ReadLastRec()
t.IQ.put(1)
while t.OQ.empty(): pass
t.OQ.get()
ReadLastRec()
t.IQ.put(1)
while t.OQ.empty(): pass
t.OQ.get()
ReadLastRec()
t.IQ.put(1)
while t.OQ.empty(): pass
t.OQ.get()
ReadLastRec()
t.IQ.put(1)
while t.OQ.empty(): pass
# c.execute('end')

print("\n- MAIN: Finished")
```

执行后会输出：

```
*THREAD: Thread started
*THREAD: <<< Prepare database
*THREAD: >>> Prepare database
- MAIN: Read last record None
*THREAD: <<< Fillup 1000 values
*THREAD: Trans. started
*THREAD: >>> Fillup 1000 values
- MAIN: Read last record None
*THREAD: Trans. finished
- MAIN: Read last record (999,)
*THREAD: <<< Fillup 1000 values
Trans. started
*THREAD: >>> Fillup 1000 values
*THREAD: Trans. finished
- MAIN: Read last record (1999,)
*THREAD: Thread end

- MAIN: Finished
```

7.2 操作 MySQL 数据库

Python 3.x 使用库 PyMySQL 来连接 MySQL 数据库服务器，Python 2 则使用库 mysqldb。

实例 07-19：显示 PyMySQL 数据库版本号

在连接数据库之前，请按照如下步骤进行操作。

（1）安装 MySQL 数据库和 PyMySQL。
（2）在 MySQL 数据库中创建数据库 TESTDB。
（3）在 TESTDB 数据库中创建表 EMPLOYEE。
（4）在表 EMPLOYEE 中添加 5 个字段，分别是 FIRST_NAME、LAST_NAME、AGE、SEX 和 INCOME。在 MySQL 数据库，表 EMPLOYEE 的界面效果如图 7-1 所示。

图 7-1 表 EMPLOYEE 的界面效果

(5) 假设本地 MySQL 数据库的登录用户名为 "root", 密码为 "66688888"。例如, 下面的实例文件 mysql.py 演示了显示 PyMySQL 数据库版本号的过程。

源码路径: daima\7\07-19\mysql.py

```
import pymysql
#打开数据库连接
db = pymysql.connect("localhost","root","66688888","TESTDB" )
#使用cursor()方法创建一个游标对象cursor
cursor = db.cursor()
#使用execute()方法执行SQL查询
cursor.execute("SELECT VERSION()")
#使用fetchone()方法获取单条数据.
data = cursor.fetchone()
print ("Database version : %s " % data)
#关闭数据库连接
db.close()
```

执行后会输出:
Database version : 5.7.17-log

实例 07-20: 创建新表

在 Python 程序中, 可以使用方法 execute() 在数据库中创建一个新表。例如, 下面的实例文件 new.py 演示了在 PyMySQL 数据库中创建新表 EMPLOYEE 的过程。

源码路径: daima\7\07-20\new.py

```
import pymysql
#打开数据库连接
db = pymysql.connect("localhost","root","66688888","TESTDB" )
#使用cursor()方法创建一个游标对象cursor
cursor = db.cursor()
#使用execute()方法执行SQL语句, 如果表存在则删除
cursor.execute("DROP TABLE IF EXISTS EMPLOYEE")
#使用预处理语句创建表
sql = """CREATE TABLE EMPLOYEE (
         FIRST_NAME  CHAR(20) NOT NULL,
         LAST_NAME   CHAR(20),
         AGE INT,
         SEX CHAR(1),
         INCOME FLOAT )"""
cursor.execute(sql)
#关闭数据库连接
db.close()
```

执行上述代码后, 将在 MySQL 数据库中创建一个名为 "EMPLOYEE" 的新表, 执行效果如图 7-2 所示。

图 7-2 执行效果

实例 07-21: 向数据库中插入数据

在 Python 程序中, 可以使用 SQL 语句向数据库中插入数据。例如, 下面的实例文件 cha.py 演示了使用 INSERT 语句向表 EMPLOYEE 中插入数据的过程。

源码路径：daima\7\07-21\cha.py

```python
import pymysql
#打开数据库连接
db = pymysql.connect("localhost","root","66688888","TESTDB" )
#使用cursor()方法获取操作游标
cursor = db.cursor()
# SQL插入语句
sql = """INSERT INTO EMPLOYEE(FIRST_NAME,
         LAST_NAME, AGE, SEX, INCOME)
         VALUES ('Mac', 'Mohan', 20, 'M', 2000)"""
try:
    #执行SQL语句
    cursor.execute(sql)
    #提交到数据库执行
    db.commit()
except:
    #如果发生错误则回滚
    db.rollback()
# 关闭数据库连接
db.close()
```

执行上述代码后，打开 MySQL 数据库中的表 EMPLOYEE，会发现在里面插入了一条新的数据。执行效果如图 7-3 所示。

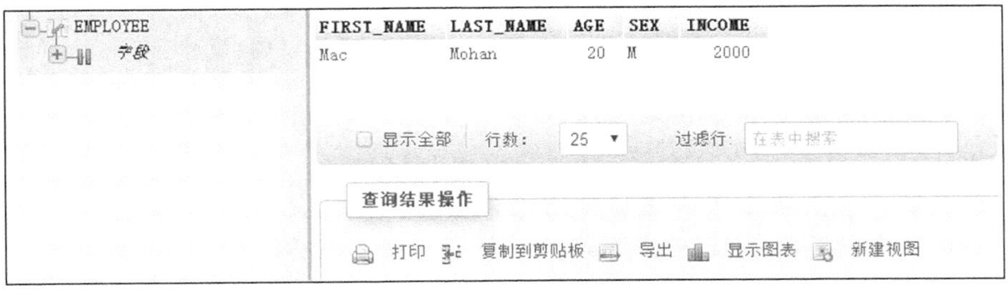

图 7-3 执行效果

实例 07-22：查询数据库中的数据

在 Python 程序中，可以使用 fetchone()方法获取 MySQL 数据库中的单条数据，使用 fetchall()方法获取 MySQL 数据库中的多条数据。例如，下面的实例文件 fi.py 演示了查询并显示表 EMPLOYEE 中 INCOME（工资）大于 1000 的所有数据的过程。

源码路径：daima\7\07-22\fi.py

```python
import pymysql
#打开数据库连接
db = pymysql.connect("localhost","root","66688888","TESTDB" )
#使用cursor()方法获取操作游标
cursor = db.cursor()
# SQL查询语句
sql = "SELECT * FROM EMPLOYEE \
       WHERE INCOME > '%d'" % (1000)
try:
    #执行SQL语句
    cursor.execute(sql)
    #获取所有记录列表
    results = cursor.fetchall()
    for row in results:
        fname = row[0]
        lname = row[1]
        age = row[2]
        sex = row[3]
        income = row[4]
        #输出结果
        print ("fname=%s,lname=%s,age=%d,sex=%s,income=%d" % \
               (fname, lname, age, sex, income ))
except:
    print ("Error: unable to fetch data")
```

```
#关闭数据库连接
db.close()
```
执行后会输出：
```
fname=Mac,lname=Mohan,age=20,sex=M,income=2000
```

实例07-23：更新数据库中的数据

在 Python 程序中，可以使用 UPDATE 语句更新数据库中的数据。例如，下面的实例文件 xiu.py 演示了将表 EMPLOYEE 中 SEX 字段为"M"的 AGE 字段递增 1 的过程。

源码路径：daima\7\07-23\xiu.py

```python
import pymysql
#打开数据库连接
db = pymysql.connect("localhost","root","66688888","TESTDB" )
#使用cursor()方法获取操作游标
cursor = db.cursor()
# SQL更新语句
sql = "UPDATE EMPLOYEE SET AGE = AGE + 1 WHERE SEX = '%c'" % ('M')
try:
    #执行SQL语句
    cursor.execute(sql)
    #提交到数据库执行
    db.commit()
except:
    #发生错误时回滚
    db.rollback()
#关闭数据库连接
db.close()
```

执行效果如图 7-4 所示。

(a) 修改前　　　　　　　　　　　　(b) 修改后

图 7-4　执行效果

实例07-24：删除数据库中的数据

在 Python 程序中，可以使用 DELETE 语句删除数据库中的数据。例如，下面的实例文件 del.py 演示了删除了表 EMPLOYEE 中所有 AGE 大于 20 的数据的过程。

源码路径：daima\7\07-24\del.py

```python
import pymysql
#打开数据库连接
db = pymysql.connect("localhost","root","66688888","TESTDB" )
#使用cursor()方法获取操作游标
cursor = db.cursor()
# SQL删除语句
sql = "DELETE FROM EMPLOYEE WHERE AGE > '%d'" % (20)
try:
    #执行SQL语句
    cursor.execute(sql)
    #提交修改
    db.commit()
except:
    #发生错误时回滚
    db.rollback()

#关闭连接
db.close()
```

执行后，会先删除表 EMPLOYEE 中所有 AGE 大于 20 的数据，然后执行查询语句会发现此时表 EMPLOYEE 中的内容为空。效果如图 7-5 所示。

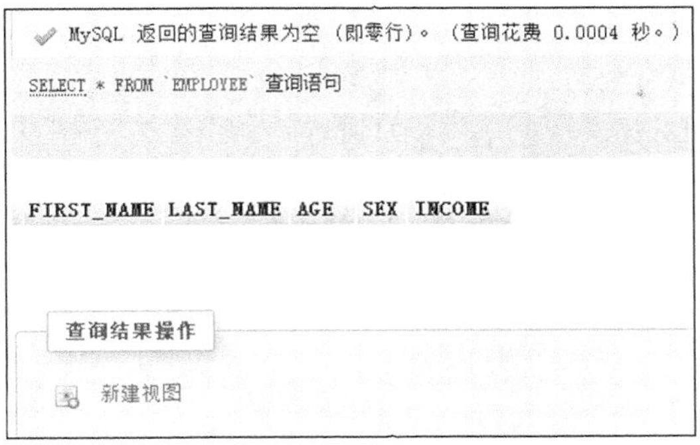

图 7-5　执行效果（表 EMPLOYEE 中的数据已经为空）

实例 07-25：通过执行事务删除表中的数据

在 Python 程序中，使用事务机制可以确保数据的一致性。通常来说，事务应该具有 4 个属性，即原子性、一致性、隔离性、持久性，这 4 个属性通常被称为 ACID 特性。Python DataBase API Specification v2.0 的事务机制提供了两个处理方法，分别是 commit()和 rollback()。例如，下面的实例文件 shi.py 通过执行事务删除表 EMPLOYEE 中所有 AGE 大于 19 的数据的过程。

源码路径：daima\7\07-25\shi.py

```
import pymysql
# 打开数据库连接
db = pymysql.connect("localhost","root","66688888","TESTDB" )
# 使用cursor()方法获取操作游标
cursor = db.cursor()
# SQL删除记录语句
sql = "DELETE FROM EMPLOYEE WHERE AGE > '%d'" % (19)
try:
    #执行SQL语句
    cursor.execute(sql)
    #向数据库提交
    db.commit()
except:
    #发生错误时回滚
    db.rollback()
```

上述代码执行后将删除表 EMPLOYEE 中所有 AGE 大于 19 的数据。

实例 07-26：足球俱乐部球员管理系统

下面的实例模拟实现了足球游戏中的足球俱乐部球员管理系统。通过这个实例，综合演练了在 Python 中操作 MySQL 数据库的过程。

（1）创建一个新的 MySQL 数据库，如 TESTDB，然后将光盘中的 league.sql 文件导入 MySQL 数据库中生成数据表。

（2）在实例文件 main.py 中编写菜单界面代码，通过菜单提示可以实现对数据库的操作处理。文件 main.py 的具体实现代码如下。

源码路径：daima\7\07-26\main.py

```
import CRUD

option = 0

print("\n欢迎光临，请选择数据库操作！")

while option != 6:
    print("\n---------------------------")
    print("选择操作：\n"
          "0. 读取显示所有数据\n"
          "1. 创建数据库\n"
          "2. 插入新的数据\n"
```

```
            "3. 注册新球员\n"
            "4. 预约\n"
            "5. 删除所有的表\n"
            "6. 退出")
    print("----------------------------")

    option = int(input())

    if option == 0:
        CRUD.readAll()

    elif option == 1:
        CRUD.createDB()

    elif option == 2:
        CRUD.insertRows()

    elif option == 3:
        CRUD.insertRow(0)

    elif option == 4:
        print("\nConsultas: \n"
            "0. 查看目前的冠军球队?\n"
            "1. 游戏角色的比较?\n"
            "2. 玩家ID的特点?\n"
            "3. 其中最有特点的球员?\n"
            "4. 你感觉每个队最多的球员？\n"
            "5. 让所有的球员\n"
            "6. 退出")
        readOpt = int(input())

        if readOpt is 0:
            CRUD.readTables(0)
        elif readOpt == 1:
            CRUD.readTables(1)
        elif readOpt == 2:
            CRUD.readTables(2)
        elif readOpt == 3:
            CRUD.readTables(3)
        elif readOpt == 4:
            CRUD.readTables(4)
        elif readOpt == 5:
            CRUD.readTables(5)

    elif option == 5:
        CRUD.deleteTables()

    elif option == 6:
        quit()
```

（3）编写实例文件 CRUD.py，其功能是根据用户选择的菜单指令进行对应的操作。具体实现流程如下。

源码路径：daima\7\07-26\CURD.py

- 建立和指定数据库的连接，填写正确的连接参数。
- 当用户在菜单界面输入"0"指令后执行函数 readAll()，功能是获取数据库内所有表的数据。
- 当用户在菜单界面输入"1"指令后执行函数 createDB()，功能是重新创建新的表。
- 编写函数 readTables()，功能是读取指定表的信息，根据用户在菜单界面中输入的指令执行指定的查询操作。

- 编写函数 insertRows()，功能是当用户在菜单界面中输入指令"2"后，执行插入新的数据操作。
- 当用户在菜单界面中输入指令"3"后，执行插入指定新数据的操作。
- 编写函数 insertNewPlayerNamed()，功能是插入指定名字的新球员信息。
- 编写函数 insertNewTeamNamed()，功能是插入指定名字的新球队信息。
- 编写函数 getTeamID()，功能是获取球队 ID 信息。
- 编写函数 valueExistsInTable()，功能是统计表的信息。
- 编写函数 deleteTables()，功能是当用户在菜单界面中输入指令"5"后执行删除数据库数据操作。

用户输入不同的指令，程序会返回不同的结果。例如，在作者计算机中执行代码后会输出：

```
欢迎光临，请选择数据库操作！
-------------------------
选择操作:
0. 读取显示所有数据
1. 创建数据库
2. 插入新的数据
3. 注册新球员
4. 预约
5. 删除所有的表
6. 退出
-------------------------
0
联赛表:
employee
jogador
partida
personagem
personagem_comprado
testtable
time
torneio

Tabela - employee
('Mac', 'Mohan', 21, 'M', 2000.0)

Tabela - jogador
(1, 1, 'Italus', 5, 2)
(2, 1, 'Thiago', 10, 7)
(3, 1, 'Gabriela', 1, 3)
(4, 1, 'Macabeus', 0, 3)
(5, 1, 'Trabson', 14, 0)
(6, 2, 'Douglas', 8, 8)
(7, 2, 'Estela', 2, 2)
(8, 2, 'Elias', 25, 15)
(9, 2, 'Designer', 3, 1)
(10, 2, 'Pistela', 2, 9)
(12, 3, 'Moskito', 3, 5)
(13, 3, 'Kira', 1, 9)
(16, 3, 'Alsaher', 7, 1)
(17, 3, 'Rodrix', 2, 20)
(18, 3, 'Lalitax', 7, 5)

Tabela - partida
(1, 1, 2, 20, 10, 1, '1')
(2, 2, 3, 30, 15, 2, '1')
(3, 1, 3, 10, 5, 1, '1')

Tabela - personagem
(1, 'Ashe', 1200)
(2, 'Vayne', 2000)
(3, 'Tryndamere', 1200)
(4, 'Blitzcrank', 2000)
(5, 'Amumu', 2000)
(6, 'Fiora', 4500)
(7, 'Sona', 6300)
```

```
(10, 'Alistar', 900)
(8, 'Morgana', 6300)
(9, 'Kayle', 4500)

Tabela - personagem_comprado
(1, 1)
(2, 1)
(3, 1)
(4, 1)
(5, 1)
(1, 2)
(2, 2)
(1, 3)
(4, 3)

Tabela - testtable

Tabela - time
(1, 'BepidPower', 30, 15)
(2, 'AlbusNox', 40, 35)
(3, 'Invocados', 20, 40)

Tabela - torneio
(1, 'Brasil')

-----------------------------
选择操作：
0. 读取显示所有数据
1. 创建数据库
2. 插入新的数据
3. 注册新球员
4. 预约
5. 删除所有的表
6. 退出
-----------------------------
```

7.3 使用 MariaDB 数据库

作为一款经典的关系数据库产品，MariaDB 是一种开源数据库，是 MySQL 数据库的一个分支。因为某些历史原因，不少用户担心 MySQL 数据库会停止开源，所以 MariaDB 逐步发展成为替代 MySQL 的数据库工具之一。

实例 07-27：搭建 MariaDB 数据库环境

搭建 MariaDB 数据库环境的基本流程如下。

（1）登录 MariaDB 官网下载页面，如图 7-6 所示。

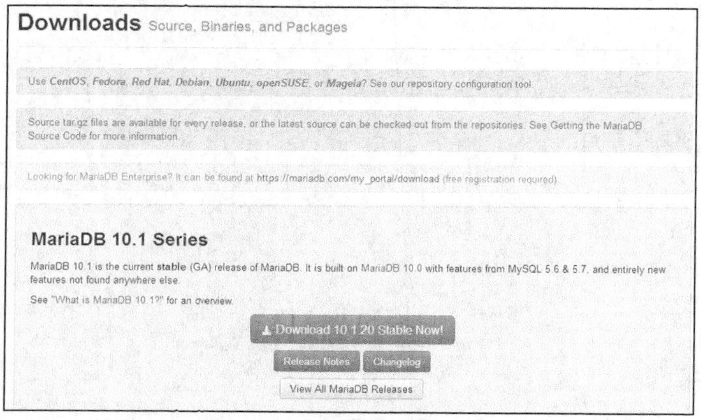

图 7-6　MariaDB 官网下载页面

7.3 使用 MariaDB 数据库

（2）单击 Download 10.1.20 Stable Now!按钮进入具体下载页面，如图 7-7 所示。在此需要根据计算机操作系统进行下载。例如，因为作者的计算机使用 64 位的 Windows 10，所以选择 mariadb-10.1.20-winx64.msi 进行下载。

图 7-7 具体下载页面

（3）下载完成后会得到一个安装文件 mariadb-10.1.20-winx64.msi，双击这个文件后进入欢迎安装界面，如图 7-8 所示。

（4）单击 Next 按钮后进入用户协议界面，在此勾选 I accept the terms in the License Agreement 复选框，如图 7-9 所示。

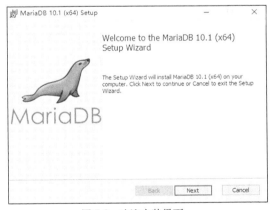
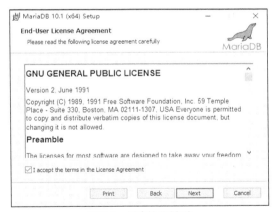

图 7-8 欢迎安装界面　　　　　　　　图 7-9 用户协议界面

（5）单击 Next 按钮后进入典型设置界面，在此设置程序文件的安装路径，如图 7-10 所示。

（6）单击 Next 按钮后进入设置密码界面，在此设置管理员用户 root 的密码，如图 7-11 所示。

（7）单击 Next 按钮后进入默认实例属性界面，在此设置服务器名字和 TCP 端口，如图 7-12 所示。

（8）单击 Next 按钮进入准备安装界面，如图 7-13 所示。

（9）单击 Install 按钮后进入安装进度条界面，开始安装 MariaDB，如图 7-14（a）所示。

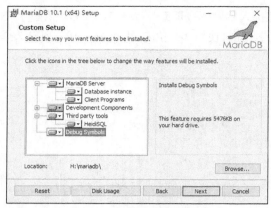

图 7-10　典型设置界面　　　　　　　　图 7-11　设置密码界面

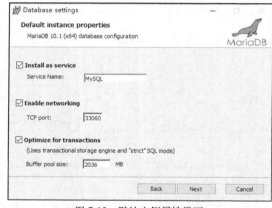
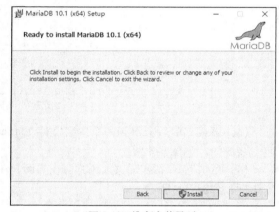

图 7-12　默认实例属性界面　　　　　　图 7-13　准备安装界面

（10）安装完成后进入完成安装界面，单击 Finish 按钮后完成安装，如图 7-14（b）所示。

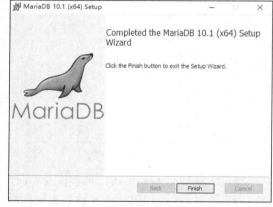

（a）安装进度条界面　　　　　　　　　（b）完成安装界面

图 7-14　安装进度条界面和完成安装界面

实例 07-28：在 Python 程序中使用 MariaDB 数据库

当在 Python 程序中使用 MariaDB 数据库时，需要在程序中加载 Python 的第三方库——MySQL Connector Python。在使用这个第三方库操作 MariaDB 数据库之前，需要先下载并安装这个第三方库。下载并安装第三方库的过程非常简单,只需在 CMD 控制台中执行如下命令即可。
pip install mysql-connector
安装成功时的界面效果如图 7-15 所示。

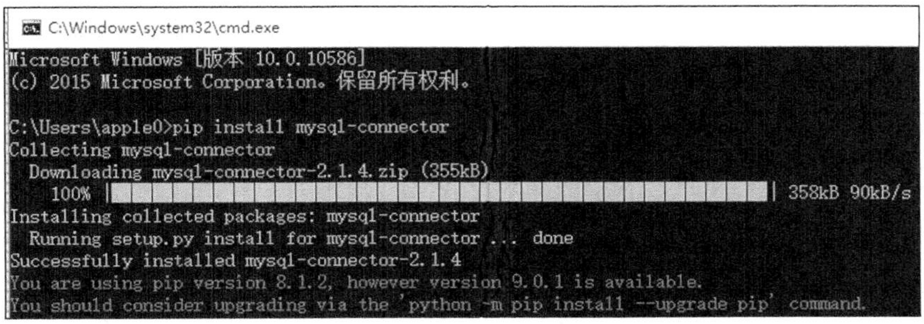

图 7-15 安装成功时的界面效果

例如,下面的实例文件 md.py 演示了在 Python 程序中使用 MariaDB 数据库的过程。

源码路径:daima\7\07-28\md.py

```
from mysql import connector
import random                        #导入内置模块
#省略部分代码
if __name__ == '__main__':
    print("建立连接...")              #输出提示信息
    #建立数据库连接
    con = connector.connect(user='root',password=
                            '66688888',database='md')
    print("建立游标...")              #输出提示信息
    cur = con.cursor()                #建立游标
    print('创建一张表mdd...')         #输出提示信息
    #创建表mdd
    cur.execute('create table mdd(id int primary key auto_increment not null,name text,passwd text)')
    #在表mdd中插入一条记录
    print('插入一条记录...')          #输出提示信息
    cur.execute('insert into mdd (name,passwd)values(%s,%s)',(get_str(2,4),get_str(8,12),))
    print('显示所有记录...')          #输出提示信息
    output()                          #输出数据库中的记录
    print('批量插入多条记录...')      #输出提示信息
    #在表mdd中插入多条记录
    cur.executemany('insert into mdd (name,passwd)values(%s,%s)',get_data_list(3))
    print("显示所有记录...")          #输出提示信息
    output_all()                      #输出数据库中的记录
    print('更新一条记录...')          #输出提示信息
    #更新表mdd中的一条记录
    cur.execute('update mdd set name=%s where id=%s',('aaa',1))
    print('显示所有记录...')          #输出提示信息
    output()                          #输出数据库中的记录
    print('删除一条记录...')          #输出提示信息
    #删除表mdd中的一条记录
    cur.execute('delete from mdd where id=%s',(3,))
    print('显示所有记录: ')           #输出提示信息
    output()                          #输出数据库中的记录
```

在上述实例代码中,使用 mysql.connector 中的函数 connect()建立了和 MariaDB 数据库的连接。连接函数 connect()在 mysql.connector 中定义,此函数的语法格式如下。

connect(host, port,user, password, database,charset)

- host:访问数据库的服务器主机(默认为本机)。
- port:访问数据库的服务端口(默认为 3306)。
- user:访问数据库的用户名。
- password:访问数据库用户名的密码。
- database:访问数据库名称。
- charset:字符编码(默认为 UTF-8)。

实例文件 md.py 执行后将显示创建表并实现数据插入、更新和删除操作的过程。
执行后会输出：

```
建立连接...
建立游标...
创建一张表mdd...
插入一条记录...
显示所有记录...
1    kpv    lrdupdsuh
批量插入多条记录...
显示所有记录...
(1, 'kpv', 'lrdupdsuh')
(2, 'hsue', 'ilrleakcoh')
(3, 'hb', 'dzmcajvm')
(4, 'll', 'ngjhixta')
更新一条记录...
显示所有记录...
1    aaa    lrdupdsuh
2    hsue   ilrleakcoh
3    hb     dzmcajvm
4    ll     ngjhixta
删除一条记录...
显示所有记录：
1    aaa    lrdupdsuh
2    hsue   ilrleakcoh
4    ll     ngjhixta
```

> **注意**：在操作 MariaDB 数据库时，与操作 SQLite3 的 SQL 语句不同的是，SQL 语句中的占位符不是 "?"，而是 "%s"。

实例07-29：使用 MariaDB 创建 MySQL 数据库

请看下面的实例文件 123.py，其功能是使用 MariaDB 创建 MySQL 数据库。

源码路径：daima\7\07-29\123.py

```python
DB_NAME='mariadb'

TABLES = {}

TABLES['location'] = (
    "CREATE TABLE IF NOT EXISTS 'location' ("
    "  'id' int(255) NOT NULL AUTO_INCREMENT,"
    "  'latitud' varchar(15) NOT NULL,"
    "  'longitud' varchar(15) NOT NULL,"
    "  'Fecha' varchar(22) NOT NULL,"
    "  'Hora' varchar(22) NOT NULL,"
    "   PRIMARY KEY ('id'), UNIQUE KEY 'Hora' ('Hora')"
    ") ENGINE=InnoDB")

cnx = mariadb.connect(host='localhost', user='root', password='66688888')
cursor = cnx.cursor()

def generate_database(curs):
    try:
        curs.execute(
            "CREATE DATABASE IF NOT EXISTS {} DEFAULT CHARACTER SET 'utf8'".format(DB_NAME))
    except mariadb.Error as err:
        print("Failed creating database: {}".format(err))
        exit(1)
    else:
        print("Database OK")

try:
    generate_database(cursor)
except mariadb.Error as err:
    print("Error: {}".format(err))

cursor.execute("USE {}".format(DB_NAME))
for name, ddl in TABLES.items():
    try:
        print("Creating table {}: ".format(name), end="")
```

```
            cursor.execute(ddl)
        except mariadb.Error as err:
            print("Failed creating table: {}".format(err))
            exit(1)
        else:
            print("Table OK")

cnx.commit()
cnx.close()
```

上述代码执行后会创建 MySQL 数据库 mariadb，并在此数据库中创建一个名为"location"的表。

执行后会输出：

```
Database OK
Creating table location: Table OK
Inicializando en Host IPV4 192.168.1.102 Puerto 10
Connected
```

7.4 使用 MongoDB 数据库

MongoDB 是一个基于分布式文件存储的数据库，使用 C++编写，旨在为 Web 应用提供可扩展的高性能数据存储解决方案。

实例 07-30：搭建 MongoDB 环境

（1）MongoDB 官网提供了可用于 32 位操作系统和 64 位操作系统的预编译二进制包，读者可以从 MongoDB 官网下载安装包，如图 7-16 所示。

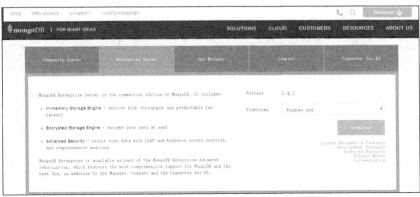

图 7-16 MongoDB 官网

（2）根据当前计算机的操作系统选择下载安装包，因为作者的计算机操作系统是 64 位的 Windows 操作系统，所以这里选择 Windows x64，然后单击 Download 按钮。在弹出的界面中选择 msi 选项，如图 7-17 所示。

（3）下载完成后得到一个扩展名为.msi 的文件，双击这个文件，然后按照操作提示进行安装即可。安装界面如图 7-18 所示。

在 Python 程序中使用 MongoDB 数据库时，必须首先确保安装了 PyMongo 第三方库。如果下载的安装文件扩展名是.exe，则可以直接执行安装。如果下载的安装文件是压缩包，则可以使用如下命令进行安装。

```
pip install pymongo
```

如果没有下载安装文件，则可以通过如下命令进行在线安装。

```
easy_install pymongo
```

图 7-17 选择 msi

图 7-18 安装界面

安装完成后的界面效果如图 7-19 所示。

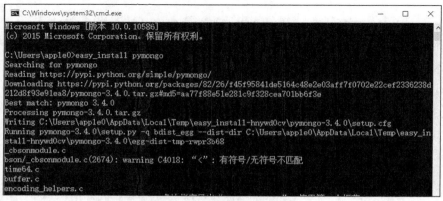
图 7-19 安装完成后的界面效果

实例 07-31：使用 PyMongo 操作 MongoDB 数据库

下面的实例文件 mdb.py 演示了在 Python 程序中使用 PyMongo 操作 MongoDB 数据库的过程。

源码路径：daima\7\07-31\mdb.py

```python
from pymongo import MongoClient
import random
#省略部分代码
if __name__ == '__main__':
    print("建立连接...")              #输出提示信息
    stus = MongoClient().test.stu     #建立连接
    print('插入一条记录...')          #输出提示信息
    #向表stus中插入一条记录
    stus.insert({'name':get_str(2,4),'passwd':get_str(8,12)})
    print("显示所有记录...")          #输出提示信息
    stu = stus.find_one()             #获取数据库记录
    print(stu)                        #输出数据库中的记录
    print('批量插入多条记录...')      #输出提示信息
    stus.insert(get_data_list(3))     #向表stus中插入多条记录
    print('显示所有记录...')          #输出提示信息
    for stu in stus.find():           #遍历记录
        print(stu)                    #输出数据库中的记录
    print('更新一条记录...')          #输出提示信息
    name = input('请输入记录的name:')  #提示输入要修改的记录
    #修改表stus中的一条记录
    stus.update({'name':name},{'$set':{'name':'langchao'}})
    print('显示所有记录...')          #输出提示信息
```

```
    for stu in stus.find():              #遍历记录
        print(stu)                        #输出数据库中的记录
    print('删除一条记录...')               #输出提示信息
    name = input('请输入记录的name:')    #提示输入要删除的记录
    stus.remove({'name':name})           #删除表中的记录
    print('显示所有记录...')               #输出提示信息
    for stu in stus.find():              #遍历记录
        print(stu)                        #输出数据库中的记录
```

上述实例代码中使用了两个生成字符串的函数。在主程序中首先连接集合，然后使用集合对象的方法对集合中的记录进行插入、更新和删除操作。每当数据被修改后，会显示集合中所有记录，以验证操作结果的正确性。

在执行本实例时，初学者很容易遇到如下 MongoDB 执行错误。

Failed to connect 127.0.0.1:27017,reason:errno:10061由于目标计算机积极拒绝，无法连接...

发生上述错误的原因是没有开启 MongoDB 服务，下面是开启 MongoDB 服务的命令。

mongod --dbpath "h:\data"

在上述命令中，h:\data 是一个保存 MongoDB 数据库数据的目录，读者可以随意在本地计算机硬盘中创建目录，并且还可以自定义目录名字。在 CMD 控制台中，开启 MongoDB 服务成功时的界面效果如图 7-20 所示。

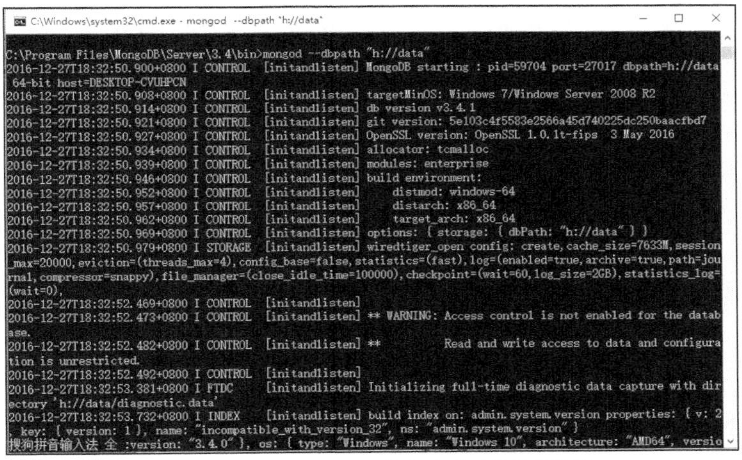

图 7-20 开启 MongoDB 服务成功时的界面效果

在执行本实例程序时，必须在 CMD 控制台中启动 MongoDB 服务，并且确保上述 CMD 控制台界面处于打开状态。

本实例执行后会输出：

```
建立连接...
插入一条记录...
显示所有记录...
{'_id': ObjectId('586243795cd071f570ed3b39'), 'name': 'vvtj', 'passwd': 'iigbddauwj'}
批量插入多条记录...
显示所有记录...
{'_id': ObjectId('586243795cd071f570ed3b39'), 'name': 'vvtj', 'passwd': 'iigbddauwj'}
{'_id': ObjectId('5862437a5cd071f570ed3b3a'), 'name': 'nh', 'passwd': 'upyufzknzgdc'}
{'_id': ObjectId('5862437a5cd071f570ed3b3b'), 'name': 'rgf', 'passwd': 'iqdlyjhztq'}
{'_id': ObjectId('5862437a5cd071f570ed3b3c'), 'name': 'dh', 'passwd': 'rgupzruqb'}
{'_id': ObjectId('586243e45cd071f570ed3b3e'), 'name': 'hcq', 'passwd': 'chiwwvxs'}
{'_id': ObjectId('586243e45cd071f570ed3b3f'), 'name': 'yrp', 'passwd': 'kiocdmeerneb'}
{'_id': ObjectId('586243e45cd071f570ed3b40'), 'name': 'hu', 'passwd': 'pknqgfnm'}
{'_id': ObjectId('5862440d5cd071f570ed3b43'), 'name': 'tlh', 'passwd': 'cikouuladgqn'}
{'_id': ObjectId('5862440d5cd071f570ed3b44'), 'name': 'qxf', 'passwd': 'jlsealrqeeel'}
{'_id': ObjectId('5862440d5cd071f570ed3b45'), 'name': 'vlzp', 'passwd': 'wolypmej'}
{'_id': ObjectId('58632e6c5cd07155543cc27a'), 'sid': 2, 'name': 'sgu', 'passwd': 'ogzvdq'}
{'_id': ObjectId('58632e6c5cd07155543cc27b'), 'sid': 3, 'name': 'jiyl', 'passwd': 'atgmhmxr'}
{'_id': ObjectId('58632e6c5cd07155543cc27c'), 'sid': 4, 'name': 'dbb', 'passwd': 'wmwoeua'}
{'_id': ObjectId('5863305b5cd07155543cc27d'), 'sid': 27, 'name': 'langchao', 'passwd': '123123'}
```

```
{'_id': ObjectId('5863305b5cd07155543cc27e'), 'sid': 28, 'name': 'oxp', 'passwd': 'acgjph'}
{'_id': ObjectId('5863305b5cd07155543cc27f'), 'sid': 29, 'name': 'sukj', 'passwd': 'hjtcjf'}
{'_id': ObjectId('5863305b5cd07155543cc280'), 'sid': 30, 'name': 'bf', 'passwd': 'cqerluvk'}
{'_id': ObjectId('5988087533fda81adc0d332f'), 'name': 'hg', 'passwd': 'gmflqxfaxxnv'}
{'_id': ObjectId('5988087533fda81adc0d3330'), 'name': 'ojb', 'passwd': 'rgxodvkprm'}
{'_id': ObjectId('5988087533fda81adc0d3331'), 'name': 'gtdj', 'passwd': 'zigavkysc'}
{'_id': ObjectId('5988087533fda81adc0d3332'), 'name': 'smgt', 'passwd': 'sizvlhdll'}
{'_id': ObjectId('5a33c1cb33fda859b82399d0'), 'name': 'dbu', 'passwd': 'ypdxtqjjafsm'}
{'_id': ObjectId('5a33c1cb33fda859b82399d1'), 'name': 'qg', 'passwd': 'frnoypez'}
{'_id': ObjectId('5a33c1cb33fda859b82399d2'), 'name': 'ky', 'passwd': 'jvzjtcfs'}
{'_id': ObjectId('5a33c1cb33fda859b82399d3'), 'name': 'glnt', 'passwd': 'ejrerztki'}
更新一条记录...
请输入记录的name:
```

实例07-32：使用 mongoengine 操作 MongoDB 数据库

在 Python 程序中，MongoDB 数据库的 ORM 框架是 MongoEngine。在使用 MongoEngine 框架之前需要先安装 MongoEngine，具体安装命令如下。

```
easy_install mongoengine
```

mongoengine 安装成功后的界面效果如图 7-21 所示。

图 7-21　mongoengine 安装成功后的界面效果

在执行上述命令之前，必须先确保使用如下命令安装了 PyMongo。

```
pip install pymongo
```

下面的实例文件 orm.py 演示了在 Python 程序中使用 mongoengine 操作 MongoDB 数据库的过程。

源码路径：daima\7\07-32\orm.py

```python
import random                              #导入内置模块
from mongoengine import *
connect('test')                            #连接数据库对象test
class Stu(Document):                       #定义ORM框架类Stu
    sid = SequenceField()                  # "序号" 属性表示用户ID
    name = StringField()                   # "用户名" 属性
    passwd = StringField()                 # "密码" 属性
    def introduce(self):                   #定义函数introduce()用于显示自己的介绍信息
        print('序号:',self.sid,end=" ")    #输出ID
        print('姓名:',self.name,end=' ')   #输出姓名
        print('密码:',self.passwd)         #输出密码
    def set_pw(self,pw):                   #定义函数set_pw()用于修改密码
        if pw:
            self.passwd = pw               #修改密码
            self.save()                    #保存修改的密码
#省略部分代码
if __name__ == '__main__':
    print('插入一个文档:')
    stu = Stu(name='langchao',passwd='123123')#创建文档类对象实例stu，设置用户名和密码
    stu.save()                             #持久化保存文档
    stu = Stu.objects(name='lilei').first()#查询数据并对类进行初始化
```

```
    if stu:
        stu.introduce()                        #输出文档信息
    print('插入多个文档')                       #输出提示信息
    for i in range(3):                         #遍历操作
        Stu(name=get_str(2,4),passwd=get_str(6,8)).save() #插入3个文档
    stus = Stu.objects()                       #文档类对象实例stus
    for stu in stus:                           #遍历所有的文档信息
        stu.introduce()                        #输出所有的遍历文档
    print('修改一个文档')                       #输出提示信息
    stu = Stu.objects(name='langchao').first() #查询某个要操作的文档
    if stu:
        stu.name='daxie'                       #修改用户名属性
        stu.save()                             #保存修改
        stu.set_pw('bbbbbbbb')                 #修改密码属性
        stu.introduce()                        #输出修改后结果
    print('删除一个文档')                       #输出提示信息
    stu = Stu.objects(name='daxie').first()    #查询某个要操作的文档
    stu.delete()                               #删除这个文档
    stus = Stu.objects()
    for stu in stus:                           #遍历所有的文档
        stu.introduce()                        #输出删除后结果
```

在上述实例代码中，在导入 MongoEngine 和连接 MongoDB 数据库后，定义了一个继承于类 Document 的子类 Stu。在主程序中通过创建类的实例，并调用其方法 save()将类持久化到数据库；通过类 Stu 中的方法 objects()来查询数据库并映射为类 Stu 的实例，并调用其自定义方法 introduce()来显示载入的信息。然后插入 3 个文档信息，并调用方法 save()将其持久化存入数据库，通过调用类中的自定义方法 set_pw()修改数据并存入数据库。最后通过调用类中的方法 delete()从数据库中删除一个文档。

开始测试程序，在执行本实例程序时，必须在 CMD 控制台中启动 MongoDB 服务，并且确保 CMD 控制台界面处于打开状态。下面是开启 MongoDB 服务的命令。

```
mongod --dbpath "h:\data"
```

在上述命令中，h:\data 是一个保存 MongoDB 数据库数据的目录。

本实例的执行效果如图 7-22 所示。

图 7-22　执行效果

7.5 使用 ORM 操作数据库

对象关系映射（Object Relational Mapping，ORM），用于实现面向对象编程语言中不同类型的数据之间的转换。

实例 07-33：使用 SQLAlchemy 操作两种数据库

在 Python 程序中，SQLAlchemy 是一种经典的 ORM。在使用 SQLAlchemy 之前需要先安装 SQLAlchemy，安装命令如下。

```
easy_install SQLAlchemy
```

SQLAlchemy 安装成功后的界面效果如图 7-23 所示。

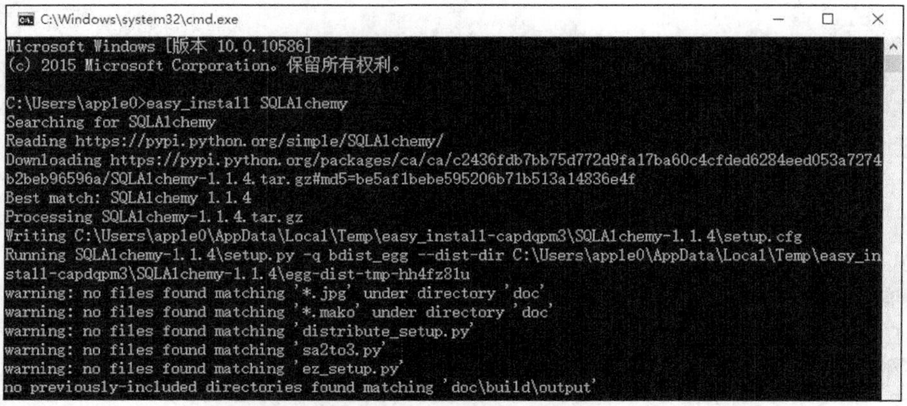

图 7-23　SQLAlchemy 安装成功后的界面效果

例如，下面的实例文件 SQLAlchemy.py 演示了在 Python 程序中使用 SQLAlchemy 操作两种数据库的过程。

源码路径：daima\7\07-33\SQLAlchemy.py

```python
from distutils.log import warn as printf
from os.path import dirname
from random import randrange as rand
from sqlalchemy import Column, Integer, String, create_engine, exc, orm
from sqlalchemy.ext.declarative import declarative_base
from db import DBNAME, NAMELEN, randName, FIELDS, tformat, cformat, setup
DSNs = {
    'mysql': 'mysql://root@localhost/%s' % DBNAME,
    'sqlite': 'sqlite:///:memory:',
}
Base = declarative_base()
class Users(Base):
    __tablename__ = 'users'
    login  = Column(String(NAMELEN))
    userid = Column(Integer, primary_key=True)
    projid = Column(Integer)
    def __str__(self):
        return ''.join(map(tformat,
            (self.login, self.userid, self.projid)))
class SQLAlchemyTest(object):
    def __init__(self, dsn):
        try:
            eng = create_engine(dsn)
        except ImportError:
            raise RuntimeError()
        try:
            eng.connect()
        except exc.OperationalError:
```

```python
                    eng = create_engine(dirname(dsn))
                    eng.execute('CREATE DATABASE %s' % DBNAME).close()
                    eng = create_engine(dsn)
            Session = orm.sessionmaker(bind=eng)
            self.ses = Session()
            self.users = Users.__table__
            self.eng = self.users.metadata.bind = eng
        def insert(self):
            self.ses.add_all(
                Users(login=who, userid=userid, projid=rand(1,5)) \
                    for who, userid in randName()
            )
            self.ses.commit()
        def update(self):
            fr = rand(1,5)
            to = rand(1,5)
            i = -1
            users = self.ses.query(
                Users).filter_by(projid=fr).all()
            for i, user in enumerate(users):
                user.projid = to
            self.ses.commit()
            return fr, to, i+1
        def delete(self):
            rm = rand(1,5)
            i = -1
            users = self.ses.query(
                Users).filter_by(projid=rm).all()
            for i, user in enumerate(users):
                self.ses.delete(user)
            self.ses.commit()
            return rm, i+1
        def dbDump(self):
            printf('\n%s' % " ".join(map(cformat, FIELDS)))
            users = self.ses.query(Users).all()
            for user in users:
                printf(user)
            self.ses.commit()
        def __getattr__(self, attr):      # use for drop/create
            return getattr(self.users, attr)
        def finish(self):
            self.ses.connection().close()
def main():
    printf('*** Connect to %r database' % DBNAME)
    db = setup()
    if db not in DSNs:
        printf('\nERROR: %r not supported, exit' % db)
        return
    try:
        orm = SQLAlchemyTest(DSNs[db])
    except RuntimeError:
        printf('\nERROR: %r not supported, exit' % db)
        return
    printf('\n*** Create users table (drop old one if appl.)')
    orm.drop(checkfirst=True)
    orm.create()
    printf('\n*** Insert names into table')
    orm.insert()
    orm.dbDump()
    printf('\n*** Move users to a random group')
    fr, to, num = orm.update()
    printf('\t(%d users moved) from (%d) to (%d)' % (num, fr, to))
    orm.dbDump()
    printf('\n*** Randomly delete group')
    rm, num = orm.delete()
    printf('\t(group #%d; %d users removed)' % (rm, num))
    orm.dbDump()
    printf('\n*** Drop users table')
    orm.drop()
    printf('\n*** Close cxns')
    orm.finish()
if __name__ == '__main__':
    main()
```

- 在上述实例代码中，首先导入了 Python 标准库中的模块（distutils、os.path、random），然后导入了第三方或外部模块（sqlalchemy），最后导入了应用的本地模块（db），该模块会给用户提供主要的常量和工具函数。
- 使用了 SQLAlchemy 的声明层，在使用前必须先导入 sqlalchemy.ext.declarative.declarative_base，然后使用它创建一个 Base 类，最后让数据子类继承 Base 类。类定义的下一个部分包含了一个 __tablename__ 属性，它定义了映射的表名。也可以显式地定义一个低级别的 sqlalchemy.Table 对象，在这种情况下需要将其写为 __table__。在大多数情况下，使用对象进行数据行的访问，不过也会使用表级别的行为保存（创建或删除）表。接下来是"列"属性，可以通过查阅文档来获取所有支持的数据类型。最后定义了一个 __str__() 方法，用来返回易于阅读的数据行的字符串。因为该输出是定制化的（通过 tformat 的协助），所以不推荐在开发过程中这样使用。
- 通过自定义函数分别实现行的插入、更新和删除操作。插入操作使用 add_all() 方法实现，使用迭代的方式产生一系列的插入操作。最后，还可以决定是进行提交还是进行回滚。update() 和 delete() 方法都具有会话查询的功能，它们使用 filter_by() 方法进行查找。进行随机更新会选择一个项目，通过改变 ID 的方法，将其从一个项目组（fr）移动到另一个项目组（to）。计数器（i）会记录有多少用户会受到影响。删除操作则是根据 ID（rm）随机选择一个项目并假设已将其取消，因此项目中的所有用户都将被删除。当要执行操作时，需要通过会话对象进行提交。
- 函数 dbDump() 负责向屏幕上显示正确的输出信息。该方法从数据库中获取数据行，并按照 db.py 中相似的样式输出数据。

执行后会输出：

```
Choose a database system:

(M)ySQL
(G)adfly
(S)SQLite

Enter choice: S

*** Create users table (drop old one if appl.)

*** Insert names into table

LOGIN      USERID     PROJID
Faye       6812       4
Serena     7003       1
Amy        7209       2
Dave       7306       3
Larry      7311       3
Mona       7404       3
Ernie      7410       3
Jim        7512       3
Angela     7603       3
Stan       7607       3
Jennifer   7608       1
Pat        7711       1
Leslie     7808       4
Davina     7902       4
Elliot     7911       1
Jess       7912       4
Aaron      8312       3
Melissa    8602       4

*** Move users to a random group
    (1 users moved) from (2) to (1)

LOGIN      USERID     PROJID
Faye       6812       4
```

```
Serena      7003     1
Amy         7209     1
Dave        7306     3
Larry       7311     3
Mona        7404     3
Ernie       7410     3
Jim         7512     3
Angela      7603     3
Stan        7607     3
Jennifer    7608     1
Pat         7711     1
Leslie      7808     4
Davina      7902     4
Elliot      7911     1
Jess        7912     4
Aaron       8312     3
Melissa     8602     4

*** Randomly delete group
        (group #1; 5 users removed)

LOGIN       USERID   PROJID
Faye        6812     4
Dave        7306     3
Larry       7311     3
Mona        7404     3
Ernie       7410     3
Jim         7512     3
Angela      7603     3
Stan        7607     3
Leslie      7808     4
Davina      7902     4
Jess        7912     4
Aaron       8312     3
Melissa     8602     4

*** Drop users table

*** Close cxns
```

实例 07-34：使用 Peewee 操作 SQLite 数据库

Peewee 是一款简单的、轻巧的 Python ORM，支持的数据库有 SQLite、MySQL 和 PostgreSQL。本实例将详细讲解使用 Peewee 连接数据库的基本知识。开发者可以通过如下命令安装 Peewee。

```
pip install Peewee
```

在使用 Peewee 连接数据库时，推荐使用 playhouse 中的 db_url 模块。db_url 的 connect()方法可以通过传入的 URL 字符串，生成数据库连接。方法 connect()的功能是通过传入的 URL 字符串，创建一个数据库实例。例如，下面的实例文件 Peewee01.py 演示了在 Python 程序中使用 Peewee 操作 SQLite 数据库的过程。

源码路径：daima\7\07-34\Peewee01.py

```python
from datetime import date
from peewee import *

db = SqliteDatabase('people.db')

'''模型定义'''
class Person(Model):
    name = CharField()
    birthday = DateField()
    is_relative = BooleanField()

    class Meta:
        database = db #这个模型使用了people.db数据库

class Pet(Model):
    owner = ForeignKeyField(Person, related_name='pets')
```

```
        name = CharField()
        animal_type = CharField()

    class Meta:
        database = db  #这个模型使用了people.db数据库

"""连接数据库"""
db.connect()
"""创建Person表和Pet表"""
db.create_tables([Person, Pet])

uncle_bob = Person(name='Bob', birthday=date(1960, 1, 15), is_relative=True)
uncle_bob.save()

"""数据库连接关闭"""
db.close()
```

通过上述代码创建了一个名为"people.db"的 SQLite 数据库，并且在里面创建了两个表——Person 和 Pet。

实例 07-35：更新和删除指定数据库中数据

通过前面的实例 07-34 创建了一个名为"people.db"数据库，下面的实例文件 Peewee02.py 演示了在 Python 程序中使用 Peewee 更新和删除数据库 people.db 中数据的过程。

源码路径：daima\7\07-35\Peewee02.py

```python
from datetime import date
from peewee import *

db = SqliteDatabase('people.db')

class Person(Model):
    name = CharField()
    birthday = DateField()
    is_relative = BooleanField()
    class Meta:
        database = db       #用了people.db数据库

class Pet(Model):
    owner = ForeignKeyField(Person, related_name='pets')
    name = CharField()
    animal_type = CharField()
    class Meta:
        database = db       #用了people.db数据库

"""-------------------------------------------------------------------------------------"""
uncle_bob = Person(name='Bob', birthday=date(1960, 1, 15), is_relative=True)
uncle_bob.save()

grandma = Person.create(name='Grandma', birthday=date(1935, 3, 1), is_relative=True)
herb = Person.create(name='Herb', birthday=date(1950, 5, 5), is_relative=False)
grandma.name = 'Grandma L.'
grandma.save()                    #更新数据库中的grandma的名字

bob_kitty = Pet.create(owner=uncle_bob, name='Kitty', animal_type='cat')
herb_fido = Pet.create(owner=herb, name='Fido', animal_type='dog')
herb_mittens = Pet.create(owner=herb, name='Mittens', animal_type='cat')
herb_mittens_jr = Pet.create(owner=herb, name='Mittens Jr', animal_type='cat')
"""-------------------------------------------------"""
herb_mittens.delete_instance()    #删除
"""""
herb_fido.owner = uncle_bob
herb_fido.save()
bob_fido = herb_fido
```

实例 07-36：查询数据库中指定范围内的数据

下面的实例文件 Peewee03.py 演示了在 Python 程序中使用 Peewee 查询 people.db 数据库中指定范围内数据的过程。

源码路径：daima\7\07-36\Peewee03.py

```python
from datetime import date
from peewee import *

db = SqliteDatabase('people.db')

class Person(Model):
    name = CharField()
    birthday = DateField()
    is_relative = BooleanField()
    class Meta:
        database = db         #用了people.db数据库

class Pet(Model):
    owner = ForeignKeyField(Person, related_name='pets')
    name = CharField()
    animal_type = CharField()
    class Meta:
         database = db       #用了people.db数据库
"""-------------------------------------------------------------------------------"""

#获取单条数据
grandma = Person.select().where(Person.name == 'Grandma L.').get()
"""同上"""
grandma = Person.get(Person.name == 'Grandma L.')

#获取数据列表

for person in Person.select():
    print("人名:",person.name, person.is_relative)

query = Pet.select().where(Pet.animal_type == 'cat')
for pet in query:
    print("宠物名:",pet.name, "主人名:",pet.owner.name)


#连接查询

query = (Pet.select(Pet, Person)
         .join(Person)
          .where(Pet.animal_type == 'cat'))
for pet in query:
    print(pet.name, pet.owner.name)



#让我们获取Bob拥有的所有宠物
for pet in Pet.select().join(Person).where(Person.name == 'Bob'):
     print(pet.name)

"""同上"""
uncle_bob = Person(name='Bob', birthday=date(1960, 1, 15), is_relative=True)
for pet in Pet.select().where(Pet.owner == uncle_bob).order_by(Pet.name):
     print(pet.name)

#日期排序
for person in Person.select().order_by(Person.birthday.desc()):
     print(person.name, person.birthday)
"""-------------------------------------------------------------------------------"""
for person in Person.select():
    print(person.name, person.pets.count(), 'pets')
    for pet in person.pets:
         print ('    ', pet.name, pet.animal_type)

"""-------------------------------------------------------------------------------"""
#日期条件查询
d1940 = date(1940, 1, 1)
d1960 = date(1960, 1, 1)

#查询生日大于1960年、小于1940年的人
query = (Person.select()
          .where((Person.birthday < d1940) | (Person.birthday > d1960)))
for person in query:
```

```
        print(person.name, person.birthday)
#查询生日在1940年与1960年之间的人
query = (Person
        .select()
        .where((Person.birthday > d1940) & (Person.birthday < d1960)))
for person in query:
    print(person.name, person.birthday)
"""-----------------------------------------------------"""
#查询以g开头的人名
expression = (fn.Lower(fn.Substr(Person.name, 1, 1)) == 'g')
for person in Person.select().where(expression):
    print(person.name)

"""连接数据库"""
db.close()
```

执行后会输出各种条件的查询结果：

```
人名: Bob True
人名: Bob True
人名: Grandma L. True
人名: Herb False
宠物名: Kitty  主人名: Bob
宠物名: Mittens Jr  主人名: Herb
Kitty Bob
Mittens Jr Herb
Kitty
Fido
Bob 1960-01-15
Bob 1960-01-15
Herb 1950-07-05
Grandma L. 1937-03-01
Bob 0 pets
Bob 2 pets
        Kitty cat
        Fido dog
Grandma L. 0 pets
Herb 1 pets
        Mittens Jr cat
Bob 1960-01-15
Bob 1960-01-15
Grandma L. 1937-03-01
Herb 1950-07-05
Grandma L.
```

实例07-37：使用 Peewee 在 MySQL 数据库中创建两个表

下面的实例文件 Peewee04.py 演示了在 Python 程序中使用 Peewee 在指定 MySQL 数据库中创建两个表（User 和 Tweet）的过程。

源码路径：daima\7\07-37\Peewee04.py

```python
from peewee import *
import datetime

db = MySQLDatabase("test", host="127.0.0.1", port=3306, user="root", passwd="66688888")
db.connect()

class BaseModel(Model):

    class Meta:
        database = db

class User(BaseModel):
    username = CharField(unique=True)

class Tweet(BaseModel):
    user = ForeignKeyField(User, related_name='tweets')
    message = TextField()
    created_date = DateTimeField(default=datetime.datetime.now)
    is_published = BooleanField(default=True)

if __name__ == "__main__":
    # 创建表
```

```
User.create_table()      #创建User表
Tweet.create_table()     #创建Tweet表
```

在上述代码中，首先创建 User 类和 Tweet 类作为表名。在类下面定义的变量为字段名，如 username、message、created_date 等。CharField、DateTimeField、BooleanField 表示字段的类型。通过 ForeignKeyField 建立外键，Peewee 会默认为我们加上 id 并且将其设置为主键。最后，执行 create_table()方法在 MySQL 数据库 test 中创建两个表 User 和 Tweet，如图 7-24 所示。并且自动创建 id 并将其设置为主键，如图 7-25 所示。

图 7-24　创建的两个表 User 和 Tweet

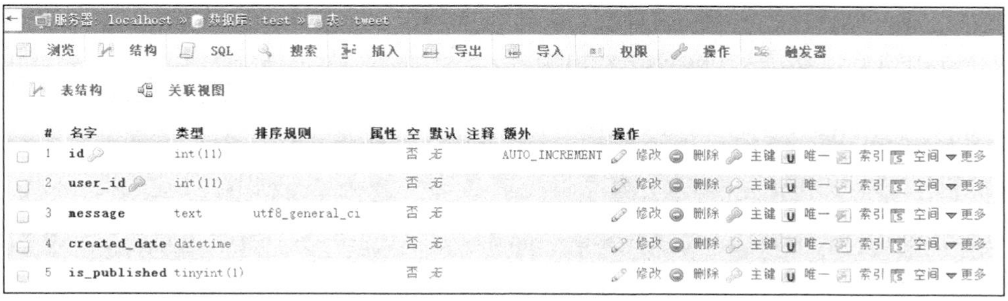

图 7-25　自动创建 id 并将其设置为主键

实例 07-38：使用 Pony 创建一个 SQLite 数据库

Pony 是 Python 中的一种 ORM，它允许使用生成器表达式来构造查询，将生成器表达式的抽象语法树解析成 SQL 语句。Pony 的功能强大，提供了在线 ER 图编辑器工具，可以帮助开发者创建模型。

在使用 Pony 之前需要先进行安装，具体安装命令如下。

```
pip install pony
```

例如，下面的实例文件 pony01.py 演示了使用 Pony 创建一个 SQLite 数据库的过程。首先创建了一个名为"music.sqlite"的数据库，然后在里面创建了两个表 Artist 和 Album。文件 pony01.py 的具体实现代码如下。

源码路径：daima\7\07-38\pony01.py

```
import datetime
import pony.orm as pny

database = pny.Database("sqlite",
                        "music.sqlite",
                        create_db=True)
########################################################################
class Artist(database.Entity):
    """
    使用Pony创建表Artist
    """
    name = pny.Required(str)
    albums = pny.Set("Album")
########################################################################
class Album(database.Entity):
    """
    使用Pony创建表Album
```

```
    """
    artist = pny.Required(Artist)
    title = pny.Required(str)
    release_date = pny.Required(datetime.date)
    publisher = pny.Required(str)
    media_type = pny.Required(str)

# 打开调试模式
pny.sql_debug(True)
# 映射模型数据库
# 如果它们不存在则创建表
database.generate_mapping(create_tables=True)
```

上述代码执行后会输出如下创建数据库和表的等效功能的 SQL 语句。多次执行上述代码，不会重新创建数据库和表，并且会自动设置主键。

```
GET CONNECTION FROM THE LOCAL POOL
PRAGMA foreign_keys = false
BEGIN IMMEDIATE TRANSACTION
CREATE TABLE "Artist" (
  "id" INTEGER PRIMARY KEY AUTOINCREMENT,
  "name" TEXT NOT NULL
)

CREATE TABLE "Album" (
  "id" INTEGER PRIMARY KEY AUTOINCREMENT,
  "artist" INTEGER NOT NULL REFERENCES "Artist" ("id"),
  "title" TEXT NOT NULL,
  "release_date" DATE NOT NULL,
  "publisher" TEXT NOT NULL,
  "media_type" TEXT NOT NULL
)

CREATE INDEX "idx_album__artist" ON "Album" ("artist")

SELECT "Album"."id", "Album"."artist", "Album"."title", "Album"."release_date", "Album"."publisher", "Album"."media_type"
FROM "Album" "Album"
WHERE 0 = 1

SELECT "Artist"."id", "Artist"."name"
FROM "Artist" "Artist"
WHERE 0 = 1

COMMIT
PRAGMA foreign_keys = true
CLOSE CONNECTION
```

实例 07-39：使用 Pony 向数据库的指定表中添加新数据

下面的实例文件 pony02.py 演示了使用 Pony 向数据库 music.sqlite 的指定表中添加新数据的过程。

源码路径：daima\7\07-39\pony02.py

```python
@pny.db_session
def add_data():

    new_artist = Artist(name=u"Newsboys")
    bands = [u"MXPX", u"Kutless", u"Thousand Foot Krutch"]
    for band in bands:
        artist = Artist(name=band)

    album = Album(artist=new_artist,
                  title=u"Read All About It",
                  release_date=datetime.date(1988, 12, 1),
                  publisher=u"Refuge",
                  media_type=u"CD")

    albums = [{"artist": new_artist,
               "title": "Hell is for Wimps",
               "release_date": datetime.date(1990, 7, 31),
               "publisher": "Sparrow",
               "media_type": "CD"
```

```python
            },
            {"artist": new_artist,
             "title": "Love Liberty Disco",
             "release_date": datetime.date(1999, 11, 16),
             "publisher": "Sparrow",
             "media_type": "CD"
            },
            {"artist": new_artist,
             "title": "Thrive",
             "release_date": datetime.date(2002, 3, 26),
             "publisher": "Sparrow",
             "media_type": "CD"}
        ]
        for album in albums:
            a = Album(**album)

if __name__ == "__main__":
    add_data()

    with pny.db_session:
        a = Artist(name="Skillet")
```

在上述代码中使用一种名为"db_session"的装饰器来操作数据库,它负责打开连接、提交数据并关闭连接,并且还可以作为上下文管理器使用。上述代码执行后会输出和添加数据功能等效的 SQL 语句:

```
GET CONNECTION FROM THE LOCAL POOL
PRAGMA foreign_keys = false
BEGIN IMMEDIATE TRANSACTION
SELECT "Album"."id", "Album"."artist", "Album"."title", "Album"."release_date", "Album"."publisher", "Album"."media_type"
FROM "Album" "Album"
WHERE 0 = 1

SELECT "Artist"."id", "Artist"."name"
FROM "Artist" "Artist"
WHERE 0 = 1

COMMIT
PRAGMA foreign_keys = true
CLOSE CONNECTION
GET NEW CONNECTION
BEGIN IMMEDIATE TRANSACTION
INSERT INTO "Artist" ("name") VALUES (?)
['Newsboys']

INSERT INTO "Artist" ("name") VALUES (?)
['MXPX']

INSERT INTO "Artist" ("name") VALUES (?)
['Kutless']

INSERT INTO "Artist" ("name") VALUES (?)
['Thousand Foot Krutch']

INSERT INTO "Album" ("artist", "title", "release_date", "publisher", "media_type") VALUES (?, ?, ?, ?, ?)
[1, 'Read All About It', '1988-12-01', 'Refuge', 'CD']

INSERT INTO "Album" ("artist", "title", "release_date", "publisher", "media_type") VALUES (?, ?, ?, ?, ?)
[1, 'Hell is for Wimps', '1990-07-31', 'Sparrow', 'CD']

INSERT INTO "Album" ("artist", "title", "release_date", "publisher", "media_type") VALUES (?, ?, ?, ?, ?)
[1, 'Love Liberty Disco', '1996-11-16', 'Sparrow', 'CD']

INSERT INTO "Album" ("artist", "title", "release_date", "publisher", "media_type") VALUES (?, ?, ?, ?, ?)
[1, 'Thrive', '2002-03-26', 'Sparrow', 'CD']

COMMIT
RELEASE CONNECTION
GET CONNECTION FROM THE LOCAL POOL
BEGIN IMMEDIATE TRANSACTION
INSERT INTO "Artist" ("name") VALUES (?)
```

```
['Skillet']

COMMIT
RELEASE CONNECTION
```

实例07-40：使用 Pony 查询并修改数据库中指定数据

下面的实例文件 pony03.py 演示了使用 Pony 查询并修改 music.sqlite 数据库中指定数据的过程。

源码路径：daima\7\07-40\pony03.py

```python
import pony.orm as pny

from pony01 import Artist, Album

with pny.db_session:
    band = Artist.get(name="Newsboys")
    print(band.name)
    for record in band.albums:
        print(record.title)
    #修改数据
    band_name = Artist.get(name="Kutless")
    band_name.name = "Beach Boys"

result = pny.select(i.name for i in Artist)
```

在上述代码中，使用 db_session 作为上下文管理器，在进行查询时从数据库中获取表 Artist 并输出其名称。然后循环表 Artist 的 Album，这些 Album 也包含在返回的对象中。最后修改指定 Artist 的名字。上述代码执行后会输出和修改、查询数据功能等效的 SQL 语句：

```
GET CONNECTION FROM THE LOCAL POOL
PRAGMA foreign_keys = false
BEGIN IMMEDIATE TRANSACTION
SELECT "Album"."id", "Album"."artist", "Album"."title", "Album"."release_date", "Album"."publisher", "Album"."media_type"
FROM "Album" "Album"
WHERE 0 = 1

SELECT "Artist"."id", "Artist"."name"
FROM "Artist" "Artist"
WHERE 0 = 1

COMMIT
PRAGMA foreign_keys = true
CLOSE CONNECTION
GET NEW CONNECTION
SWITCH TO AUTOCOMMIT MODE
SELECT "id", "name"
FROM "Artist"
WHERE "name" = ?
LIMIT 2
['Newsboys']

Newsboys
SELECT "id", "artist", "title", "release_date", "publisher", "media_type"
FROM "Album"
WHERE "artist" = ?
[1]

Thrive
Read All About It
Hell is for Wimps
Love Liberty Disco
SELECT "id", "name"
FROM "Artist"
WHERE "name" = ?
LIMIT 2
['Kutless']

BEGIN IMMEDIATE TRANSACTION
UPDATE "Artist"
SET "name" = ?
```

```
      WHERE "id" = ?
        AND "name" = ?
    ['Beach Boys', 3, 'Kutless']

    COMMIT
    RELEASE CONNECTION
```

实例07-41：使用 Pony 删除数据库的表中的某条数据

下面的实例文件 pony04.py 演示了使用 Pony 删除 music.sqlite 数据库的表 Artist 中 name 为 "MXPX" 的这条数据的过程。

源码路径：daima\7\07-41\pony04.py

```python
import pony.orm as pny
from pony01 import Artist
with pny.db_session:
    band = Artist.get(name="MXPX")
    band.delete()
```

上述代码执行后会输出和删除数据功能等效的 SQL 语句：

```
GET CONNECTION FROM THE LOCAL POOL
PRAGMA foreign_keys = false
BEGIN IMMEDIATE TRANSACTION
SELECT "Album"."id", "Album"."artist", "Album"."title", "Album"."release_date", "Album"."publisher", "Album"."media_type"
FROM "Album" "Album"
WHERE 0 = 1

SELECT "Artist"."id", "Artist"."name"
FROM "Artist" "Artist"
WHERE 0 = 1

COMMIT
PRAGMA foreign_keys = true
CLOSE CONNECTION
GET NEW CONNECTION
SWITCH TO AUTOCOMMIT MODE
SELECT "id", "name"
FROM "Artist"
WHERE "name" = ?
LIMIT 2
['MXPX']

SELECT "id", "artist", "title", "release_date", "publisher", "media_type"
FROM "Album"
WHERE "artist" = ?
[2]

BEGIN IMMEDIATE TRANSACTION
DELETE FROM "Artist"
WHERE "id" = ?
  AND "name" = ?
[2, 'MXPX']

COMMIT
RELEASE CONNECTION
```

实例07-42：在指定 MySQL 数据库中创建指定的表

例如，下面的实例文件 pony05.py 演示了使用 Pony 在指定 MySQL 数据库中创建指定的表的过程。

源码路径：daima\7\07-42\pony05.py

```python
from pony.orm import *

db = Database("mysql", host="localhost",
              user="root",
              passwd="66688888",
              db="test")
```

```python
class Person(db.Entity):
    name = Required(str)
    age = Required(int)

sql_debug(True)#显示调试信息（SQL语句）
db.generate_mapping(create_tables=True)#如果数据库没有创建表

@db_session
def create_persons():
    p1 = Person(name="Person1", age=20)
    p2 = Person(name="Person2", age=22)
    p3 = Person(name="Person3", age=12)
    print(p1.id) #这里得不到id，没提交
    commit()
    print(p1.id) #这里得到已经有的id
create_persons()
```

上述代码执行后会创建指定的表 person，并且向表中插入了 3 条数据。上述代码执行后会输出和创建表功能等效的 SQL 语句：

```
GET CONNECTION FROM THE LOCAL POOL
cursor.execute("SHOW VARIABLES LIKE 'foreign_key_checks'")
SET foreign_key_checks = 0
SELECT 'person'. 'id', 'person'. 'name', 'person'. 'age'
FROM 'person' 'person'
WHERE 0 = 1

COMMIT
SET foreign_key_checks = 1
CLOSE CONNECTION
None
GET NEW CONNECTION
INSERT INTO 'person' ('name', 'age') VALUES (%s, %s)
['Person1', 20]

INSERT INTO 'person' ('name', 'age') VALUES (%s, %s)
['Person2', 22]

INSERT INTO 'person' ('name', 'age') VALUES (%s, %s)
['Person3', 12]

COMMIT
7
RELEASE CONNECTION
```

实例 07-43：在 MySQL 数据库中实现一对多和继承操作

在使用 Pony 操作 MySQL 数据库时，外键必须设置为一对多中两边的字段；对于 Django，可以只设置多的一边的关系。例如，下面是一对一的代码：

```python
class User(db.Entity):
    name = Required(str)
    cart = Optional("Cart")    #必须是Optional-Required或Optional-Optional

class Cart(db.Entity):
    user = Required("User")
```

而下面是多对一的代码：

```python
class Product(db.Entity):
    tags = Set("Tag")

class Tag(db.Entity):
    products = Set(Product)
```

为了更好地说明对应关系，下面举一个通俗易懂的例子。例如，学生和教授继承 Person，教授有多个学生（一对多的关系），Person 包含了所有子类的属性，可以用一个 classtype 来区分其属于哪个对象，默认是 class name，可以添加_discriminator_属性修改 classtype。例如，下面的实例文件 pony06.py 演示了使用 Pony 在 MySQL 数据库中实现一对多和继承操作的过程。

源码路径：daima\7\07-43\pony06.py

```python
from pony.orm import *
from decimal import Decimal

db = Database("mysql", host="localhost",
              user="root",
              passwd="66688888",
              db="test")
db.drop_table("person", with_all_data=True)

class Person(db.Entity):
    _discriminator_ = 1
    name = Required(str)
    age = Required(int)

class Student(Person):
    _discriminator_ = 3
    gpa = Optional(Decimal)
    mentor = Optional("Professor")

class Professor(Person):
    _discriminator_ = 2
    degree = Required(str)
    students = Set("Student")

sql_debug(True)                              # 显示调试信息（SQL语句）
db.generate_mapping(create_tables=True)      # 如果数据库没有创建表

@db_session
def create_persons():
    p1 = Person(name="Person", age=20)
    s = Student(name="Student", age=22, gpa=1.2)
    p2 = Professor(name="Professor", age=12, degree="aaaaaa", students=[s])
    commit()

create_persons()
```

上述代码执行后会输出和相应数据操作功能等效的 SQL 语句：

```
GET NEW CONNECTION
SET foreign_key_checks = 0
CREATE TABLE 'person' (
  'id' INTEGER PRIMARY KEY AUTO_INCREMENT,
  'name' VARCHAR(255) NOT NULL,
  'age' INTEGER NOT NULL,
  'classtype' VARCHAR(255) NOT NULL,
  'gpa' DECIMAL(12, 2),
  'mentor' INTEGER,
  'degree' VARCHAR(255)
)

CREATE INDEX 'idx_person__mentor' ON 'person' ('mentor')

ALTER TABLE 'person' ADD CONSTRAINT 'fk_person__mentor' FOREIGN KEY ('mentor') REFERENCES 'person' ('id')

SELECT 'person'.'id', 'person'.'name', 'person'.'age', 'person'.'classtype', 'person'.'gpa', 'person'.'mentor', 'person'.'degree'
FROM 'person' 'person'
WHERE 0 = 1

COMMIT
SET foreign_key_checks = 1
CLOSE CONNECTION
GET NEW CONNECTION
INSERT INTO 'person' ('name', 'age', 'classtype') VALUES (%s, %s, %s)
['Person', 20, '1']

INSERT INTO 'person' ('name', 'age', 'classtype', 'degree') VALUES (%s, %s, %s, %s)
['Professor', 12, '2', 'aaaaaa']

INSERT INTO 'person' ('name', 'age', 'classtype', 'gpa', 'mentor') VALUES (%s, %s, %s, %s, %s)
['Student', 22, '3', Decimal('1.2'), 2]

COMMIT
RELEASE CONNECTION
```

实例 07-44：下载并安装 PostgreSQL

PostgreSQL 是一款开源的数据库，其优势在于具有 SQL 标准的完备性，支持事务、事务隔离级别，并对数据类型、内置函数、索引具有很好的扩展性。

在 Windows 操作系统下，下载并安装 PostgreSQL 的具体流程如下。

（1）在 PostgreSQL 官网首页单击 Download 链接进入下载界面，如图 7-26 所示。

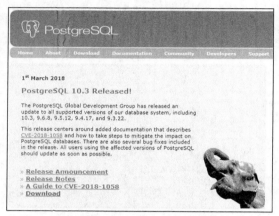

图 7-26　单击 Download 链接

（2）单击下方的 Windows 链接进入 Windows 操作系统的下载界面。

（3）单击 Download the installer 链接，在弹出的界面中选择安装的 PostgreSQL 版本和安装的操作系统类型，单击 DOWNLOAD NOW 按钮开始下载。

（4）下载完成后得到一个扩展名为.exe 的可执行文件，双击该文件后即可开始安装。注：需要先安装 Microsoft Visual C++。

（5）最后连续单击 Next 按钮即可，一直到弹出安装成功界面，如图 7-27 所示。

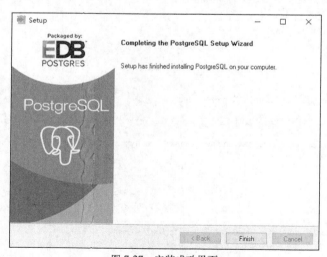

图 7-27　安装成功界面

实例 07-45：连接指定 PostgreSQL 数据库

与 MySQL 数据库不同，PostgreSQL 至少包含 3 种 Python 适配器驱动：psycopg、PyPgSQL 和 PyGreSQL。另外还有一种适配器——PoPy，目前已废弃，并且该项目在 2003 年与 PyGreSQL

7.5 使用 ORM 操作数据库

进行了合并。目前剩下的这 3 种适配器都有其特性和优缺点，建议开发者根据实际项目进行选择。需要注意的是，PyPgSQL 自 2006 年起就不再开发了，而 PyGreSQL 则是在 2009 年发布了最新版本（4.0）。这两种适配器不再活跃，这使得 psycopg 成为 PostgreSQL 适配器的唯一引领者，因此本书的示例均基于该适配器进行讲解。psycopg 目前已进入到第二个版本 psycopg2，所以本书将以 psycopg2 模块为主来介绍连接 PostgreSQL 数据库的知识。

在使用 psycopg2 之前需要先进行安装，具体安装命令如下。

```
pip install python-psycopg2
```

例如，下面的实例文件 PostgreSQL01.py 演示了使用 Python 程序连接指定 PostgreSQL 数据库的过程。

源码路径：daima\7\07-45\PostgreSQL01.py

```python
import psycopg2

def connectPostgreSQL():
    conn = psycopg2.connect(database="mydb", user="postgres", password="66688888", host="127.0.0.1", port="5432")
    print('connect successful!')

if __name__ == '__main__':
    connectPostgreSQL()
```

执行后会输出：

```
connect successful!
```

实例 07-46：在 PostgreSQL 数据库中创建指定表

下面的实例文件 PostgreSQL02.py 演示了使用 Python 程序在 PostgreSQL 数据库中创建指定表的过程。

源码路径：daima\7\07-46\PostgreSQL02.py

```python
import os
import sys
import psycopg2

def connectPostgreSQL():
    conn = psycopg2.connect(database="mydb", user="postgres", password="66688888", host="127.0.0.1", port="5432")
    print('connect successful!')
    cursor = conn.cursor()
    cursor.execute('''
        create table public.member(
        id integer not null primary key,
        name varchar(32) not null,
        password varchar(32) not null,
        singal varchar(128)
        )''')
    conn.commit()
    conn.close()
    print('table public.member is created!')

if __name__ == '__main__':
    connectPostgreSQL()
```

执行后会输出：

```
connect successful!
table public.member is created!
```

实例 07-47：创建 PostgreSQL 表并插入新数据

下面的实例文件 PostgreSQL03.py 演示了使用 Python 程序创建 PostgreSQL 表并插入新数据的过程。

源码路径：daima\7\07-47\PostgreSQL03.py

```python
def connectPostgreSQL():
    conn = psycopg2.connect(database="mydb", user="postgres", password="66688888", host="127.0.0.1", port="5432")
```

```
        print('connect successful!')
        cursor = conn.cursor()
        cursor.execute('''create table public.member(
id integer not null primary key,
name varchar(32) not null,
password varchar(32) not null,
singal varchar(128)
)''')
        conn.commit()
        conn.close()
        print('table public.member is created!')

def insertOperate():
        conn = psycopg2.connect(database="mydb", user="postgres", password="66688888", host="127.0.0.1", port="5432")
        cursor = conn.cursor()
        cursor.execute("insert into public.member(id,name,password,singal)\
values(1,'member0','password0','signal0')")
        cursor.execute("insert into public.member(id,name,password,singal)\
values(2,'member1','password1','signal1')")
        cursor.execute("insert into public.member(id,name,password,singal)\
values(3,'member2','password2','signal2')")
        cursor.execute("insert into public.member(id,name,password,singal)\
values(4,'member3','password3','signal3')")
        conn.commit()
        conn.close()

        print('insert records into public.memmber successfully')

if __name__ == '__main__':
        insertOperate()
```

上述代码执行后会在表 public.member 中插入指定的数据，并输出如下提示信息。

```
insert records into public.memmber successfully
```

实例07-48：查询显示指定表中的数据

下面的实例文件 PostgreSQL04.py 演示了使用 Python 程序先创建 PostgreSQL 表并插入新数据，然后查询显示指定表中的数据的过程。

源码路径：daima\7\07-48\PostgreSQL04.py

```
def connectPostgreSQL():
        conn = psycopg2.connect(database="mydb", user="postgres", password="66688888", host="127.0.0.1", port="5432")
        print
'connect successful!'
        cursor = conn.cursor()
        cursor.execute('''create table public.member(
id integer not null primary key,
name varchar(32) not null,
password varchar(32) not null,
singal varchar(128)
)''')
        conn.commit()
        conn.close()
        print('table public.member is created!')

def insertOperate():
        conn = psycopg2.connect(database="mydb", user="postgres", password="66688888", host="127.0.0.1", port="5432")
        cursor = conn.cursor()
        cursor.execute("insert into public.member(id,name,password,singal)\
values(1,'member0','password0','signal0')")
        cursor.execute("insert into public.member(id,name,password,singal)\
values(2,'member1','password1','signal1')")
        cursor.execute("insert into public.member(id,name,password,singal)\
values(3,'member2','password2','signal2')")
        cursor.execute("insert into public.member(id,name,password,singal)\
values(4,'member3','password3','signal3')")
        conn.commit()
        conn.close()

        print('insert records into public.memmber successfully')
```

```python
def selectOperate():
    conn = psycopg2.connect(database="mydb", user="postgres", password="66688888", host="127.0.0.1", port="5432")
    cursor = conn.cursor()
    cursor.execute("select id,name,password,singal from public.member where id>2")
    rows = cursor.fetchall()
    for row in rows:
        print('id=', row[0], ',name=', row[1], ',pwd=', row[2], ',singal=', row[3], '\n')
    conn.close()

if __name__ == '__main__':
    selectOperate()
```

上述代码执行后会输出指定表中的数据：

id= 3 ,name= member2 ,pwd= password2 ,singal= signal2

id= 4 ,name= member3 ,pwd= password3 ,singal= signal3

实例 07-49：向 PostgreSQL 数据库中插入新数据并更新数据

下面的实例文件 PostgreSQL05.py 演示了向 PostgreSQL 数据库中插入新数据并更新数据的过程。

源码路径：daima\7\07-49\PostgreSQL05.py

```python
def connectPostgreSQL():
    conn = psycopg2.connect(database="mydb", user="postgres", password="66688888", host="127.0.0.1", port="5432")
    print('connect successful!')
    cursor = conn.cursor()
    cursor.execute('''create table public.member(
id integer not null primary key,
name varchar(32) not null,
password varchar(32) not null,
singal varchar(128)
)''')
    conn.commit()
    conn.close()
    print('table public.member is created!')

def insertOperate():
    conn = psycopg2.connect(database="mydb", user="postgres", password="66688888", host="127.0.0.1", port="5432")
    cursor = conn.cursor()
    cursor.execute("insert into public.member(id,name,password,singal)\
 values(1,'member0','password0','signal0')")
    cursor.execute("insert into public.member(id,name,password,singal)\
 values(2,'member1','password1','signal1')")
    cursor.execute("insert into public.member(id,name,password,singal)\
 values(3,'member2','password2','signal2')")
    cursor.execute("insert into public.member(id,name,password,singal)\
 values(4,'member3','password3','signal3')")
    conn.commit()
    conn.close()
    print('insert records into public.memmber successfully')

def selectOperate():
    conn = psycopg2.connect(database="mydb", user="postgres", password="66688888", host="127.0.0.1", port="5432")
    cursor = conn.cursor()
    cursor.execute("select id,name,password,singal from public.member where id>2")
    rows = cursor.fetchall()
    for row in rows:
        print('id=', row[0], ',name=', row[1], ',pwd=', row[2], ',singal=', row[3], '\n')
    conn.close()

def updateOperate():
    conn = psycopg2.connect(database="mydb", user="postgres", password="66688888", host="127.0.0.1", port="5432")
    cursor = conn.cursor()
    cursor.execute("update public.member set name='update ...' where id=2")
    conn.commit()
    print("Total number of rows updated :", cursor.rowcount)
    cursor.execute("select id,name,password,singal from public.member")
    rows = cursor.fetchall()
    for row in rows:
        print('id=', row[0], ',name=', row[1], ',pwd=', row[2], ',singal=', row[3], '\n')
```

```
        conn.close()

if __name__ == '__main__':
    updateOperate()
```

上述代码执行后会显示更新数据表后的数据：

```
Total number of rows updated : 1
id= 1 ,name= member0 ,pwd= password0 ,singal= signal0

id= 3 ,name= member2 ,pwd= password2 ,singal= signal2

id= 4 ,name= member3 ,pwd= password3 ,singal= signal3

id= 2 ,name= update ... ,pwd= password1 ,singal= signal1
```

实例 07-50：删除 PostgreSQL 数据库中的指定数据

下面的实例文件 PostgreSQL06.py 演示了删除 PostgreSQL 数据库表中 id 为 2 的数据的过程。

源码路径：daima\7\07-50\PostgreSQL06.py

```python
def connectPostgreSQL():
    conn = psycopg2.connect(database="mydb", user="postgres", password="66688888", host="127.0.0.1", port="5432")
    print('connect successful!')
    cursor = conn.cursor()
    cursor.execute('''
        create table public.member(
        id integer not null primary key,
        name varchar(32) not null,
        password varchar(32) not null,
        singal varchar(128)
    )''')
    conn.commit()
    conn.close()
    print( 'table public.member is created!')

def insertOperate():
    conn = psycopg2.connect(database="mydb", user="postgres", password="66688888", host="127.0.0.1", port="5432")
    cursor = conn.cursor()
    cursor.execute("insert into public.member(id,name,password,singal)\
        values(1,'member0','password0','signal0')")
    cursor.execute("insert into public.member(id,name,password,singal)\
        values(2,'member1','password1','signal1')")
    cursor.execute("insert into public.member(id,name,password,singal)\
        values(3,'member2','password2','signal2')")
    cursor.execute("insert into public.member(id,name,password,singal)\
        values(4,'member3','password3','signal3')")
    conn.commit()
    conn.close()
    print('insert records into public.memmber successfully')

def selectOperate():
    conn = psycopg2.connect(database="mydb", user="postgres", password="66688888", host="127.0.0.1", port="5432")
    cursor = conn.cursor()
    cursor.execute("select id,name,password,singal from public.member where id>2")
    rows = cursor.fetchall()
    for row in rows:
        print('id=', row[0], ',name=', row[1], ',pwd=', row[2], ',singal=', row[3], '\n')
    conn.close()

def updateOperate():
    conn = psycopg2.connect(database="mydb", user="postgres", password="66688888", host="127.0.0.1", port="5432")
    cursor = conn.cursor()
    cursor.execute("update public.member set name='update ...' where id=2")
    conn.commit()
    print("Total number of rows updated :", cursor.rowcount)
    cursor.execute("select id,name,password,singal from public.member")
    rows = cursor.fetchall()
    for row in rows:
        print('id=', row[0], ',name=', row[1], ',pwd=', row[2], ',singal=', row[3], '\n')
    conn.close()
```

```python
def deleteOperate():
    conn = psycopg2.connect(database="mydb", user="postgres", password="66688888", host="127.0.0.1", port="5432")
    cursor = conn.cursor()

    cursor.execute("select id,name,password,singal from public.member")
    rows = cursor.fetchall()
    for row in rows:
        print('id=', row[0], ',name=', row[1], ',pwd=', row[2], ',singal=', row[3], '\n')
    print('begin delete')
    cursor.execute("delete from public.member where id=2")
    conn.commit()
    print('end delete')
    print("Total number of rows deleted :", cursor.rowcount)
    cursor.execute("select id,name,password,singal from public.member")
    rows = cursor.fetchall()
    for row in rows:
        print('id=', row[0], ',name=', row[1], ',pwd=', row[2], ',singal=', row[3], '\n')
    conn.close()

if __name__ == '__main__':
    deleteOperate()
```

上述代码执行后会输出删除前和删除后的数据库信息以进行对比：

```
id= 1 ,name= member0 ,pwd= password0 ,singal= signal0

id= 3 ,name= member2 ,pwd= password2 ,singal= signal2

id= 4 ,name= member3 ,pwd= password3 ,singal= signal3

id= 2 ,name= update ... ,pwd= password1 ,singal= signal1

begin delete
end delete
Total number of rows deleted : 1
id= 1 ,name= member0 ,pwd= password0 ,singal= signal0

id= 3 ,name= member2 ,pwd= password2 ,singal= signal2

id= 4 ,name= member3 ,pwd= password3 ,singal= signal3
```

实例07-51：创建 PostgreSQL 表并实现插入、查询、更新和删除数据

下面的实例文件 PostgreSQL.py 演示了创建 PostgreSQL 表并实现插入、查询、更新和删除数据的过程。

源码路径：daima\7\07-51\PostgreSQL.py

```python
import psycopg2

#连接数据库
conn = psycopg2.connect(dbname="mydb", user="postgres",
        password="6668888", host="127.0.0.1", port="5432")

#创建游标以访问数据库
cur = conn.cursor()

#创建表
cur.execute(
        'CREATE TABLE stu ('
        'name     varchar(80),'
        'address varchar(80),'
        'age      int,'
        'date     date'
        ')'
)

#插入数据
cur.execute("INSERT INTO stu "
        "VALUES('Gopher', 'China Beijing', 100, '2018-07-27')")

#查询数据
cur.execute("SELECT * FROM stu")
```

```
    rows = cur.fetchall()
    for row in rows:
        print('name=' + str(row[0]) + ' address=' + str(row[1]) +
              ' age=' + str(row[2]) + ' date=' + str(row[3]))

    #更新数据
    cur.execute("UPDATE stu SET age=12 WHERE name='Gopher'")

    #删除数据
    cur.execute("DELETE FROM stu WHERE name='Gopher'")

    #提交事务
    conn.commit()

    #关闭连接
    conn.close()
```

执行后会输出：

```
name=Gopher address=China Beijing age=100 date=2018-07-27
```

实例 07-52：使用模块 queries 查询 PostgreSQL 数据库中的数据

模块 queries 是对 psycopg2 库的封装。通过使用模块 queries，Python 程序可以更加灵活地操作 PostgreSQL 数据库。安装模块 queries 的具体命令如下。

```
pip install queries
```

例如，下面的实例文件 simple-tornado.py 演示了使用模块 queries 查询 PostgreSQL 数据库中的数据的过程。

源码路径：daima\7\07-52\simple-tornado.py

```
import queries
with queries.Session('postgresql://postgres:66688888@127.0.0.1:5432/mydb') as session:
    for row in session.query('SELECT * FROM public.member'):
        print(row)
```

上述代码执行后会输出表 public.member 中的数据：

```
{'id': 1, 'name': 'member0', 'password': 'password0', 'singal': 'signal0'}
{'id': 3, 'name': 'member2', 'password': 'password2', 'singal': 'signal2'}
{'id': 4, 'name': 'member3', 'password': 'password3', 'singal': 'signal3'}
```

实例 07-53：查询并显示 PostgreSQL 数据库中的数据

下面的实例文件 simple.py 是一个 Tornado Web 程序，演示了使用模块 queries 查询并显示 PostgreSQL 数据库中的数据的过程。

源码路径：daima\7\07-53\simple.py

```
class ExampleHandler(web.RequestHandler):
    queries.Session('postgresql://postgres:66688888@127.0.0.1:5432/mydb')
    SQL = 'SELECT * FROM public.member'

    @gen.coroutine
    def get(self):
        try:
            result = yield self.application.session.query(self.SQL)
        except queries.OperationalError as error:
            logging.error('Error connecting to the database: %s', error)
            raise web.HTTPError(503)

        rows = []
        for row in result.items():
            row = dict([(k, v.isoformat()
                         if isinstance(v, datetime.datetime) else v)
                        for k, v in row.items()])
            rows.append(row)
        result.free()
        self.finish({'pg_stat_activity': rows})

class ReportHandler(web.RequestHandler):
```

```
        @gen.coroutine
        def get(self):
            self.finish(pool.PoolManager.report())


if __name__ == '__main__':
    logging.basicConfig(level=logging.DEBUG)
    application = web.Application([
        (r'/', ExampleHandler),
        (r'/report', ReportHandler),
    ], debug=True)
    application.session = queries.TornadoSession()
    application.listen(8000)
    ioloop.IOLoop.instance().start()
```

7.6 连接 SQL Server 数据库

在开发 Python 程序的过程中，可以使用模块 pymssql 实现与 SQL Server 数据库的连接和操作功能。开发者可以通过如下命令安装 pymssql。

```
pip install pymssql
```

实例 07-54：连接并操作 SQL Server 数据库

下面的实例演示了使用模块 pymssql 连接并操作 SQL Server 数据库数据的过程。

(1) 编写 SQL 文件 23.sql，然后在 SQL Server 数据库中打开 SQL 文件，单击执行按钮后会生成一个名为"tset"的数据库，并在里面添加数据。

(2) 编写程序文件 mssql01.py，使用模块 pymssql 连接并操作 SQL Server 数据库数据，具体实现代码如下。

源码路径：daima\7\07-54\mssql01.py

```python
import pymssql
# 数据库连接
conn=pymssql.connect(host='DESKTOP-VMVTB06',user='sa',password='guanxijing',database='test')
# 打开游标
cur=conn.cursor();
if not cur:
    raise Exception('数据库连接失败！')
sSQL = 'SELECT * FROM TB'
# 执行SQL语句，获取所有数据
cur.execute(sSQL)
result=cur.fetchall()
# result是列表，而其中的每个元素是元组
print(type(result),type(result[0]))
print('\n\n总行数：'+ str(cur.rowcount))
# 通过enumerate返回行号
for i,(id,name,v) in enumerate(result):
    print('第 '+str(i+1)+' 行记录->>> '+ str(id) +':'+ name+ ':' + str(v) )
# 更新数据
cur.execute("insert into tb(id,name,score) values(9,'历史',75)")
cur.execute("update tb set score=95 where id=7")
conn.commit() # 更新数据后提交事务
# 再查一次
cur.execute(sSQL)
# 一次取一条数据，cur.rowcount为-1
r=cur.fetchone()
i=1
print('\n')
while r:
    id,name,v =r #r是一个元组
    print('第 '+str(i)+' 行记录->>> '+ str(id) +':'+ name+ ':' + str(v) )
    r=cur.fetchone()
    i+= 1
conn.close()
```

上述代码执行后会建立和指定 SQL Server 数据库的连接，并分别实现查询、更新和统计表

TB 中数据的功能，最后输出整个操作过程如下。

```
<class 'list'> <class 'tuple'>

总行数：8
第 1 行记录->>> 1:语文:100.00
第 2 行记录->>> 2:数学:80.00
第 3 行记录->>> 3:英语:900.00
第 4 行记录->>> 4:政治:65.00
第 5 行记录->>> 5:物理:65.00
第 6 行记录->>> 6:化学:85.00
第 7 行记录->>> 7:生物:55.00
第 8 行记录->>> 8:地理:100.00

第 1 行记录->>> 1:语文:100.00
第 2 行记录->>> 2:数学:80.00
第 3 行记录->>> 3:英语:900.00
第 4 行记录->>> 4:政治:65.00
第 5 行记录->>> 5:物理:65.00
第 6 行记录->>> 6:化学:85.00
第 7 行记录->>> 7:生物:95.00
第 8 行记录->>> 8:地理:100.00
第 9 行记录->>> 9:历史:75.00
```

实例 07-55：创建 SQL Server 表并查询其数据

下面的实例文件 mssql02.py 演示了使用模块 pymssq 创建 SQL Server 表并查询其数据的过程。在编写 Python 程序文件之前，需要先在 SQL Server 数据库中创建一个名为"dyt"的数据库。程序文件 mssql02.py 的具体实现代码如下。

源码路径：daima\7\07-55\mssql02.py

```python
import pymssql

conn = pymssql.connect('DESKTOP-VMVTB06', 'sa', 'guanxijing', 'dyt')
cursor = conn.cursor()
#新建、插入操作
cursor.execute("""
IF OBJECT_ID('persons', 'U') IS NOT NULL
    DROP TABLE persons
CREATE TABLE persons (
    id INT NOT NULL,
    name VARCHAR(100),
    salesrep VARCHAR(100),
    PRIMARY KEY(id)
)
""")
cursor.executemany(
    "INSERT INTO persons VALUES (%d, %s, %s)",
    [(1, 'John Smith', 'John Doe'),
     (2, 'Jane Doe', 'Joe Dog'),
     (3, 'Mike T.', 'Sarah H.')])
#如果没有指定autocommit属性为True的话就需要调用commit()方法
conn.commit()

#查询操作
cursor.execute('SELECT * FROM persons WHERE salesrep=%s', 'John Doe')
row = cursor.fetchone()
while row:
    print("ID=%d, Name=%s" % (row[0], row[1]))
    row = cursor.fetchone()
#也可以使用for循环来迭代查询结果
for row in cursor:print("ID=%d, Name=%s" % (row[0], row[1]))
#关闭连接
conn.close()
```

执行后会输出：

```
D=1, Name=John Smith
ID=1, Name=John Smith
ID=1, Name=John Smith
```

```
ID=1, Name=John Smith
ID=1, Name=John Smith
ID=1, Name=John Smith
ID=1, Name=John Smith
ID=1, Name=John Smith
ID=1, Name=John Smith
ID=1, Name=John Smith
```

7.7 使用 Redis 存储

Redis 是一个 "key-value"（键-值）类型的存储系统，为开发者提供了丰富的数据结构，包括 lists、sets、ordered sets 和 hashes，还包括对这些数据结构的丰富操作。

实例 07-56：使用 Redis 连接服务器

Redis 提供了两个类 Redis 和 StrictRedis 用于实现 Redis 的命令。其中，StrictRedis 用于实现大部分官方的命令，并使用官方的语法和命令，Redis 是 StrictRedis 的子类。例如，下面的实例文件 r01.py 演示了使用 Redis 连接服务器的过程。

源码路径：daima\7\07-56\r01.py

```python
import redis

r = redis.Redis(host='127.0.0.1', port=6379, db=0)
r.set('name', 'zhangsan')    # 添加
print(r.get('name'))    # 获取
```

执行后会输出：

```
b'zhangsan'
```

需要注意，在执行上述实例代码之前，必须确保已经启动了 Redis。启动命令如下。

```
redis-server redis.windows.conf
```

另外，官方还提供了针对 Windows 操作系统的开源版本，读者下载后用 Visual Studio 执行即可。成功启动 Redis 后的界面效果如图 7-28 所示。

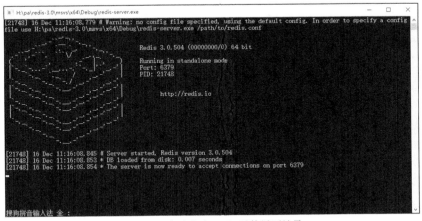

图 7-28　成功启动 Redis 后的界面效果

实例 07-57：使用 ConnectionPool 创建连接池

Redis 使用 ConnectionPool 来管理 Redis 服务器的所有连接，减少每次建立、释放连接的开销。在默认情况下，每个 Redis 实例都会维护自己的连接池。我们可以直接建立一个连接池，然后将其作为参数 Redis，这样就可以实现多个 Redis 实例共享一个连接池。例如，下面的实例文件 r02.py 演示了使用 ConnectionPool 创建连接池的过程。

源码路径：daima\7\07-57\r02.py

```python
import redis

pool = redis.ConnectionPool(host='127.0.0.1', port=6379)
r = redis.Redis(connection_pool=pool)
r.set('name', 'zhangsan')    #添加
print(r.get('name'))         #获取
```

执行后会输出：

b'zhangsan'

实例07-58：实现"发布-订阅"模式

Redis 也可以实现"发布-订阅"模式，定义一个 RedisHelper 类，连接 Redis，定义频道为 monitor，定义发布（publish）方法及订阅（subscribe）方法。具体实现代码如下。

源码路径：daima\7\07-58\RedisHelper.py

```python
import redis

class RedisHelper(object):
    def __init__(self):
        self.__conn = redis.Redis(host='127.0.0.1', port=6379)  #连接Redis
        self.channel = 'monitor'    #定义名称

    def publish(self, msg):         #定义发布方法
        self.__conn.publish(self.channel, msg)
        return True

    def subscribe(self):            #定义订阅方法
        pub = self.__conn.pubsub()
        pub.subscribe(self.channel)
        pub.parse_response()
        return pub
```

下面是发布者的实现代码。

源码路径：daima\7\07-58\r03.py

```python
# 发布
from RedisHelper import RedisHelper

obj = RedisHelper()
obj.publish('hello')   # 调用发布方法
```

下面是订阅者的实现代码。

源码路径：daima\7\07-58\r04.py

```python
#订阅
from RedisHelper import RedisHelper

obj = RedisHelper()
redis_sub = obj.subscribe()    #调用订阅方法

while True:
    msg = redis_sub.parse_response()
    print(msg)
```

实例07-59：在 Redis 中使用 delete 命令和 exists 命令

在 Redis 应用中，delete 命令用于删除已存在的键，不存在的键会被忽略。使用 exists 命令可以判断键是否存在。例如，下面的实例文件 r05.py 演示了使用 delete 命令和 exists 命令的过程。

源码路径：daima\7\07-59\r05.py

```python
import redis
#这个Redis连接不能用，请根据自己的需要修改
r =redis.Redis(host="127.0.0.1",port=6379)
print(r.set('1', '4028b2883d3f5a8b013d57228d760a93'))
print(r.get('1'))
print(r.delete('1'))
print(r.get('1'))
```

```
#设定键为2的值是4028b2883d3f5a8b013d57228d760a93
r.set('2', '4028b2883d3f5a8b013d57228d760a93')
# 存在就返回True，不存在就返回False
print(r.exists('2'))      #返回True
print(r.exists('33'))     #返回False
```

执行后会输出：

```
True
b'4028b2883d3f5a8b013d57228d760a93'
1
None
True
False
```

实例 07-60：使用 expire 命令和 expireat 命令

在 Redis 应用中，expire 命令用于设置键的过期时间，键过期后将不再可用。expireat 命令用于以 UNIX 时间戳格式设置键的过期时间，键过期后将不再可用。需要注意，时间精确到秒，时间戳为数字。例如，下面的实例文件 r06.py 演示了使用 expire 命令和 expireat 命令的过程。

源码路径：daima\7\07-60\r06.py

```
import redis
# 这个Redis连接不能用，请根据自己的需要修改
r = redis.Redis(host="127.0.0.1", port=6379)
r.set('2', '4028b2883d3f5a8b013d57228d760a93')
#成功就返回True。失败就返回False，下面的20表示20s
print(r.expire('2',20))
#如果键没有过期，则我们能得到键2的值，否则是None
print(r.get('2'))
r.set('2', '4028b2883d3f5a8b013d57228d760a93')
#成功就返回True，失败就返回False，下面的1598033936表示在2020-08-22 02:18:56时键2过期
print(r.expireat('2',1598033936))
print(r.get('2'))
```

执行后会输出：

```
True
b'4028b2883d3f5a8b013d57228d760a93'
True
b'4028b2883d3f5a8b013d57228d760a93'
```

实例 07-61：使用 persist 命令、keys 命令和 move 命令

在 Redis 应用中，persist 命令用于移除给定键的过期时间，使得键永不过期。keys 命令用于查找所有符合给定模式（pattern）的键。move 命令用于将当前数据库的键移动到给定的数据库中，select 可以设定当前的数据库。例如，下面的实例文件 r07.py 演示了使用 persist、keys 命令和 move 命令的过程。

源码路径：daima\7\07-61\r07.py

```
import redis
#这个Redis连接不能用，请根据自己的需要修改
r = redis.Redis(host="127.0.0.1", port=6379)
print (r.set('111', '11'))
print (r.set('122', '12'))
print (r.set('113', '13'))
print (r.keys(pattern='11*'))
#输出的结果是 ['111', '113']，因为键122不和 11* 匹配
r.move(2,1)#将键2移动到数据库1中去
#设定键1的值为11
print (r.set('1', '11'))
#设定键1的过期时间为100s
print (r.expire(1,100))
#查看键1的过期时间还剩下多少
print (r.ttl('1'))
#目的是13s后移除键1的过期时间
import time
time.sleep(3)
#查看键1的过期时间还剩下多少
```

```
print (r.ttl('1'))
#移除键1的过期时间
r.persist(1)
#查看键1的过期时间还剩下多少，输出的结果是None，可以通过redis desktop manager查看键1的过期时间
print (r.ttl('1'))
```

执行后会输出：

```
True
True
True
[b'111', b'113']
True
True
100
97
None
```

在 Redis 应用中有很多重要的命令，具体说明如表 7-1 所示。

表 7-1　　　　　　　　　　　　Redis 的重要命令

命　　令	描　　述
del 命令	该命令用于在键存在时删除键
dump 命令	序列化给定键，并返回被序列化的值
exists 命令	检查给定键是否存在
expire 命令	为给定键设置过期时间
expireat 命令	作用和 expire 命令类似，都用于为键设置过期时间，不同之处在于 expireat 命令接受的时间参数是 UNIX 时间戳
pexpireat 命令	设置键的过期时间，以毫秒计
pexpireat 命令	设置键过期时间的 UNIX 时间戳，以毫秒计
keys 命令	查找所有符合给定模式的键
move 命令	将当前数据库的键移动到给定的数据库当中
persist 命令	移除键的过期时间，键将持久保持
pttl 命令	以毫秒为单位返回键的剩余过期时间
ttl 命令	以秒为单位，返回给定键的剩余生存时间（Time To Live，TTL）
randomkey 命令	从当前数据库中随机返回一个键
rename 命令	修改键的名称
renamenx 命令	仅当 newkey 键不存在时，将键改名为 newkey
type 命令	返回键所储存的值的类型

第 8 章

特殊文本格式处理实战

在开发 Python 应用程序的过程中,开发人员经常需要处理一些数据并将其保存成不同的文本格式,如 Office 对应的文本格式、PDF 和 CSV 等文本格式。本章将通过具体实例的实现过程,详细讲解将数据处理成特殊文本格式的方法。

8.1　Tablib 模块实战演练

在 Python 程序中，可以使用模块 Tablib 将数据导出为各种不同的格式，包括 Excel、JSON、HTML、YAML、CSV 和 TSV 等格式。

实例 08-01：操作数据集中的指定行和列

在使用 Tablib 之前需要先安装 Tablib，安装命令如下。

```
pip install tablib
```

下面的实例文件 Tablib01.py 演示了使用 Tablib 模块操作数据集中的指定行和列的过程。

源码路径：daima\8\08-01\Tablib01.py

```python
import tablib
names = ['Kenneth Reitz', 'Bessie Monke']
data = tablib.Dataset()
for name in names:
    fname, lname = name.split()
    data.append([fname, lname])

data.headers = ['First Name', 'Last Name']
data.append_col([22, 20], header='Age')
#显示某条数据信息
print(data[0])
#显示某列的值
print(data['First Name'])
#使用索引访问列
print(data.headers)
print(data.get_col(1))
#计算平均年龄
ages = data['Age']
print(float(sum(ages)) / len(ages))
```

执行后会输出：

```
('Kenneth', 'Reitz', 22)
['Kenneth', 'Bessie']
['First Name', 'Last Name', 'Age']
['Reitz', 'Monke']
21.0
```

实例 08-02：删除指定数据并导出不同文本格式的数据

下面的实例文件 Tablib02.py 演示了使用 Tablib 模块删除数据集中指定数据，并将数据导出为不同文本格式的过程。

源码路径：daima\8\08-02\Tablib02.py

```python
import tablib
headers = ('area', 'user', 'recharge')
data = [
    ('1', 'Rooney', 20),
    ('2', 'John', 30),
]
data = tablib.Dataset(*data, headers=headers)

#可以通过下面的方式得到各种格式的数据
print(data.csv)
print(data.html)
print(data.xls)
print(data.ods)
print(data.json)
print(data.yaml)
print(data.tsv)

#增加行
data.append(['3', 'Keven',18])
#增加列
data.append_col([22, 20,13], header='Age')
```

```
print(data.csv)

#删除行
del data[1:3]
#删除列
del data['Age']
print(data.csv)
```

执行后会输出：

```
area,user,recharge
1,Rooney,20
2,John,30

<table>
<thead>
<tr><th>area</th>
<th>user</th>
<th>recharge</th></tr>
</thead>
<tr><td>1</td>
<td>Rooney</td>
<td>20</td></tr>
<tr><td>2</td>
<td>John</td>
<td>30</td></tr>
</table>

#省略其他文本格式

[{"area": "1", "user": "Rooney", "recharge": 20}, {"area": "2", "user": "John", "recharge": 30}]
- {area: '1', recharge: 20, user: Rooney}
- {area: '2', recharge: 30, user: John}

area   user      recharge
1      Rooney    20
2      John      30

area,user,recharge,Age
1,Rooney,20,22
2,John,30,20
3,Keven,18,13

area,user,recharge
1,Rooney,20
```

实例08-03：将 Tablib 数据集导出到新建 Excel 文件

下面的实例文件 Tablib03.py 演示了将 Tablib 数据集导出到新建 Excel 文件的过程。

源码路径：daima\8\08-03\Tablib03.py

```
import tablib
headers = ('lie1', 'lie2', 'lie3', 'lie4', 'lie5')
mylist = [('23','23','34','23','34'),('sadf','23','sdf','23','fsad')]
mylist = tablib.Dataset(*mylist, headers=headers)
with open('excel.xls', 'wb') as f:
    f.write(mylist.xls)
```

上述代码执行后会新建一个 Excel 文件 excel.xls，里面填充的是数据集中的数据，如图 8-1 所示。

	A	B	C	D	E
1	lie1	lie2	lie3	lie4	lie5
2	23	23	34	23	34
3	sadf	23	sdf	23	fsad

图 8-1 新建一个 Excel 文件

实例 08-04：将多个 Tablib 数据集导出到 Excel 文件

在现实应用中，有时需要在表格中处理多个数据集。例如，将多个数据集数据导出到一个 Excel 文件中，这时候可以使用 Tablib 模块中的 Databook 实现。例如，下面的实例文件 Tablib04.py 不仅演示了增加、删除数据集数据的方法，而且演示了将多个 Tablib 数据集导出到 Excel 文件的过程。

源码路径：daima\8\08-04\Tablib04.py

```python
import tablib
import os

#创建数据集，方法1
dataset1 = tablib.Dataset()
header1 = ('ID', 'Name', 'Tel', 'Age')
dataset1.headers = header1
dataset1.append([1, 'zhangsan', 13711111111, 16])
dataset1.append([2, 'lisi',     13811111111, 18])
dataset1.append([3, 'wangwu',   13911111111, 20])
dataset1.append([4, 'zhaoliu',  15811111111, 25])
print('dataset1:', os.linesep, dataset1, os.linesep)

#创建数据集，方法2
header2 = ('ID', 'Name', 'Tel', 'Age')
data2 = [
    [1, 'zhangsan', 13711111111, 16],
    [2, 'lisi',     13811111111, 18],
    [3, 'wangwu',   13911111111, 20],
    [4, 'zhaoliu',  15811111111, 25]
]
dataset2 = tablib.Dataset(*data2, headers = header2)
print('dataset2: ', os.linesep, dataset2, os.linesep)

#增加行
dataset1.append([5, 'sunqi', 15911111111, 30])          #添加到最后一行的下面
dataset1.insert(0, [0, 'liuyi', 18211111111, 35])       #在指定位置添加行
print('增加行后的dataset1: ', os.linesep, dataset1, os.linesep)

#删除行
dataset1.pop()                                          #删除最后一行
dataset1.lpop()                                         #删除第1行
del dataset1[0:2]                                       #删除第[0,2]行数据
print('删除行后的dataset1:', os.linesep, dataset1, os.linesep)

#增加列
#现在dataset1就剩两行数据了
dataset1.append_col(('beijing', 'shenzhen'), header='city')     #增加列到最后一列
dataset1.insert_col(2, ('male', 'female'), header='sex')        #在指定位置添加列
print('增加列后的dataset1: ', os.linesep, dataset1, os.linesep)

#删除列
del dataset1['Tel']
print('删除列后的dataset1: ', os.linesep, dataset1, os.linesep)

#获取各种格式的数据
print('yaml format: ', os.linesep ,dataset1.yaml, os.linesep)
print('csv format: ', os.linesep ,dataset1.csv , os.linesep)
print('tsv format: ', os.linesep ,dataset1.tsv , os.linesep)

#导出到Excel文件中
dataset1.title = 'dataset1'                             #设置Excel文件中表单的名称
dataset2.title = 'dataset2'
myfile = open('mydata.xls', 'wb')
myfile.write(dataset1.xls)
myfile.close()

#如果有多个表单，则使用Databook就可以了
myDataBook = tablib.Databook((dataset1, dataset2))
myfile = open(myfile.name, 'wb')
```

```
myfile.write(myDataBook.xls)
myfile.close()
```

执行后会输出：

```
dataset1:
 ID|Name    |Tel        |Age
--|--------|-----------|---
1 |zhangsan|13711111111|16
2 |lisi    |13811111111|18
3 |wangwu  |13911111111|20
4 |zhaoliu |15811111111|25

dataset2:
 ID|Name    |Tel        |Age
--|--------|-----------|---
1 |zhangsan|13711111111|16
2 |lisi    |13811111111|18
3 |wangwu  |13911111111|20
4 |zhaoliu |15811111111|25

增加行后的dataset1:
 ID|Name    |Tel        |Age
--|--------|-----------|---
0 |liuyi   |18211111111|35
1 |zhangsan|13711111111|16
2 |lisi    |13811111111|18
3 |wangwu  |13911111111|20
4 |zhaoliu |15811111111|25
5 |sunqi   |15911111111|30

删除行后的dataset1:
 ID|Name   |Tel        |Age
--|-------|-----------|---
3 |wangwu |13911111111|20
4 |zhaoliu|15811111111|25

增加列后的dataset1:
 ID|Name   |sex   |Tel        |Age|city
--|-------|------|-----------|---|--------
3 |wangwu |male  |13911111111|20 |beijing
4 |zhaoliu|female|15811111111|25 |shenzhen

删除列后的dataset1:
 ID|Name   |sex   |Age|city
--|-------|------|---|--------
3 |wangwu |male  |20 |beijing
4 |zhaoliu|female|25 |shenzhen

yaml format:
- {Age: 20, ID: 3, Name: wangwu, city: beijing, sex: male}
- {Age: 25, ID: 4, Name: zhaoliu, city: shenzhen, sex: female}

csv format:
ID,Name,sex,Age,city
3,wangwu,male,20,beijing
4,zhaoliu,female,25,shenzhen

tsv format:
ID   Name    sex     Age   city
3    wangwu  male    20    beijing
4    zhaoliu female  25    shenzhen
```

上述代码执行后会创建 Excel 文件 mydata.xls，里面保存了从数据集中导出的数据，如图 8-2 所示。

图 8-2　从数据集中导出的数据

实例 08-05：使用标签过滤 Tablib 数据集

在使用 Tablib 数据集时，可以添加一个标签到指定的行作为参数。这样在后面的程序中，可以通过这个标签筛选数据集基于任意条件的数据。例如，下面的实例文件 Tablib05.py 演示了使用标签过滤 Tablib 数据集的过程。

源码路径：daima\8\08-05\Tablib05.py

```
import tablib
students = tablib.Dataset()
students.headers = ['first', 'last']
students.rpush(['Kenneth', 'Reitz'], tags=['male', 'technical'])
students.rpush(['Bessie', 'Monke'], tags=['female', 'creative'])
print(students.filter(['male']).yaml)
```

执行后会输出：

```
- {first: Kenneth, last: Reitz}
```

实例 08-06：将两组数据分离导入 Excel 文件

在将 Tablib 数据导出到某个格式的文件中时，有时需要将多种数据集对象进行分类。例如，下面的实例文件 Tablib06.py 演示了将两组数据分类导入 Excel 文件的过程。

源码路径：daima\8\08-06\Tablib06.py

```
import tablib
daniel_tests = [
    ('11/24/09', 'Math 101 Mid-term Exam', 56.),
    ('05/24/10', 'Math 101 Final Exam', 62.)
]

suzie_tests = [
    ('11/24/09', 'Math 101 Mid-term Exam', 56.),
    ('05/24/10', 'Math 101 Final Exam', 62.)
]
tests = tablib.Dataset()
tests.headers = ['Date', 'Test Name', 'Grade']

# Daniel数据测试
tests.append_separator('Daniel的得分')

for test_row in daniel_tests:
    tests.append(test_row)

# Susie数据测试
tests.append_separator('Susie的得分')

for test_row in suzie_tests:
    tests.append(test_row)

# 写入Eccel文件
with open('grades.xls', 'wb') as f:
    f.write(tests.export('xls'))
```

上述代码执行后会将 Tablib 数据分类导入 Excel 文件中，如图 8-3 所示。

	A	B	C
1	Date	Test Name	Grade
2	Daniel的得分		
3	11/24/09	Math 101 Mid-term Exam	56
4	05/24/10	Math 101 Mid-term Exam	62
5	Susie的得分		
6	11/24/09	Math 101 Mid-term Exam	56
7	05/24/10	Math 101 Mid-term Exam	62

图 8-3 分类的数据

8.2 Office 处理实战

在 Python 程序中，可以使用专用模块将数据转换成 Office（Word、Excel 和 PowerPoint）文件对应的格式。

实例 08-07：使用 openpyxl 读取 Excel 文件

使用模块 openpyxl 可以读写 Excel 文件，包括.xlsx、.xlsm、.xltx 和.xltm 文件。在使用 openpyxl 之前需要先安装它，安装命令如下。

```
pip install openpyxl
```

在 openpyxl 中主要用到如下 3 个概念。

- Workbook：代表一个 Excel 工作表。
- Worksheet：代表工作表中的一张表单。
- Cell：代表最简单的一个单元格。

下面的实例文件 office01.py 演示了使用 openpyxl 读取 Excel 文件的过程。

源码路径：daima\8\08-07\office01.py

```python
from openpyxl import load_workbook
wb = load_workbook("template.xlsx")        #打开一个.xlsx文件
print(wb.sheetnames)
sheet = wb.get_sheet_by_name("Sheet3")     #看看打开的Excel文件里面有哪些表单
#下面读取到指定的Sheet页
print(sheet["C"])
print(sheet["4"])
print(sheet["C4"].value)                   # c4，即第C4格的值
print(sheet.max_row)                       # 10，即最大行数
print(sheet.max_column)                    # 5，即最大列数
for i in sheet["C"]:
    print(i.value, end=" ")                # c1 c2 c3 c4 c5 c6 c7 c8 c9 c10，即C列中的所有值
```

执行后会输出：

```
['Sheet1', 'Sheet2', 'Sheet3']
    sheet = wb.get_sheet_by_name("Sheet3")
(<Cell 'Sheet3'.C1>, <Cell 'Sheet3'.C2>, <Cell 'Sheet3'.C3>, <Cell 'Sheet3'.C4>, <Cell 'Sheet3'.C5>, <Cell 'Sheet3'.C6>, <Cell 'Sheet3'.C7>, <Cell 'Sheet3'.C8>, <Cell 'Sheet3'.C9>, <Cell 'Sheet3'.C10>)
(<Cell 'Sheet3'.A4>, <Cell 'Sheet3'.B4>, <Cell 'Sheet3'.C4>, <Cell 'Sheet3'.D4>, <Cell 'Sheet3'.E4>)
c4
10
5
c1 c2 c3 c4 c5 c6 c7 c8 c9 c10
```

实例 08-08：将 4 组数据导入 Excel 文件中

下面的实例文件 office02.py 演示了将 4 组数据导入 Excel 文件中的过程。

源码路径：daima\8\08-08\office02.py

```python
import openpyxl
import time

ls = [['马坡','接入交换','192.168.1.1','G0/3','AAAA-AAAA-AAAA'],
    ['马坡','接入交换','192.168.1.2','G0/8','BBBB-BBBB-BBBB'],
    ['马坡','接入交换','192.168.1.2','G0/8','CCCC-CCCC-CCCC'],
    ['马坡','接入交换','192.168.1.2','G0/8','DDDD-DDDD-DDDD']]

#定义数据

time_format = '%Y-%m-%d__%H:%M:%S'
time_current = time.strftime(time_format)
#定义时间格式
```

```
def savetoexcel(data,sheetname,wbname):
    print("写入Excel文件：")
    wb=openpyxl.load_workbook(filename=wbname)
    #打开Excel文件

    sheet=wb.active                    #关联Excel文件中活动的表单（这里关联的是Sheet1）
    max_row = sheet.max_row            #获取Sheet1中当前数据最大的行数
    row = max_row + 3                  #将新数据写入最大行数+3行的位置
    data_len=row+len(data)             #计算当前数据长度

    for data_row in range(row,data_len):       #写入数据
    #轮询每一行进行数据写入
        for data_col1 in range(2,7):
        #针对每一行还要进行for循环来写入列的数据
            _ =sheet.cell(row=data_row, column=1, value=str(time_current))
            #每行第1列写入时间
            _ =sheet.cell(row=data_row,column=data_col1,value=str(data[data_row-data_len][data_col1-2]))
            #从第2列开始写入数据

    wb.save(filename=wbname)           #保存数据
    print("保存成功")

savetoexcel(ls,"Sheet1","template.xlsx")
```

上述代码执行后会在指定文件 template.xlsx 中导入 4 组数据，如图 8-4 所示。

	A	B	C	D	E	F	G
1	a1	b1	c1	d1	e1		
2	a2	b2	c2	d2	e2		
3	a3	b3	c3	d3	e3		
4	a4	b4	c4	d4	e4		
5	a5	b5	c5	d5	e5		
6	a6	b6	c6	d6	e6		
7	a7	b7	c7	d7	e7		
8	a8	b8	c8	d8	e8		
9	a9	b9	c9	d9	e9		
10	a10	b10	c10	d10	e10		
11							
12							
13	2018-04-04	马坡	接入交换	192.168.1.	G0/3	AAAA-AAAA-AAAA	
14	2018-04-04	马坡	接入交换	192.168.1.	G0/8	BBBB-BBBB-BBBB	
15	2018-04-04	马坡	接入交换	192.168.1.	G0/8	CCCC-CCCC-CCCC	
16	2018-04-04	马坡	接入交换	192.168.1.	G0/8	DDDD-DDDD-DDDD	

图 8-4 导入的 4 组数据

实例 08-09：在 Excel 文件中检索某关键字

下面的实例文件 office03.py 演示了在 Excel 文件中检索某关键字的过程。

源码路径：daima\8\08-09\office03.py

```
import openpyxl

wb=openpyxl.load_workbook("template.xlsx")
the_list =[]

while True:
    info = input('请输入关键字查找：').upper().strip()
    if len(info) == 0:                    #输入的关键字不能为空，否则继续循环
        continue
    count = 0
    for line1 in wb['Sheet3'].values:     #轮询列表
        if None not in line1:
        #Excel文件中空行的数据表示None，当这里匹配None时就不会再进行for循环，所以需要匹配非None的数据才能
        #进行下面的for循环
            for line2 in line1:           #由于列表中还存在元组，所以需要将元组的内容也轮询一遍
                if info in line2:
```

```
                count += 1       #统计关键字被匹配了多少次
                print(line1)     #匹配关键字后输出元组信息

    else:
        print('匹配"%s"的数量统计：%s个条目被匹配' % (info, count))    #输出查找的关键字被匹配了多少次
```

上述代码执行后可以通过输入关键字的方式快速查询 Excel 文件中的数据，例如下面的检索过程。

```
请输入关键字查找：马坡
('2018-04-04__16:28:45', '马坡', '接入交换', '192.168.1.1', 'G0/3', 'AAAA-AAAA-AAAA')
('2018-04-04__16:28:45', '马坡', '接入交换', '192.168.1.2', 'G0/8', 'BBBB-BBBB-BBBB')
('2018-04-04__16:28:45', '马坡', '接入交换', '192.168.1.2', 'G0/8', 'CCCC-CCCC-CCCC')
('2018-04-04__16:28:45', '马坡', '接入交换', '192.168.1.2', 'G0/8', 'DDDD-DDDD-DDDD')
匹配"马坡"的数量统计：4个条目被匹配
请输入关键字查找：192.168.1.1
('2018-04-04__16:28:45', '马坡', '接入交换', '192.168.1.1', 'G0/3', 'AAAA-AAAA-AAAA')
匹配"192.168.1.1"的数量统计：1个条目被匹配
```

实例 08-10：将数据导入 Excel 文件并生成图表

下面的实例文件 office04.py 演示了将数据导入 Excel 文件中，并根据导入的数据在 Excel 文件中生成图表的过程。

源码路径：daima\8\08-10\office04.py

```python
from openpyxl import Workbook
from openpyxl.chart import (
    AreaChart,
    Reference,
    Series,
)

wb = Workbook()
ws = wb.active

rows = [
    ['Number', 'Batch 1', 'Batch 2'],
    [2, 40, 30],
    [3, 40, 25],
    [4, 50, 30],
    [5, 30, 10],
    [6, 25, 5],
    [7, 50, 10],
]

for row in rows:
    ws.append(row)

chart = AreaChart()
chart.title = "Area Chart"
chart.style = 13
chart.x_axis.title = 'Test'
chart.y_axis.title = 'Percentage'

cats = Reference(ws, min_col=1, min_row=1, max_row=7)
data = Reference(ws, min_col=2, min_row=1, max_col=3, max_row=7)
chart.add_data(data, titles_from_data=True)
chart.set_categories(cats)

ws.add_chart(chart, "A10")

wb.save("area.xlsx")
```

上述代码执行后会将 rows 中的数据导入文件 area.xlsx 中，并在文件 area.xlsx 中根据数据生成图表，如图 8-5 所示。

第 8 章 特殊文本格式处理实战

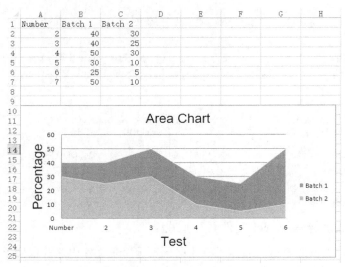

图 8-5 导入数据并生成图表

实例 08-11：使用 pyexcel 读取并写入 CSV 文件

使用模块 pyexcel 可以操作 Excel 文件和 CSV 文件，在使用 pyexcel 之前需要先安装 pyexcel_xls，安装命令如下。

```
pip install pyexcel_xls
```

然后通过如下命令安装 pyexcel。

```
pip install pyexcel
```

例如，下面的实例文件 office05.py 演示了使用 pyexcel 读取并写入 CSV 文件的过程。

源码路径：daima\8\08-11\office05.py

```python
import pyexcel as p
sheet = p.get_sheet(file_name="example.csv")
print(sheet)

with open('tab_example.csv', 'w') as f:
    unused = f.write('I\tam\ttab\tseparated\tcsv\n')
    unused = f.write('You\tneed\tdelimiter\tparameter\n')
sheet = p.get_sheet(file_name="tab_example.csv", delimiter='\t')
print(sheet)
```

在上述代码中，执行顺序是，首先读取了文件 example.csv 中的内容，然后将指定的数据写入新建 CSV 文件 tab_example.csv 中。执行效果如图 8-6 所示。

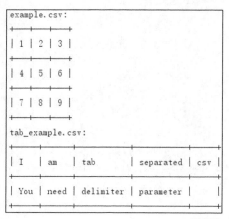

图 8-6 执行效果

实例 08-12：使用 pyexcel 读取 Excel 文件中的每个单元格内容

下面的实例文件 read_cell_by_cell.py 演示了使用 pyexcel 读取 Excel 文件中每个单元格内容的过程。

源码路径：daima\8\08-12\read_cell_by_cell.py

```
import os
import pyexcel as pe
def main(base_dir):
    #读取的文件example.xlsm
    spreadsheet = pe.get_sheet(file_name=os.path.join(base_dir, "example.csv"))

    #遍历每一行
    for r in spreadsheet.row_range():
        #遍历每一列
        for c in spreadsheet.column_range():
            print(spreadsheet.cell_value(r, c))

if __name__ == '__main__':
    main(os.getcwd())
```

执行效果如图 8-7 所示。

(a) Excel 文件中的内容　　　　(b) 输出结果

图 8-7　执行效果

实例 08-13：按列读取并显示 Excel 文件中的每个单元格内容

下面的实例文件 read_column_by_column.py 演示了使用 pyexcel 按列读取并显示 Excel 文件中的每个单元格内容的过程。

源码路径：daima\8\08-13\read_column_by_column.py

```
def main(base_dir):
    spreadsheet = pe.get_sheet(file_name=os.path.join(base_dir, "example.xlsx"))
    for value in spreadsheet.columns():
        print(value)

if __name__ == '__main__':
    main(os.getcwd())
```

执行效果如图 8-8 所示。

(a) Excel 文件中的内容　　　　(b) 读取后的输出结果

图 8-8　执行效果

实例 08-14：读取并显示 Excel 文件中的所有数据

一个 Excel 文件中可能有多个表单，例如，文件 multiple-sheets-example.xls 中有 3 个表单，其数据如图 8-9 所示。通过下面的实例文件 read_excel_book.py，可以读取 multiple-sheets-example.xls 文件中的所有数据。

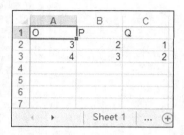 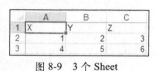 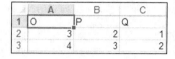

图 8-9 3 个 Sheet

源码路径：daima\8\08-14\read_excel_book.py

```python
def main(base_dir):
    book = pe.get_book(file_name=os.path.join(base_dir,"multiple-sheets-example.xls"))

    # 默认的迭代器为工作表实例
    for sheet in book:
        #每张表单都有名字
        print("sheet: %s" % sheet.name)
        #一旦拥有了一个表单实例，就可以将其视为一个读取器实例。可以按照想要的方式迭代其成员
        for row in sheet:
            print(row)

if __name__ == '__main__':
    main(os.getcwd())
```

执行后会输出：

```
sheet: Sheet 1
[1, 2, 3]
[4, 5, 6]
[7, 8, 9]
sheet: Sheet 2
['X', 'Y', 'Z']
[1, 2, 3]
[4, 5, 6]
sheet: Sheet 3
['O', 'P', 'Q']
[3, 2, 1]
[4, 3, 2]
```

实例 08-15：将 3 组数据导入新建的 Excel 文件中

通过下面的实例文件 write_excel_book.py，可以将 3 组数据导入新建的 multiple-sheets1.xls 文件中，3 组数据分别对应其中的 3 个表单。

源码路径：daima\8\08-15\write_excel_book.py

```python
def main(base_dir):
    data = {
        "Sheet 1": [[1, 2, 3], [4, 5, 6], [7, 8, 9]],
        "Sheet 2": [['X', 'Y', 'Z'], [1, 2, 3], [4, 5, 6]],
        "Sheet 3": [['O', 'P', 'Q'], [3, 2, 1], [4, 3, 2]]
    }
    pe.save_book_as(bookdict=data, dest_file_name="multiple-sheets1.xls")

if __name__ == '__main__':
    main(os.getcwd())
```

上述代码执行后会创建拥有 3 个表单的 Excel 文件，如图 8-10 所示。

图 8-10　创建拥有 3 个表单的 Excel 文件

实例 08-16：使用 pyexcel 以多种方式获取 Excel 数据

下面的实例文件 series.py 演示了使用 pyexcel 以多种方式获取 Excel 数据的过程。

源码路径：daima\8\08-16\series.py

```python
def main(base_dir):
    sheet = pe.get_sheet(file_name=os.path.join(base_dir,"example_series.xls"),
                        name_columns_by_row=0)
    print(json.dumps(sheet.to_dict()))
    #获取列标题
    print(sheet.colnames)
    #在一维数组中获取内容
    data = list(sheet.enumerate())
    print(data)
    #逆序获取一维数组中的内容
    data = list(sheet.reverse())
    print(data)

    #在一维数组中获取内容，但垂直地迭代它
    data = list(sheet.vertical())
    print(data)
    #获取一维数组中的内容，遍历垂直相反的顺序
    data = list(sheet.rvertical())
    print(data)

    #获取二维数组数据
    data = list(sheet.rows())
    print(data)

    #以相反的顺序获取二维数组
    data = list(sheet.rrows())
    print(data)

    #获取二维数组，堆栈列
    data = list(sheet.columns())
    print(data)

    #获取一个二维数组，以相反的顺序堆栈列
    data = list(sheet.rcolumns())
    print(data)

    #可以把结果写入一个文件中
    sheet.save_as("example_series.xls")

if __name__ == '__main__':
    main(os.getcwd())
```

通过上述代码以多种方式获取了 Excel 数据，包括一维数组顺序和逆序、二维数组顺序和逆序。

执行后会输出：
```
{"Column 1": [1, 2, 3], "Column 2": [4, 5, 6], "Column 3": [7, 8, 9]}
['Column 1', 'Column 2', 'Column 3']
[1, 4, 7, 2, 5, 8, 3, 6, 9]
[9, 6, 3, 8, 5, 2, 7, 4, 1]
[1, 2, 3, 4, 5, 6, 7, 8, 9]
[9, 8, 7, 6, 5, 4, 3, 2, 1]
[[1, 4, 7], [2, 5, 8], [3, 6, 9]]
[[3, 6, 9], [2, 5, 8], [1, 4, 7]]
[[1, 2, 3], [4, 5, 6], [7, 8, 9]]
[[7, 8, 9], [4, 5, 6], [1, 2, 3]]
```

实例 08-17：将数据分别导入 Excel 文件和 SQLite 数据库

下面的实例文件 import_xls_into_database_via_sqlalchemy.py 演示了使用 pyexcel 将数据分别导入 Excel 文件和 SQLite 数据库的过程。

源码路径：daima\8\08-17\import_xls_into_database_via_sqlalchemy.py

```python
engine = create_engine("sqlite:///birth.db")
Base = declarative_base()
Session = sessionmaker(bind=engine)


#目标表
class BirthRegister(Base):
    __tablename__ = 'birth'
    id = Column(Integer, primary_key=True)
    name = Column(String)
    weight = Column(Float)
    birth = Column(Date)

Base.metadata.create_all(engine)
#创建数据
data = [
    ["name", "weight", "birth"],
    ["Adam", 3.4, datetime.date(2017, 2, 3)],
    ["Smith", 4.2, datetime.date(2014, 11, 12)]
]
pyexcel.save_as(array=data,
                dest_file_name="birth.xls")

#导入Excel文件
session = Session()  # obtain a sql session
pyexcel.save_as(file_name="birth.xls",
                name_columns_by_row=0,
                dest_session=session,
                dest_table=BirthRegister)

#验证结果
sheet = pyexcel.get_sheet(session=session, table=BirthRegister)
print(sheet)
session.close()
```

执行后会输出：
```
birth:
+------------+----+-------+--------+
| birth      | id | name  | weight |
+------------+----+-------+--------+
| 2018-02-03 | 1  | Adam  | 3.4    |
+------------+----+-------+--------+
| 2014-11-12 | 2  | Smith | 4.2    |
+------------+----+-------+--------+
```

实例 08-18：使用 python-docx 创建 Word 文档

使用模块 python-docx 可以读取、查询及修改 Office 文件。其安装命令如下。

```
pip install python-docx
```

下面的实例文件 python-docx01.py 演示了使用 python-docx 创建 Word 文档的过程。

源码路径：daima\8\08-18\python-docx01.py

```
from docx import Document
document = Document()
document.add_paragraph('Hello,Word!')
document.save('demo.docx')
```

在上述代码中，第 1 行引入 Python-docx 库和 Document 类，类 Document 即代表"文档"。第 2 行创建了类 Document 的实例 document，相当于"这篇文档"。然后我们在文档中利用函数 add_paragraph() 添加了一个段落，段落的内容是"Hello,Word!"。最后，使用函数 save() 将文档保存在磁盘上。上述代码执行后会创建一个名为"demo.docx"的文件，其内容如图 8-11 所示。

图 8-11　文件 demo.docx 的内容

实例 08-19：在 Word 文档中插入 20 个实心图形

下面的实例文件 python-docx02.py 演示了使用 python-docx 向 Word 文档中插入 20 个实心圆形的过程。

源码路径：daima\8\08-19\python-docx02.py

```
from docx import Document
from PIL import Image,ImageDraw
from io import BytesIO

document = Document()                    #新建文档
p = document.add_paragraph()             #添加一个段落
r = p.add_run()                          #添加一个游程
img_size = 20
for x in range(20):
    im = Image.new("RGB", (img_size,img_size), "white")
    draw_obj = ImageDraw.Draw(im)
    draw_obj.ellipse((0,0,img_size-1,img_size-1), fill=255-x)#画圆
    fake_buf_file = BytesIO()            #用BytesIO将图片保存在内存里，减少磁盘操作
    im.save(fake_buf_file,"png")
    r.add_picture(fake_buf_file)         #在当前游程中插入图片
    fake_buf_file.close()
document.save("demo.docx")
```

执行上述代码后，会在 Word 文档 demo.docx 中添加 20 个实心圆形，颜色为红色且由浅入深。文件 demo.docx 的内容如图 8-12 所示。

第 8 章 特殊文本格式处理实战

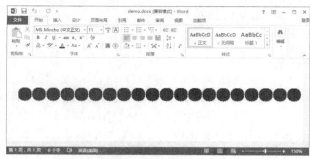

图 8-12 文件 demo.docx 的内容

实例 08-20：向 Word 文档中添加指定段落样式的内容

创建结构文档，就是创建具有不同样式的段落。在类 Document 的函数 add_paragraph()中，第 1 个参数表示段落的文字，第 2 个可选参数表示段落的样式。通过这个样式参数即可设置所添加段落的样式。如果不指定这个参数，则默认样式为"正文"。函数 add_paragraph()的返回值是一个段落对象，可以通过这个对象的 style 属性得到该段落的样式，也可以设置 Style 属性以设置该段落的样式。

下面的实例文件 python-docx03.py 演示了向 Word 文档中添加指定段落样式的内容的过程。

源码路径：daima\8\08-20\python-docx03.py

```
from docx import Document

doc = Document()
doc.add_paragraph(u'Python为什么这么受欢迎？','Title')
doc.add_paragraph(u'作者','Subtitle')
doc.add_paragraph(u'摘要：本文阐明了Python的优势...','Body Text 2')
doc.add_paragraph(u'简单','Heading 1')
doc.add_paragraph(u'易学')
doc.add_paragraph(u'易用','Heading 2')
doc.add_paragraph(u'功能强')
p = doc.add_paragraph(u'贴合大家的风格')
p.style = 'Heading 2'
doc.save('demo.docx')
```

执行上述代码后，会在 Word 文档 demo.docx 中添加一段指定段落样式的内容，打开后的文件 demo.docx 的内容如图 8-13 所示。

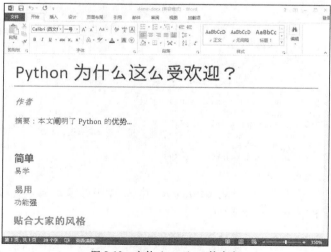

图 8-13 文件 demo.docx 的内容

在上述代码中，Title、Heading 1 等都是 Word 的内建样式。启动 Word 后，在"样式"窗格中看到的样式图标就是 Word 的内置样式，如图 8-14 所示。

图 8-14　Word 的内置样式

实例 08-21：得到英文的样式名称

对于英文版的 Word 来说，在样式图标下标注的样式名称就是在 Python 代码可以使用的样式名称。而对于中文版的 Word 来说，内建样式仍然使用英文名称。如果我们没有安装英文版的 Word，可以用下面的实例文件 python-docx04.py 得到样式的英文名称。

源码路径：daima\8\08-21\python-docx04.py

```
from docx import Document
from docx.enum.style import WD_STYLE_TYPE
doc = Document()
styles = doc.styles
print("\n".join([s.name for s in styles if s.type == WD_STYLE_TYPE.PARAGRAPH]))
```

执行后会输出：

```
Normal
Heading 1
Heading 2
Heading 3
Heading 4
Heading 5
Heading 6
Heading 7
Heading 8
Heading 9
No Spacing
Title
Subtitle
List Paragraph
Body Text
Body Text 2
Body Text 3
List
List 2
List 3
List Bullet
List Bullet 2
List Bullet 3
List Number
List Number 2
List Number 3
List Continue
List Continue 2
List Continue 3
macro
Quote
Caption
Intense Quote
TOC Heading
```

在上述样式列表中，文档结构常用的样式有以下几种。

- Title：文档的标题，样式窗格里显示为"标题"。
- Subtitle：副标题。
- Heading n：n 级标题，样式窗格里显示为"标题 n"。
- Normal：正文。

文档标题及 n 级标题样式可以使用类 Document 中的函数 add_heading() 来设置。这个函数的第 1 个参数表示文本内容。如果第 2 个参数设置为 0，则等价于 add_paragraph(text,'Title')；如果第 2 个参数设置为大于 0 的整数 *n*，则等价于 add_paragraph(text,'Heading %d' % n)。

实例 08-22：获取 Word 文档中的文本样式名称

下面的实例文件 python-docx05.py 演示了获取 Word 文档中的文本样式名称和每个样式的文字数目的过程。

源码路径：daima\8\08-22\python-docx05.py

```
from docx import Document
import sys
path = "demo.docx"
document = Document(path)
for p in document.paragraphs:
    print(len(p.text))
    print(p.style.name)
```

对于中文文本来说，len()得到的是汉字个数，这和 Python 默认的多语言处理方法是一致的。读者可以拿现成 Word 文档试试（需要.docx 格式），如果文档是使用 Word 默认的样式创建的，则程序会输出 Title、Normal、Heading *n* 之类的样式名称。如果文档对默认样式进行了修改，那么程序依然会输出原有样式名称，不受影响。如果文档创建了新样式，则使用新样式的段落会显示新样式的名称。在作者计算机中执行后会输出：

```
15
Title
2
Subtitle
20
Body Text 2
2
Heading 1
2
Normal
2
Heading 2
3
Normal
5
Heading 2
```

实例 08-23：获取 Word 文档中的文本内容

下面的实例文件 python-docx06.py 演示了获取 Word 文档中的文本内容的过程。

源码路径：daima\8\08-23\python-docx06.py

```
from docx import Document
path = "demo.docx"
document = Document(path)
for paragraph in document.paragraphs:
    print(paragraph.text)
```

执行后会输出文件 demo.docx 的内容：

```
Python为什么这么受欢迎？
作者
摘要：本文阐明了Python的优势…
简单
易学
易用
功能强
贴合大家的风格
```

实例 08-24：在 Word 文档中创建表格

Word 文档中有两种表格，具体说明如下。

❏ 和段落同级的顶级表格：我们可以使用类 Document 中函数的 add_table()创建一个新的顶级表格对象，也可以使用类 Document 中的 tables 得到文档中所有的顶级表格。

❏ 表格里嵌套的表格。

本书将主要讨论顶级表格，下面内容提及的表格均指顶级表格。

我们可以把一个表格看成 M 行（rows）N 列（cols）的矩阵。利用类 Table 中的_Cell 对象的 text 属性，可以设置、获取表格中任一单元格的文本。下面的实例文件 python-docx07.py 演示了在 Word 文档中创建表格的过程。

源码路径：daima\8\08-24\python-docx07.py

```python
from docx import Document
import psutil

#获取当前计算机配置数据
vmem = psutil.virtual_memory()
vmem_dict = vmem._asdict()

trow = 2
tcol = len(vmem_dict.keys())
#创建表格

document = Document()
table = document.add_table(rows=trow,cols=tcol,style = 'Table Grid')
for col,info in enumerate(vmem_dict.keys()):
    table.cell(0,col).text = info
    if info == 'percent':
        table.cell(1,col).text = str(vmem_dict[info])+'%'
    else:
        table.cell(1,col).text = str(vmem_dict[info]/(1024*1024)) + 'M'
document.save('table.docx')
```

在上述代码中，使用库 psutil 获取了当前计算机的内存信息，将获取到的这些信息作为表格的填充数据。首先把从 psutil 得到的内存信息转化为一个字典，它的键是项目（物理内存总数、使用数、剩余数等），值是各个项目的数值。代码执行后会创建文件 table.docx，文件的表格中显示当前计算机的内存信息，如图 8-15 所示。

total	available	percent	used	free
16290.6875M	8248.26953125M	49.4%	8042.41796875M	8248.26953125M

图 8-15 文件的表格中显示的内存信息

实例 08-25：创建表格并合并其中的单元格

我们可以利用_Cell 对象中的函数 merge(other_cell)合并单元格。合并的方式是以当前_Cell 为左上角、other_cell 为右下角进行合并。例如，下面的实例文件 python-docx08.py 执行顺序是，首先创建一个表格，然后合并其中的单元格并保存为 Word 文档，最后读取这个 Word 文档，并把每个单元格的坐标注到单元格。

源码路径：daima\8\08-25\python-docx08.py

```python
document = Document()
table = document.add_table(rows=9,cols=10,style = 'Table Grid')
cell_1 = table.cell(1,2)
cell_2 = table.cell(4,6)
cell_1.merge(cell_2)
document.save('table-1.docx')

document = Document('table-1.docx')
table = document.tables[0]
for row,obj_row in enumerate(table.rows):
    for col,cell in enumerate(obj_row.cells):
        cell.text = cell.text + "%d,%d " % (row,col)

document.save('table-2.docx')
```

上述代码执行后会生成两个 Word 文件 table-1.docx 和 table-2.docx，内容如图 8-16 所示。

(a) 文件 table-1.docx 的内容

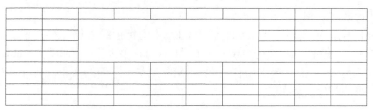

0,0	0,1	0,2	0,3	0,4	0,5	0,6	0,7	0,8	0,9
1,0	1,1	1,2 1,3 1,4 1,5 1,6 2,2 2,3 2,4 2,5 2,6 3,2 3,3					1,7	1,8	1,9
2,0	2,1	3,4 3,5 3,6 4,2 4,3 4,4 4,5 4,6					2,7	2,8	2,9
3,0	3,1						3,7	3,8	3,9
4,0	4,1						4,7	4,8	4,9
5,0	5,1	5,2	5,3	5,4	5,5	5,6	5,7	5,8	5,9
6,0	6,1	6,2	6,3	6,4	6,5	6,6	6,7	6,8	6,9
7,0	7,1	7,2	7,3	7,4	7,5	7,6	7,7	7,8	7,9
8,0	8,1	8,2	8,3	8,4	8,5	8,6	8,7	8,8	8,9

(b) 文件 table-2.docx 的内容

图 8-16 生成的两个 Word 文件的内容

由此可见，在合并单元格之后，可以利用合并区域的任何一个单元格的坐标指代这个合并区域。也就是说，单元格的合并并没有使_Cell 消失，只是这些_Cell 共享里面的内容而已。

实例 08-26：调整 Word 表格宽度

行及单元格的高度受字体限制，不能够手动调整高度，但是列和单元格的宽度可以手动调整。下面的实例文件 python-docx09.py 演示了调整 Word 表格宽度的过程。

源码路径：daima\8\08-26\python-docx09.py

```
document = Document()
for row in range(9):
    t = document.add_table(rows=1, cols=1, style='Table Grid')
    t.autofit = False    # 很重要，必须设置！
    w = float(row) / 2.0
    t.columns[0].width = Inches(w)
document.save('table-step.docx')
```

上述代码执行后会设置文件 table-step.docx 中表格的宽度，如图 8-17 所示。

图 8-17 表格的宽度

实例 08-27：获取 python-docx 内部的表格样式名称

Word 的表格设计面板中有各种样式，通过这些样式可以为表格设置不同的显示效果。在库 python-docx 中可以通过 Document.add_table 的第 3 个参数设定表格样式，也可以用 Table 的属性 style 获取和设置样式。如果要设置样式，则可以直接用样式的英文名称，如"Table Grid"；如果读取样式，那么会得到一个 Styles 对象，这个对象是可以跨文档使用的。另外，也可以使用 Styles.name 得到它的名称。

库 python-docx 使用独立于 Office 的样式命名体系。随着 Office 的更新，python-docx 内建的样式名称和较新版本的 Office 样式名称可能会不一致。下面的实例文件 python-docx10.py 演示了获取 python-docx 内部的表格样式名称的过程。

源码路径：daima\8\08-27\python-docx10.py

```
document = Document()
styles = document.styles
table_styles = [s for s in styles if s.type == WD_STYLE_TYPE.TABLE]
for style in table_styles:
    print(style.name)
```

执行后会输出：

```
Normal Table
Table Grid
Light Shading
Light Shading Accent 1
Light Shading Accent 2
Light Shading Accent 3
Light Shading Accent 4
Light Shading Accent 5
Light Shading Accent 6
#省略后面的样式名称列表
```

实例 08-28：使用指定样式修饰表格

下面的实例文件 python-docx11.py 演示了使用指定样式修饰表格的过程。

源码路径：daima\8\08-28\python-docx11.py

```
table = document.add_table(rows=trow,cols=tcol,style = 'Colorful Grid Accent 4')
```

新的表格样式如图 8-18 所示。

total	available	percent	used	free
16290.6875M	8866.4765625M	45.6%	7424.2109375M	8866.4765625M

图 8-18　新的表格样式

实例 08-29：创建样式和设置字体

字符对象、Paragraph 对象和 Table 对象都有一个成员 style，style 都有一个字体成员 font，其中包含了字体的所有设置。如果凭空地创建一个 style，则需要使用 Document 的 Styles 的 add_style()函数实现。例如，下面的实例文件 python-docx12.py 演示了添加了 10 个段落，并为每个段落创建一个新的 style 的过程。这些 style 的区别是字号依次增加。

源码路径：daima\8\08-29\python-docx12.py

```
from docx import Document
from docx.shared import Pt
from docx.enum.style import WD_STYLE_TYPE

doc = Document()
for i in range(10):
    p = doc.add_paragraph(u'段落 %d' % i)
    style = doc.styles.add_style('UserStyle%d' % i, WD_STYLE_TYPE.PARAGRAPH)
    style.font.size = Pt(i + 20)
    p.style = style

doc.save('style-1.docx')
```

在上述代码中，库 python-docx 中的 docx.shared 模块提供了以 Pt 为单位设置的字体。类似的单位有 mm、cm、in、emu（物理尺寸）及 twips。上述代码执行后会创建一个包含指定样式段落文本的 Word 文件 style-1.docx，如图 8-19 所示。

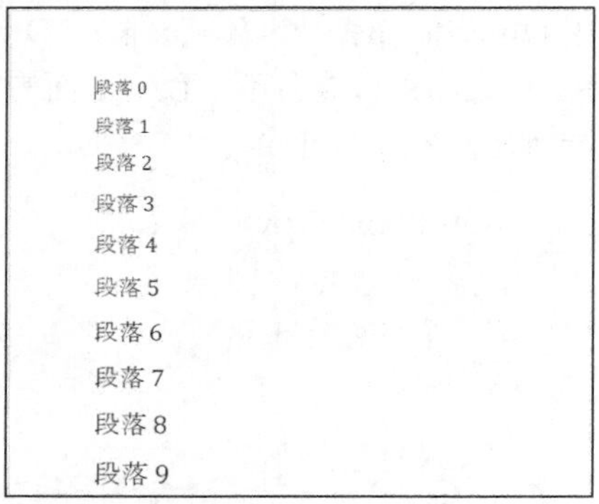

第 8 章 特殊文本格式处理实战

```
段落 0
段落 1
段落 2
段落 3
段落 4
段落 5
段落 6
段落 7
段落 8
段落 9
```

图 8-19　文件 style-1.docx 的内容

实例 08-30：使用 Run.font 设置字体样式

下面的实例文件 python-docx13.py 演示了使用 Run.font 设置字体样式的过程。

源码路径：daima\8\08-30\python-docx13.py

```
doc = Document()
p = doc.add_paragraph()
text_str = u'好好学习Python，努力做到开发专家，成为最牛的程序员。'
for i, ch in enumerate(text_str):
    run = p.add_run(ch)
    font = run.font
    font.name = u'微软雅黑'
    # python-docx的bug
    run._element.rPr.rFonts.set(qn('w:eastAsia'), u'微软雅黑')
    font.bold = (i % 2 == 0)
    font.italic = (i % 3 == 0)
    color = font.color
    color.rgb = RGBColor(i * 10 % 200 + 55, i * 20 % 200 + 55, i * 30 % 200 + 55)

doc.save('style-2.docx')
```

上述代码执行后将会创建一个包含指定样式文本的 Word 文件 style-2.docx，如图 8-20 所示。因为本书不是彩色印刷，所以书中的执行效果不够明显，建议读者在计算机中执行后观看文本颜色。

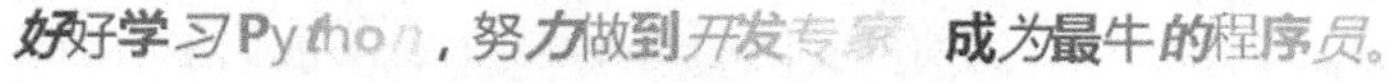

好好学习Python，努力做到开发专家 成为最牛的程序员。

图 8-20　文件 style-2.docx 的内容

实例 08-31：设置段落递进的左对齐样式

通过使用库 python-docx 中的段落格式的成员 paragraph_format，可以设置指定的段落样式和表格样式，这相当于 Word 中的"段落"面板，如图 8-21 所示。库 python-docx 的段落格式有各种成员，相当于 Word 的"段落"对话框中的各个设置要素，如图 8-22 所示。

下面的实例文件 python-docx14.py 演示了设置段落递进的左对齐样式的过程。

源码路径：daima\8\08-31\python-docx14.py

```
doc = Document()
for i in range(10):
    p = doc.add_paragraph(u'段落 %d' % i)
    style = doc.styles.add_style('UserStyle%d' % i, WD_STYLE_TYPE.PARAGRAPH)
    style.paragraph_format.left_indent = Cm(i)
    p.style = style

doc.save('style-3.docx')
```

上述代码执行后将会创建一个包含指定段落样式文本的 Word 文件 style-3.docx，如图 8-23 所示。

8.2 Office 处理实战

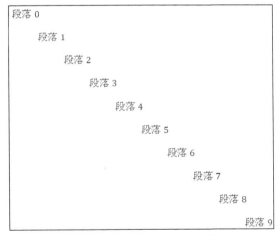

图 8-21 Word 中的"段落"面板　　　　　图 8-22 "段落"对话框

图 8-23 文件 style-3.docx 的内容

实例 08-32：自定义创建 Word 样式

Word 自带了多种样式，我们可以使用库 python-docx 中的 Document.styles 来访问 builtin 属性为 True 的自带样式。当然，开发者通过 add_style()函数增加的样式也会被放在 Document.styles 中。对于开发者自己创建的样式，如果将其属性 hidden 和 quick_style 分别设置为 False 和 True，则可以将这个自建样式添加到 Word 快速样式管理器中。例如，下面的实例文件 python-docx15.py 演示了开发者自定义创建 Word 样式的过程。

源码路径：daima\8\08-32\python-docx15.py

```
doc = Document()
for i in range(10):
    p = doc.add_paragraph(u'段落 %d' % i)
    style = doc.styles.add_style('UserStyle%d' % i, WD_STYLE_TYPE.PARAGRAPH)
    style.paragraph_format.left_indent = Cm(i)
```

```
            p.style = style
        if i == 7:
            style.hidden = False
            style.quick_style = True

for style in doc.styles:
    print(style.name, style.builtin)

doc.paragraphs[3].style = doc.styles['Subtitle']
doc.save('style-4.docx')
```

通过上述代码，在 Word 文件 style-4.docx 中自定义创建了 9 种样式（UserStyle1～UserStyle9），如图 8-24 所示。

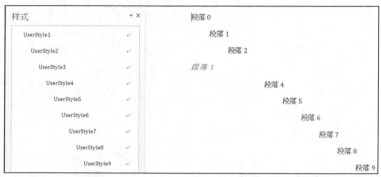

图 8-24　文件 style-4.docx 中自定义创建的 9 种样式和内容

实例 08-33：使用库 xlrd 读取 Excel 文件的内容

在 Python 程序中，可以使用库 xlrd 读取 Excel 文件的内容。安装库 xlrd 的命令如下。

```
pip install xlrd
```

例如，下面的实例文件 ex01.py 演示了使用库 xlrd 读取 Excel 文件内容的过程。

源码路径：daima\8\08-33\ex01.py

```python
import xlrd
#打开Excel文件
data = xlrd.open_workbook('example.xlsx')
#查看文件中表单的名称
data.sheet_names()
#得到第1个表单，或者通过索引或表单名称得到第1个
table = data.sheets()[0]
table = data.sheet_by_index(0)
table = data.sheet_by_name(u'Sheet1')
#获取行数和列数
nrows = table.nrows
ncols = table.ncols
print(nrows)
print(ncols)
#循环行，得到索引的列表
for rownum in range(table.nrows):
    print(table.row_values(rownum))
#分别使用行、列索引
cell_A1 = table.row(0)[0].value
cell_A2 = table.col(1)[0].value
print(cell_A1)
print(cell_A2)
```

执行后会输出：

```
3
3
[1.0, 2.0, 3.0]
[4.0, 5.0, 6.0]
[8.0, 8.0, 9.0]
1.0
6.0
1.0
```

实例 08-34：将指定内容写入 Excel 文件并创建 Excel 文件

在 Python 程序中，可以使用库 xlwt 向 Excel 文件中写入内容。安装库 xlwt 的命令如下。

```
pip install xlwt
```

例如，下面的实例文件 ex02.py 演示了使用库 xlwt 将指定内容写入 Excel 文件并创建 Excel 文件的过程。

源码路径：daima\8\08-34\ex02.py

```python
import xlwt
from datetime import datetime

style0 = xlwt.easyxf('font: name Times New Roman, color-index red, bold on',
    num_format_str='#,##0.00')
style1 = xlwt.easyxf(num_format_str='D-MMM-YY')#当前日期

wb = xlwt.Workbook()
ws = wb.add_sheet('A Test Sheet')          #表单的名字

ws.write(0, 0, 1234.56, style0)            #第1个单元格的内容
ws.write(1, 0, datetime.now(), style1)     #第2个单元格的内容
ws.write(2, 0, 1)                          #第3个单元格的内容
ws.write(2, 1, 1)                          #第4个单元格的内容
ws.write(2, 2, xlwt.Formula("A3+B3"))      #第5个单元格的内容

wb.save('example02.xls')
```

上述代码执行后会将指定内容写入文件 example02.xls 中，如图 8-25 所示。

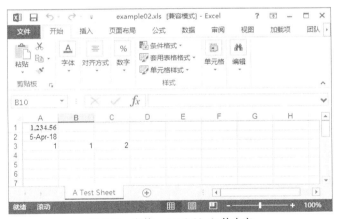

图 8-25　文件 example02.xls 的内容

实例 08-35：使用库 xlsxwriter 创建一个指定内容的 Excel 文件

在 Python 程序中，可以使用库 xlsxwriter 操作 Excel 文件。安装库 xlsxwriter 的命令如下。

```
pip install xlsxwriter
```

下面的实例文件 xlsxwriter01.py 演示了使用库 xlsxwriter 创建一个指定内容的 Excel 文件的过程。

源码路径：daima\8\08-35\xlsxwriter01.py

```python
import xlsxwriter    #导入库

workbook = xlsxwriter.Workbook('hello.xlsx')  #创建一个名为"hello.xlsx"的文件，赋值给workbook
worksheet = workbook.add_worksheet()          #创建一个默认表单，赋值给worksheet
# 表单也支持命名，如workbook.add_worksheet('hello')

worksheet.write('A1', 'Hello world')          #使用表单在A1处写入"Hello world"
workbook.close()                              #关闭文件
```

上述代码执行后会创建一个 Excel 文件 hello.xlsx，如图 8-26 所示。

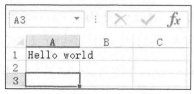

图 8-26 文件 hello.xlsx 的内容

实例 08-36：向 Excel 文件中批量写入内容

下面的实例文件 xlsxwriter02.py 演示了使用库 xlsxwriter 向 Excel 文件中批量写入内容的过程。

源码路径：daima\8\08-36\xlsxwriter02.py

```
import xlsxwriter

workbook = xlsxwriter.Workbook('Expenses01.xlsx')
worksheet = workbook.add_worksheet()

#需要写入的内容
expenses = (['Rent', 1000],
            ['Gas', 100],
            ['Food', 300],
            ['Gym', 50],
            )

#行跟列的初始位置
row = 0
col = 0

# write()方法，即write(行,列,写入的内容,样式)
for item, cost in (expenses):
    worksheet.write(row, col, item)         #在第1列的地方写入item
    worksheet.write(row, col + 1, cost)     #在第2列的地方写入cost
    row + 1                                  #每次循环中行数发生改变

worksheet.write(row, 0, 'Total')
worksheet.write(row, 1, '=SUM(B1:B4)')      #写入公式
```

上述代码执行后会创建一个 Excel 文件 Expenses01.xlsx，如图 8-27 所示。

图 8-27 文件 Expenses01.xlsx 的内容

实例 08-37：设置表格样式

表格样式包含字体、颜色、模式、边框和数字格式等。在设置表格样式时需要使用函数 add_format()，库 xlsxwriter 包含的样式信息如表 8-1 所示。

表 8-1　　　　　　　　　　　　库 xlsxwriter 包含的样式信息

类　别	描　述	属　性	方　法　名
字体	字体名称	font_name	set_font_name()
	字体大小	font_size	set_font_size()
	字体颜色	font_color	set_font_color()

续表

类别	描述	属性	方法名
字体	加粗	bold	set_bold()
	斜体	italic	set_italic()
	下划线	underline	set_underline()
	删除线	font_strikeout	set_font_strikeout()
	上标/下标	font_script	set_font_script()
数字	数字格式	num_format	set_num_format()
保护	表格锁定	locked	set_locked()
	隐藏公式	hidden	set_hidden()
对齐	水平对齐	align	set_align()
	垂直对齐	valign	set_valign()
	旋转	rotation	set_rotation()
	文本包装	text_wrap	set_text_warp()
	底端对齐	text_justlast	set_text_justlast()
	中心对齐	center_across	set_center_across()
	缩进	indent	set_indent()
	缩小填充	shrink	set_shrink()
模式	表格模式	pattern	set_pattern()
	背景颜色	bg_color	set_bg_color()
	前景颜色	fg_color	set_fg_color()
边框	表格边框	border	set_border()
	底部边框	bottom	set_bottom()
	顶部边框	top	set_top()
	右边框	right	set_right()
	边框颜色	border_color	set_border_color()
	底部边框颜色	bottom_color	set_bottom_color()
	顶部边框颜色	top_color	set_top_color()
	左边框颜色	left_color	set_left_color()
	右边框颜色	right_color	set_right_color()

下面的实例文件 xlsxwriter03.py 演示了使用库 xlsxwriter 设置表格样式的过程。

源码路径: daima\8\08-37\xlsxwriter03.py

```
#创建文件及表单
workbook = xlsxwriter.Workbook('Expenses03.xlsx')
worksheet = workbook.add_worksheet()
#设置粗体，默认是False
bold = workbook.add_format({'bold': True})
#定义数字格式
money = workbook.add_format({'num_format': '$#,##0'})
#使用自定义blod格式写表头
worksheet.write('A1', 'Item', bold)
worksheet.write('B1', 'Cost', bold)
#写入表中的数据
expenses = (
['Rent', 1000],
['Gas',    100],
['Food',   300],
```

```
        ['Gym',        50],
    )

    #从标题下面的第1个单元格开始 .
    row = 1
    col = 0

    #迭代数据并逐行地写出它
    for item, cost in (expenses):
        worksheet.write(row, col, item)               #使用默认格式写入
        worksheet.write(row, col + 1, cost, money)    #使用自定义money格式写入
        row += 1

    # 用公式计算总数
    worksheet.write(row, 0, 'Total',          bold)
    worksheet.write(row, 1, '=SUM(B2:B5)', money)

    workbook.close()
```

上述代码执行后会创建一个 Excel 文件 Expenses03.xlsx，表格中的字体样式是用户自己定义的，如图 8-28 所示。

	A	B	C
1	**Item**	**Cost**	
2	Rent	$1,000	
3	Gas	$100	
4	Food	$300	
5	Gym	$50	
6	**Total**	$1,450	
7			

图 8-28　文件 Expenses03.xlsx 的内容

实例 08-38：向 Excel 文件中插入图像

下面的实例文件 xlsxwriter04.py 演示了使用库 xlsxwriter 向 Excel 文件中插入图像的过程。

源码路径：daima\8\08-38\xlsxwriter04.py

```
#创建一个新Excel文件并添加表单
workbook = xlsxwriter.Workbook('demo.xlsx')
worksheet = workbook.add_worksheet()

#展开第1列，使正文更清楚
worksheet.set_column('A:A', 20)

#添加一个粗体格式用于突出单元格内容
bold = workbook.add_format({'bold': True})

#写一些简单的文字
worksheet.write('A1', 'Hello')

#设置文本与格式
worksheet.write('A2', 'World', bold)

#写一些数字，行/列符号
worksheet.write(2, 0, 123)
worksheet.write(3, 0, 123.456)
#插入图像
worksheet.insert_image('B5', '123.png')
workbook.close()
```

上述代码执行后会创建一个包含指定图像的 Excel 文件 demo.xlsx，如图 8-29 所示。

图 8-29 文件 demo.xlsx 的内容

实例 08-39：向 Excel 文件中插入数据并绘制柱状图

Excel 的核心功能之一是用表格内的数据绘制统计图表，使整体数据变得更加直观。通过使用库 xlsxwriter，可以用 Excel 表格内的数据绘制图表。下面的实例文件 xlsxwriter05.py 演示了使用库 xlsxwriter 向 Excel 文件中插入数据并绘制柱状图的过程。

源码路径：daima\8\08-39\xlsxwriter05.py

```
import xlsxwriter

workbook = xlsxwriter.Workbook('chart.xlsx')
worksheet = workbook.add_worksheet()

#新建图表对象
chart = workbook.add_chart({'type': 'column'})

#向Excel文件中插入数据，绘制图表时要用到
data = [
    [1, 2, 3, 4, 5],
    [2, 4, 6, 8, 10],
    [3, 6, 9, 12, 15],
]

worksheet.write_column('A1', data[0])
worksheet.write_column('B1', data[1])
worksheet.write_column('C1', data[2])

#向图表中添加数据，例如，第1行为：用A1~A5的数据绘制图表
chart.add_series({'values': '=Sheet1!$A$1:$A$5'})
chart.add_series({'values': '=Sheet1!$B$1:$B$5'})
chart.add_series({'values': '=Sheet1!$C$1:$C$5'})

#将图表插入表单中
worksheet.insert_chart('A7', chart)

workbook.close()
```

上述代码执行后会创建一个包含指定数据的 Excel 文件 chart.xlsx，并根据数据绘制了一个柱状图，如图 8-30 所示。

第 8 章 特殊文本格式处理实战

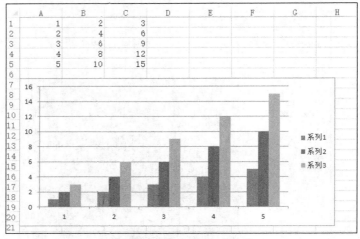

图 8-30 文件 chart.xlsx 的内容

实例 08-40：向 Excel 文件中插入数据并绘制散点图

下面的实例文件 xlsxwriter06.py 演示了使用库 xlsxwriter 向 Excel 文件中插入数据并绘制散点图的过程。

源码路径：daima\8\08-40\xlsxwriter06.py

```
import xlsxwriter

workbook = xlsxwriter.Workbook('chart_scatter.xlsx')
worksheet = workbook.add_worksheet()
bold = workbook.add_format({'bold': 1})

#添加图表将引用的数据
headings = ['Number', 'Batch 1', 'Batch 2']
data = [
    [2, 3, 4, 5, 6, 7],
    [10, 40, 50, 20, 10, 50],
    [30, 60, 70, 50, 40, 30],
]

worksheet.write_row('A1', headings, bold)
worksheet.write_column('A2', data[0])
worksheet.write_column('B2', data[1])
worksheet.write_column('C2', data[2])

#绘制一个散点图
chart1 = workbook.add_chart({'type': 'scatter'})

#配置第1个系列散点.
chart1.add_series({
    'name':       '=Sheet1!$B$1',
    'categories': '=Sheet1!$A$2:$A$7',
    'values':     '=Sheet1!$B$2:$B$7',
})

#配置第2个系列散点，注意使用替代语法来定义范围
chart1.add_series({
    'name':       ['Sheet1', 0, 2],
    'categories': ['Sheet1', 1, 0, 6, 0],
    'values':     ['Sheet1', 1, 2, 6, 2],
})

#添加图表标题和一些轴标签
chart1.set_title ({'name': 'Results of sample analysis'})
chart1.set_x_axis({'name': 'Test number'})
chart1.set_y_axis({'name': 'Sample length (mm)'})
```

```
#设置Excel图表样式
chart1.set_style(11)

#将图表插入表单（带偏移量）
worksheet.insert_chart('D2', chart1, {'x_offset': 25, 'y_offset': 10})
workbook.close()
```

上述代码执行后会创建一个包含指定数据的 Excel 文件 chart_scatter.xlsx，并根据数据内容绘制了一个散点图，如图 8-31 所示。

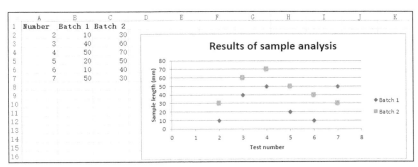

图 8-31　文件 chart_scatter.xlsx 的内容

实例 08-41：向 Excel 文件中插入数据并绘制柱状图和饼图

下面的实例文件 xlsxwriter07.py 演示了使用库 xlsxwriter 向 Excel 文件中插入数据并绘制柱状图和饼图的过程。

源码路径：daima\8\08-41\xlsxwriter07.py

```python
import xlsxwriter

#新建一个Excel文件，起名为expense01.xlsx
workbook = xlsxwriter.Workbook("123.xlsx")
#添加一个表单页，不添加名字，默认为Sheet1
worksheet = workbook.add_worksheet()
#准备数据
headings=["姓名","数学","语文"]
data=[["C罗张",78,60],["糖人李",98,89],["梅西徐",88,100]]
#样式
head_style = workbook.add_format({"bold":True,"bg_color":"yellow","align":"center","font":13})
#写数据
worksheet.write_row("A1",headings,head_style)
for i in range(0,len(data)):
    worksheet.write_row("A{}".format(i+2),data[i])
#绘制柱状图
chart1 = workbook.add_chart({"type":"column"})
chart1.add_series({
    "name":"=Sheet1!$B$1",          #图例项
    "categories":"=Sheet1!$A$2:$A$4",  #x轴名称
    "values":"=Sheet1!$B$2:$B$4"    #x轴值
})
chart1.add_series({
    "name":"=Sheet1!$C$1",
    "categories":"=Sheet1!$A$2:$A$4",
    "values":"=Sheet1!$C$2:$C$4"
})
#添加柱状图标题
chart1.set_title({"name":"柱状图"})
#y轴名称
chart1.set_y_axis({"name":"分数"})
#x轴名称
chart1.set_x_axis({"name":"人名"})
#图表样式
chart1.set_style(11)

#添加柱状图叠图子类型
chart2 = workbook.add_chart({"type":"column","subtype":"stacked"})
```

```
chart2.add_series({
    "name":"=Sheet1!$B$1",
    "categories":"=Sheet1!$A$2:$a$4",
    "values":"=Sheet1!$B$2:$B$4"
})
chart2.add_series({
    "name":"=Sheet1!$C$1",
    "categories":"=Sheet1!$A$2:$a$4",
    "values":"=Sheet1!$C$2:$C$4"
})
chart2.set_title({"name":"叠图子类型"})
chart2.set_x_axis({"name":"姓名"})
chart2.set_y_axis({"name":"成绩"})
chart2.set_style(12)

#绘制饼图
chart3 = workbook.add_chart({"type":"pie"})
chart3.add_series({
    #"name":"饼图",
    "categories":"=Sheet1!$A$2:$A$4",
    "values":"=Sheet1!$B$2:$B$4",
    #定义各饼块的颜色
    "points":[
        {"fill":{"color":"yellow"}},
        {"fill":{"color":"blue"}},
        {"fill":{"color":"red"}}
    ]
})
chart3.set_title({"name":"饼图成绩单"})
chart3.set_style(3)

#插入图表
worksheet.insert_chart("B7",chart1)
worksheet.insert_chart("B25",chart2)
worksheet.insert_chart("J2",chart3)

#关闭Excel文件
workbook.close()
```

上述代码执行后会创建一个包含指定数据的 Excel 文件 123.xlsx，并根据数据分别绘制了两个柱状图和一个饼图，如图 8-32 所示。

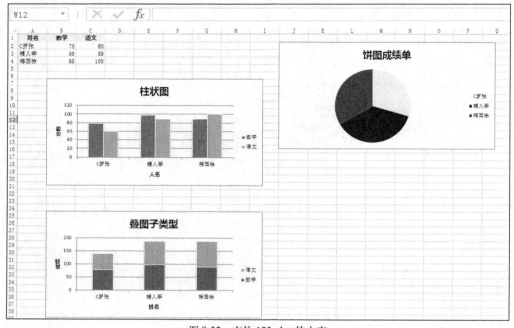

图 8-32　文件 123.xlsx 的内容

8.3 PDF 处理实战

在 Python 程序中,可以使用第三方模块处理 PDF 文件中的数据。本节将详细讲解这些模块的使用方法。

实例 08-42:将 PDF 文件中的内容转换为 TXT 文本

使用模块 PDFMiner 可以解析 PDF 文件,在使用 PDFMiner 之前需要先安装 PDFMiner。其安装命令如下。

```
pip install pdfminer3k
```

因为解析 PDF 文件是一件非常耗时和耗费内存的工作,所以 PDFMiner 使用了懒解析(lazy parsing)策略,即只在需要的时候才解析 PDF 文件,以减少时间和内存的使用。要想使用 PDFMiner 解析 PDF 文件至少用到两个类:PDFParser 和 PDFDocument。其中,类 PDFParser 用于从文件中提取数据;类 PDFDocument 用于保存数据;另外还需要 PDFPageInterprete 类,用于处理 PDF 文件中的内容;类 PDFDevice,用于将内容转换为我们所需要的结果;类 PDFResourceManager 用于保存共享内容,如字体或图片。

假设存在一个 PDF 文件"开发 Python 应用程序.pdf",其内容如图 8-33 所示。

图 8-33 文件"开发 Python 应用程序.pdf"的内容

下面的实例文件 PDFMiner01.py 演示了使用 PDFMiner 将 PDF 文件中的内容转换为 TXT 文本的过程。

源码路径:daima\8\08-42\PDFMiner01.py

```
import sys
import importlib
importlib.reload(sys)

from pdfminer.pdfparser import PDFParser,PDFDocument
from pdfminer.pdfinterp import PDFResourceManager, PDFPageInterpreter
from pdfminer.converter import PDFPageAggregator
from pdfminer.layout import LTTextBoxHorizontal,LAParams
from pdfminer.pdfinterp import PDFTextExtractionNotAllowed
```

```python
'''
解析PDF 文本，保存到TXT文件中
'''
path = r'开发Python应用程序.pdf'
def parse():
    fp = open(path, 'rb') #以二进制读方式打开
    #用文件对象来创建一个PDF文档分析器
    praser = PDFParser(fp)
    #创建一个PDF文档
    doc = PDFDocument()
    #连接分析器与文档对象
    praser.set_document(doc)
    doc.set_parser(praser)

    #提供初始化密码
    #如果没有密码，就创建一个空的字符串
    doc.initialize()

    #检测文档是否提供TXT转换，不提供就忽略
    if not doc.is_extractable:
        raise PDFTextExtractionNotAllowed
    else:
        #创建PDF资源管理器来管理共享资源
        rsrcmgr = PDFResourceManager()
        #创建一个PDF设备对象
        laparams = LAParams()
        device = PDFPageAggregator(rsrcmgr, laparams=laparams)
        #创建一个PDF解释器对象
        interpreter = PDFPageInterpreter(rsrcmgr, device)

        #循环遍历列表，每次处理一个page的内容
        for page in doc.get_pages(): # doc.get_pages()获取page列表
            interpreter.process_page(page)
            #接受该page的LTPage对象
            layout = device.get_result()
            #这里layout是一个LTPage对象，里面存放着这个page解析出的各种对象，一般包括LTTextBox、LTFigure、
            #LTImage、LTTextBoxHorizontal等。想要获取文本就要获得对象的text属性
            for x in layout:
                if (isinstance(x, LTTextBoxHorizontal)):
                    with open(r'123.txt', 'a') as f:
                        results = x.get_text()
                        print(results)
                        f.write(results + '\n')
if __name__ == '__main__':
    parse()
```

上述代码执行后后会将文件"开发 Python 应用程序.pdf"中的内容解析并保存到 TXT 文件"123.txt"中（见图 8-34），并在命令行界面中显示解析后的内容，如图 8-35 所示。

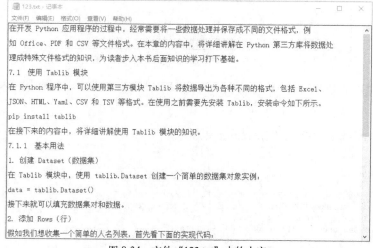

图 8-34 文件"123.txt"中的内容

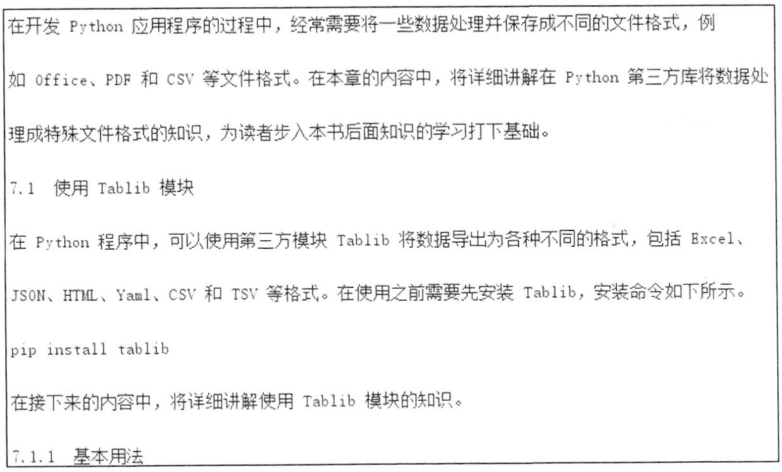

> 在开发 Python 应用程序的过程中，经常需要将一些数据处理并保存成不同的文件格式，例如 Office、PDF 和 CSV 等文件格式。在本章的内容中，将详细讲解在 Python 第三方库将数据处理成特殊文件格式的知识，为读者步入本书后面知识的学习打下基础。
>
> 7.1 使用 Tablib 模块
>
> 在 Python 程序中，可以使用第三方模块 Tablib 将数据导出为各种不同的格式，包括 Excel、JSON、HTML、Yaml、CSV 和 TSV 等格式。在使用之前需要先安装 Tablib，安装命令如下所示。
>
> pip install tablib
>
> 在接下来的内容中，将详细讲解使用 Tablib 模块的知识。
>
> 7.1.1 基本用法

图 8-35　在命令行界面中显示解析后的内容

实例 08-43：解析某个在线 PDF 文件的内容

下面的实例文件 PDFMiner02.py 演示了使用 PDFMiner 解析某个在线 PDF 文件的内容的过程。

源码路径：daima\8\08-43\PDFMiner02.py

```python
import importlib
import sys
import random
from urllib.request import urlopen
from urllib.request import Request

from pdfminer.converter import PDFPageAggregator
from pdfminer.layout import LTTextBoxHorizontal, LAParams
from pdfminer.pdfinterp import PDFResourceManager, PDFPageInterpreter
from pdfminer.pdfinterp import PDFTextExtractionNotAllowed
from pdfminer.pdfparser import PDFParser, PDFDocument

'''
解析PDF文本，保存到TXT文件中
'''
importlib.reload(sys)

user_agent = ['Mozilla/5.0 (Windows NT 10.0; WOW64)', 'Mozilla/5.0 (Windows NT 6.3; WOW64)',
              'Mozilla/5.0 (Windows NT 6.1; WOW64; rv:54.0) Gecko/20100101 Firefox/54.0',
              'Mozilla/5.0 (Windows NT 6.1) AppleWebKit/538.11 (KHTML, like Gecko) Chrome/23.0.1271.64 Safari/538.11',
              'Mozilla/5.0 (Windows NT 6.3; WOW64; Trident/8.0; rv:11.0) like Gecko',
              'Mozilla/5.0 (Windows NT 5.1) AppleWebKit/538.36 (KHTML, like Gecko) Chrome/28.0.1500.95 Safari/538.36',
              'Mozilla/5.0 (Windows NT 6.1; WOW64; Trident/8.0; SLCC2; .NET CLR 2.0.50727; .NET CLR 3.5.30729; .NET CLR 3.0.30729; Media Center PC 6.0; .NET4.0C; rv:11.0) like Gecko)',
              'Mozilla/5.0 (Windows; U; Windows NT 5.2) Gecko/2008070208 Firefox/3.0.1',
              'Mozilla/5.0 (Windows; U; Windows NT 5.1) Gecko/20070309 Firefox/2.0.0.3',
              'Mozilla/5.0 (Windows; U; Windows NT 5.1) Gecko/20070803 Firefox/1.5.0.12',
              'Opera/9.27 (Windows NT 5.2; U; zh-cn)',
              'Mozilla/5.0 (Macintosh; PPC macOS X; U; en) Opera 8.0',
              'Opera/8.0 (Macintosh; PPC macOS X; U; en)',
              'Mozilla/5.0 (Windows; U; Windows NT 5.1; en-US; rv:1.8.1.12) Gecko/20080219 Firefox/2.0.0.12 Navigator/9.0.0.6',
              'Mozilla/4.0 (compatible; MSIE 8.0; Windows NT 6.1; Win64; x64; Trident/4.0)',
              'Mozilla/4.0 (compatible; MSIE 8.0; Windows NT 6.1; Trident/4.0)',
              'Mozilla/5.0 (compatible; MSIE 10.0; Windows NT 6.1; WOW64; Trident/6.0; SLCC2; .NET CLR 2.0.50727; .NET CLR 3.5.30729; .NET CLR 3.0.30729; Media Center PC 6.0; InfoPath.2; .NET4.0C; .NET4.0E)',
              'Mozilla/5.0 (Windows NT 6.1; WOW64) AppleWebKit/538.1 (KHTML, like Gecko) Maxthon/4.0.6.2000 Chrome/26.0.1410.43 Safari/538.1 ',
              'Mozilla/5.0 (compatible; MSIE 10.0; Windows NT 6.1; WOW64; Trident/6.0; SLCC2; .NET CLR
```

第8章 特殊文本格式处理实战

```
                            2.0.50727; .NET CLR 3.5.30729; .NET CLR 3.0.30729; Media Center PC 6.0;
                            InfoPath.2; .NET4.0C; .NET4.0E; QQBrowser/8.3.9825.400)',
              'Mozilla/5.0 (Windows NT 6.1; WOW64; rv:21.0) Gecko/20100101 Firefox/21.0 ',
              'Mozilla/5.0 (Windows NT 6.1; WOW64) AppleWebKit/538.1 (KHTML, like Gecko) Chrome/21.0.1180.92
                            Safari/538.1 LBBROWSER',
              'Mozilla/5.0 (compatible; MSIE 10.0; Windows NT 6.1; WOW64; Trident/6.0; BIDUBrowser 2.x)',
              'Mozilla/5.0 (Windows NT 6.1; WOW64) AppleWebKit/536.11 (KHTML, like Gecko) Chrome/20.0.1132.11
                            TaoBrowser/3.0 Safari/536.11']

def parse(_path):
    # rb以二进制读模式打开本地PDF文件
    request = Request(url=_path, headers={'User-Agent': random.choice(user_agent)})
    #随机从user_agent列表中抽取一个元素
    fp = urlopen(request)      #打开在线PDF文档

    #用文件对象来创建一个PDF文档分析器
    praser_pdf = PDFParser(fp)

    #创建一个PDF文档
    doc = PDFDocument()

    #连接分析器与文档对象
    praser_pdf.set_document(doc)
    doc.set_parser(praser_pdf)

    #提供初始化密码doc.initialize("123456")
    #如果没有密码，则不添加参数
    doc.initialize()

    #检测文档是否提供TXT转换，不提供就忽略
    if not doc.is_extractable:
        raise PDFTextExtractionNotAllowed
    else:
        #创建PDf资源管理器来管理共享资源
        rsrcmgr = PDFResourceManager()

        #创建一个PDF参数分析器
        laparams = LAParams()

        #创建聚合器
        device = PDFPageAggregator(rsrcmgr, laparams=laparams)

        #创建一个PDF页面解释器对象
        interpreter = PDFPageInterpreter(rsrcmgr, device)

        #循环遍历列表，每次处理一页的内容
        # doc.get_pages()获取page列表
        for page in doc.get_pages():
            #使用页面解释器来读取
            interpreter.process_page(page)

            #使用聚合器获取内容
            layout = device.get_result()

            #这里layout是一个LTPage对象，里面存放着这个page解析出的各种对象,一般包括LTTextBox、LTFigure、
            #LTImage、LTTextBoxHorizontal等。想要获取文本就要获得对象的text属性
            for out in layout:
                if isinstance(out, LTTextBoxHorizontal):
                    results = out.get_text()
                    print("results: " + results)
if __name__ == '__main__':
    url = "http://www.caac.gov.cn/XXGK/XXGK/TJSJ/201708/P020170821330916187824.pdf"
    parse(url)
```

上述代码执行后将会解析并输出在线 PDF 文件的内容：

results: 中国民航2017年6月份主要生产指标统计

results: 统计指标

results: 运输总周转量

results: 国内航线

```
results:    其中：港澳台航线
results:       国际航线
results:    旅客运输量
results:       国内航线
results:    其中：港澳台航线
results:       国际航线
results:    货邮运输量
results:       国内航线
#省略剩余解析结果
```

实例 08-44：使用 PyPDF2 读取 PDF 文件

使用库 PyPDF2 可以读写、分割或合并 PDF 文件，在使用 PyPDF2 之前需要先安装 PyPDF2。其安装命令如下。

```
pip install PyPDF2
```

类 PdfFileReader 的功能是初始化一个 PdfFileReader 对象并读取 PDF 文件的内容。因为 PDF 流的交叉引用表被读入内存，所以此操作可能需要一些时间。例如，下面的实例文件 PyPDF201.py 演示了使用 PyPDF2 读取 PDF 文件的过程。

源码路径：daima\8\08-44\PyPDF201.py

```python
from PyPDF2 import PdfFileReader, PdfFileWriter

readFile = '开发Python应用程序.pdf'
#获取 PdfFileReader 对象
pdfFileReader = PdfFileReader(readFile)   # 或者这个方式：pdfFileReader = PdfFileReader(open(readFile, 'rb'))
#获取 PDF 文件的文档信息
documentInfo = pdfFileReader.getDocumentInfo()
print('documentInfo = %s' % documentInfo)
#获取页面布局
pageLayout = pdfFileReader.getPageLayout()
print('pageLayout = %s ' % pageLayout)

#获取页面模式
pageMode = pdfFileReader.getPageMode()
print('pageMode = %s' % pageMode)

xmpMetadata = pdfFileReader.getXmpMetadata()
print('xmpMetadata   = %s ' % xmpMetadata)

#获取PDF文件页数
pageCount = pdfFileReader.getNumPages()

print('pageCount = %s' % pageCount)
for index in range(0, pageCount):
    #返回指定页编号的 PageObject
    pageObj = pdfFileReader.getPage(index)
    print('index = %d , pageObj = %s' % (index, type(pageObj)))   # <class 'PyPDF2.pdf.PageObject'>
    #获取 PageObject 在 PDF 文档中处于的页码
    pageNumber = pdfFileReader.getPageNumber(pageObj)
    print('pageNumber = %s ' % pageNumber)
```

上述代码执行后将会读取并显示文件"开发 Python 应用程序.pdf"的信息：

```
documentInfo = {'/Author': 'apple', '/Creator': 'Microsoft® Word 2013', '/CreationDate': "D:20180406164456+08'00'", '/ModDate': "D:20180406164456+08'00'", '/Producer': 'Microsoft® Word 2013'}
pageLayout = None
pageMode = None
xmpMetadata    = None
pageCount = 1
index = 0 , pageObj = <class 'PyPDF2.pdf.PageObject'>
pageNumber = 0
```

实例 08-45：使用 PyPDF2 将 PDF 文件中写入另一个 PDF 文件内容

类 PdfFileWriter 提供了向 PDF 文件中写入数据的功能。例如，下面的实例文件 PyPDF202.py 演示了使用 PyPDF2 将 PDF 文件中写入另一个 PDF 文件内容的过程。

源码路径：daima\8\08-45\PyPDF202.py

```python
from PyPDF2 import PdfFileReader, PdfFileWriter

readFile = '开发Python应用程序.pdf'
outFile = 'copy.pdf'
pdfFileWriter = PdfFileWriter()

# 获取 PdfFileReader 对象
pdfFileReader = PdfFileReader(readFile)              #或者这个方式：pdfFileReader = PdfFileReader(open(readFile, 'rb'))
numPages = pdfFileReader.getNumPages()

for index in range(0, numPages):
    pageObj = pdfFileReader.getPage(index)
    pdfFileWriter.addPage(pageObj)                   #根据每页返回的 PageObject写入文件
    pdfFileWriter.write(open(outFile, 'wb'))

pdfFileWriter.addBlankPage()                         #在文件的最后一页写入一个空白页，保存文件
pdfFileWriter.write(open(outFile,'wb'))
```

上述代码执行后会将文件"开发 Python 应用程序.pdf"的内容写入文件"copy.pdf"中，新生成文件"copy.pdf"的内容如图 8-36 所示。

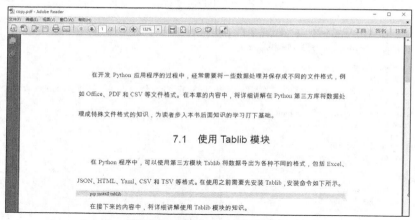

图 8-36　文件"copy.pdf"的内容

实例 08-46：将两个 PDF 文件合并为一个 PDF 文件

类 PageObject 表示 PDF 文件中的单个页面，通常这个对象是通过访问 PdfFileReader 对象的 getPage()方法得到的，也可以使用 createBlankPage()静态方法创建一个空的页面。下面的实例文件 PyPDF203.py 演示了使用 PyPDF2 将两个 PDF 文件合并为一个 PDF 文件的过程。

源码路径：daima\8\08-46\PyPDF203.py

```python
from PyPDF2 import PdfFileReader, PdfFileWriter
def mergePdf(inFileList, outFile):
    """
    合并文档
    :param inFileList:  要合并的文档的列表
    :param outFile:     合并后的输出文件
    :return:
    """
    pdfFileWriter = PdfFileWriter()
    for inFile in inFileList:
        #依次循环打开要合并的文件
        pdfReader = PdfFileReader(open(inFile, 'rb'))
```

```
            numPages = pdfReader.getNumPages()
            for index in range(0, numPages):
                pageObj = pdfReader.getPage(index)
                pdfFileWriter.addPage(pageObj)

            #最后统一写入输出文件中
            pdfFileWriter.write(open(outFile, 'wb'))

mergePdf(['copy.pdf','123.pdf'],'456.pdf')
```

上述代码执行后会将文件 copy.pdf 和 123.pdf'的内容合并到文件 456.pdf 中。

实例 08-47：分割某个指定 PDF 文件

下面的实例文件 PyPDF204.py 演示了使用 PyPDF2 分割某个指定 PDF 文件的过程。

源码路径：daima\8\08-47\PyPDF204.py

```
from PyPDF2 import PdfFileReader, PdfFileWriter
def splitPdf():
    readFile = '123.pdf'
    outFile = '789.pdf'
    pdfFileWriter = PdfFileWriter()

    #获取PdfFileReader对象
    pdfFileReader = PdfFileReader(readFile)    #或者这个方式：pdfFileReader = PdfFileReader(open(readFile, 'rb'))
    #文档总页数
    numPages = pdfFileReader.getNumPages()

    if numPages > 5:
        #从第5页之后的页面，被输出到一个新的文件中，即分割文档
        for index in range(5, numPages):
            pageObj = pdfFileReader.getPage(index)
            pdfFileWriter.addPage(pageObj)
        #添加完每页，再一起保存至文件中
        pdfFileWriter.write(open(outFile, 'wb'))

splitPdf()
```

上述执行代码后会分割文件 123.pdf 的内容，将此文件第 5 页之后的页面分离出来并保存为文件 789.pdf。

实例 08-48：合并 3 个 PDF 文件

下面的实例文件 PyPDF205.py 演示了使用 PyPDF2 合并 3 个 PDF 文件的过程。

源码路径：daima\8\08-48\PyPDF205.py

```
import PyPDF2

pdff1 = open("123.pdf", "rb")
pr = PyPDF2.PdfFileReader(pdff1)
print(pr.numPages)

pdff2 = open("456.pdf", "rb")
pr2 = PyPDF2.PdfFileReader(pdff2)

pdf3 = open("789.pdf", "rb")
pr3 = PyPDF2.PdfFileReader(pdf3)

pdfw = PyPDF2.PdfFileWriter()
pageobj = pr.getPage(0)
pdfw.addPage(pageobj)

for pageNum in range(pr2.numPages):
    pageobj2 = pr2.getPage(pageNum)
    pdfw.addPage(pageobj2)

pageobj3 = pr3.getPage(0)
pdfw.addPage(pageobj3)

pdfout = open("aaa.pdf", "wb")
pdfw.write(pdfout)
```

```
pdfout.close()
pdff1.close()
pdff2.close()
pdff3.close()
```

上述代码执行后会合并 3 个文件（123.pdf、456.pdf 和 789.pdf）的内容，并将合并后的内容保存到新建文件 aaa.pdf 中。

实例 08-49：向指定 PDF 文件中写入文本

使用库 reportlab 可以在 PDF 文件中写入文本、绘制图形或图像等，在使用 reportlab 之前需要先安装 reportlab。其安装命令如下。

```
pip install reportlab
```

例如，下面的实例文件 reportlab01.py 演示了使用 reportlab 在指定 PDF 文件中写入文本的过程。

源码路径：daima\8\08-49\reportlab01.py

```
#引入所需要的基本包
from reportlab.pdfgen import canvas
#设置绘画开始的位置
def hello(c):
    c.drawString(100, 100, "hello world!")
#定义要生成的PDF文件的名称
c=canvas.Canvas("1.pdf")
#调用函数进行绘画，并将canvas对象作为参数传递
hello(c)
#showPage()函数：保存当前页的canvas
c.showPage()
#save()函数：保存文件并关闭canvas
c.save()
```

上述代码执行后会在文件 1.pdf 中写入文本"hello world!"。

实例 08-50：向 PDF 文件中写入指定样式的文本

下面的实例文件 reportlab02.py 演示了使用 reportlab 向 PDF 文件中写入指定样式文本的过程。

源码路径：daima\8\08-50\reportlab02.py

```
from reportlab.lib.styles import getSampleStyleSheet
from reportlab.platypus import Paragraph,SimpleDocTemplate
from reportlab.lib import colors

Style=getSampleStyleSheet()

bt = Style['Normal']           #字体的样式
bt.fontSize=14                 #字号
bt.wordWrap = 'CJK'
#该属性支持自动换行，CJK是中文方式换行，用于英文中会截断单词造成阅读困难，可改为Normal
bt.firstLineIndent = 32        #该属性支持第1行以空格开头
bt.leading = 20                #该属性用于设置行距

ct=Style['Normal']
ct.fontSize=12
ct.alignment=1                 #居中

ct.textColor = colors.red

t = Paragraph('hello',bt)
pdf=SimpleDocTemplate('2.pdf')
pdf.multiBuild([t])
```

通过上述代码在文件 2.pdf 中写入了指定样式的文本。一个 PDF 文件可以定义多种字体样式，如 bt 和 ct。字体有多种属性，例如，在上述代码中，自动换行属性 wordWrap 的参数 CJK 是按照中文方式换行的（可以在字符之间换行），英文方式为 Normal（在空格处换行）。alignment 的取值有 3 个，其中 0 表示左对齐，1 表示居中，2 表示右对齐。

实例 08-51：在 PDF 文件中绘制矢量图形

下面的实例文件 reportlab03.py 演示了使用 reportlab 在 PDF 文件中绘制矢量图形的过程。

源码路径：daima\8\08-51\reportlab03.py

```python
#引入所需要的基本包
from reportlab.pdfgen import canvas
from reportlab.lib.units import inch
#设置绘画开始的位置
def hello(c):
    #设置描边色
    c.setStrokeColorRGB(0, 0, 1.0)
    #设置填充色
    c.setFillColorRGB(1,0,1)
    #画线
    c.line(0.1*inch, 0.1*inch, 0.1*inch, 1.7*inch)
    c.line(0.1*inch, 0.1*inch, 1*inch, 0.1*inch)
    #绘制一个长方形
    c.rect(0.2*inch, 0.2*inch, 1*inch, 1.5*inch, fill=1)
#定义要生成的PDF文件的名称
c=canvas.Canvas("3.pdf")
#调用函数进行绘画，并将canvas对象作为参数传递
hello(c)
#showPage()函数：保存当前页的canvas
c.showPage()
#save()函数：保存文件并关闭canvas
c.save()
```

上述代码执行后会在文件 3.pdf 中绘制一个矢量图形，如图 8-37 所示。

图 8-37　绘制的矢量图形

实例 08-52：在 PDF 文件中绘制图像

下面的实例文件 reportlab04.py 演示了使用 Reportlab 在 PDF 文件中绘制图像的过程。

源码路径：daima\8\08-52\reportlab04.py

```python
#引入所需要的基本包
from reportlab.pdfgen import canvas
from reportlab.lib.units import mm

def drawBitmap(c):
    c.drawImage("123.png", 5*mm, 5*mm, 62*mm, 88.6*mm)

#定义要生成的PDF文件的名称
c=canvas.Canvas("4.pdf")
#调用函数绘制图像，并将canvas对象作为参数传递
drawBitmap(c)
#showPage()函数：保存当前页的canvas
c.showPage()
#save()函数：保存文件并关闭canvas
c.save()
```

上述代码执行后会在文件 4.pdf 中绘制图像 123.png，如图 8-38 所示。

图 8-38 绘制的图像

实例 08-53：分别在 PDF 文件和 PNG 文件中绘制饼图

下面的实例文件 reportlab05.py 演示了分别在 PDF 文件和 PNG 文件中绘制饼图的过程。

源码路径：daima\8\08-53\reportlab05.py

```python
from reportlab.graphics.charts.piecharts import Pie
from reportlab.graphics.shapes import Drawing, _DrawingEditorMixin
from reportlab.lib.colors import Color, magenta, cyan


class pietests(_DrawingEditorMixin, Drawing):
    def __init__(self, width=400, height=200, *args, **kw):
        Drawing.__init__(self, width, height, *args, **kw)
        self._add(self, Pie(), name='pie', validate=None, desc=None)
        self.pie.sideLabels = 1
        self.pie.labels = ['Label 1', 'Label 2', 'Label 3', 'Label 4', 'Label 5']
        self.pie.data = [20, 10, 5, 5, 5]
        self.pie.width = 140
        self.pie.height = 140
        self.pie.y = 35
        self.pie.x = 125


def main():
    drawing = pietests()
    #绘制图形，并保存为PDF文件和PNG文件
    drawing.save(formats=['pdf', 'png'], outDir='.', fnRoot=None)
    return 0


if __name__ == '__main__':
    main()
```

上述代码执行后将分别在文件 pietests000.pdf 和 pietests000.png 中绘制一个饼状图，如图 8-39 所示。

图 8-39 绘制的饼状图

实例 08-54：在 PDF 文件中分别生成条形图和二维码

例如，下面的实例文件 reportlab06.py 演示了使用 reportlab 在 PDF 文件中生成条形图和二维码的过程。

源码路径：daima\8\08-54\reportlab06.py

```python
#引入所需要的基本包
from reportlab.pdfgen import canvas
from reportlab.graphics.barcode import code39, code128, code93
from reportlab.graphics.barcode import eanbc, qr, usps
from reportlab.graphics.shapes import Drawing
from reportlab.lib.units import mm
from reportlab.graphics import renderPDF

#----------------------------------------------------------------
def createBarCodes(c):
    barcode_value = "1234567890"

    barcode39 = code39.Extended39(barcode_value)
    barcode39Std = code39.Standard39(barcode_value, barHeight=20, stop=1)

    barcode93 = code93.Standard93(barcode_value)

    barcode128 = code128.Code128(barcode_value)

    barcode_usps = usps.POSTNET("50158-9999")

    codes = [barcode39, barcode39Std, barcode93, barcode128, barcode_usps]

    x = 1 * mm
    y = 285 * mm

    for code in codes:
        code.drawOn(c, x, y)
        y = y - 15 * mm

    barcode_eanbc8 = eanbc.Ean8BarcodeWidget(barcode_value)
    d = Drawing(50, 10)
    d.add(barcode_eanbc8)
    renderPDF.draw(d, c, 15, 555)

    barcode_eanbc13 = eanbc.Ean13BarcodeWidget(barcode_value)
    d = Drawing(50, 10)
    d.add(barcode_eanbc13)
    renderPDF.draw(d, c, 15, 465)

    qr_code = qr.QrCodeWidget('http://www.toppr.net')
    bounds = qr_code.getBounds()
    width = bounds[2] - bounds[0]
    height = bounds[3] - bounds[1]
    d = Drawing(45, 45, transform=[45./width,0,0,45./height,0,0])
    d.add(qr_code)
    renderPDF.draw(d, c, 15, 405)

#定义要生成的PDF文件的名称
c=canvas.Canvas("6.pdf")
#调用函数生成条形码和二维码,并将canvas对象作为参数传递
createBarCodes(c)
#showPage()函数：保存当前页的canvas
c.showPage()
#save()函数：保存文件并关闭canvas
c.save()
```

上述代码执行后将在文件 6.pdf 中生成条形图和二维码，每个条形图和二维码都有自己的具体含义，如图 8-40 所示。

图 8-40　生成条形图和二维码

第 9 章

图形化界面开发实战

图形用户界面（Graphical User Interface，GUI）又称图形用户接口，是指采用图形方式显示的计算机操作用户界面。例如，Windows 操作系统就是一个功能强大的 GUI 程序。本章将通过具体实例的实现过程，详细讲解开发图形化界面的知识。

9.1 使用内置库 tkinter

在 Python 程序中，tkinter 是 Python 的一个内置模块，可以像其他模块一样在 Python Shell 中（或 Python 程序中）被导入，tkinter 模块被导入后即可使用 tkinter 模块中的函数、方法等。开发者可以使用 tkinter 库中的文本框、按钮、标签等组件（widget）实现 GUI 开发功能。整个实现过程十分简单，例如，要实现某个界面元素，只需要调用对应的 tkinter 组件即可。

实例 09-01：创建第一个 GUI 程序

当在 Python 程序中使用 tkinter 模块时，需要先使用 tkinter.Tk()生成一个主窗口对象，然后才能使用 tkinter 模块中的其他函数和方法等元素。当生成主窗口以后才可以向里面添加组件，或者直接调用其 mainloop()方法进行消息循环。例如，下面的实例文件 first.py 演示了使用 tkinter 创建第一个 GUI 程序的过程。

源码路径：daima\9\09-01\first.py

```
import tkinter                #导入tkinter模块
top = tkinter.Tk()            #生成一个主窗口对象
# 进入消息循环
top.mainloop()
```

在上述实例代码中，首先导入了 tkinter 库，然后使用 tkinter.Tk()生成一个主窗口对象，并进入消息循环。生成的窗口具有一般应用程序窗口的基本功能，可以最小化、最大化、关闭，还具有标题栏，甚至可以使用鼠标调整其大小。执行效果如图 9-1 所示。

通过上述实例代码创建了一个简单的 GUI 窗口，在完成窗口内部组件的创建工作后，也要进入消息循环中，这样可以处理窗口及其内部组件的事件。

图 9-1　执行效果

实例 09-02：向窗口中添加组件

前面创建的窗口只是一个容器，在这个容器中还可以添加其他元素。在 Python 程序中，当使用 tkinter 创建 GUI 窗口后，接下来可以向窗口中添加组件。其实，组件与窗口一样，也是通过 tkinter 模块中相应的组件函数生成的。在生成组件以后，就可以使用 pack()、grid()或 place()等方法将其添加到窗口中。例如，下面的实例文件 zu.py 演示了使用 tkinter 向窗口中添加组件的过程。

源码路径：daima\9\09-02\zu.py

```
import tkinter                                      #导入tkinter模块
root = tkinter.Tk()                                 #生成一个主窗口对象
#实例化标签组件
label= tkinter.Label(root, text="Python, tkinter!")
label.pack()                                        #将标签添加到窗口
button1 = tkinter.Button(root, text="按钮1")        #创建按钮1
button1.pack(side=tkinter.LEFT)                     #将按钮1添加到窗口
button2 = tkinter.Button(root, text="按钮2")        #创建按钮2
button2.pack(side=tkinter.RIGHT)                    #将按钮2添加到窗口
root.mainloop()                                     #进入消息循环
```

在上述实例代码中，分别实例化了库 tkinter 中的 1 个标签组件（Label）和 2 个按钮组件（Button），然后调用 pack()方法将这 3 个组件添加到主窗口中。执行效果如图 9-2 所示。

图 9-2 执行效果

实例 09-03：使用 Frame() 布局窗体界面

在实例文件 zu.py 中，曾经使用组件的 pack() 方法将组件添加到窗口中，而没有设置组件的位置，实例中的组件位置都是由 tkinter 模块自动确定的。对于一个包含多个组件的窗口，为了让组件布局更加合理，可以通过向方法 pack() 传递参数来设置组件在窗口中的具体位置。除了组件的 pack() 方法以外，还可以通过方法 grid() 和方法 place() 来设置组件的位置。例如，下面的实例文件 Frame.py 演示了使用 Frame() 布局窗体界面的过程。

源码路径：daima\9\09-03\Frame.py

```
from tkinter import *
root = Tk()
root.title("hello world")
root.geometry('300x200')

Label(root, text='校训', font=('Arial', 20)).pack()

frm = Frame(root)
#左
frm_L = Frame(frm)
Label(frm_L, text='厚德', font=('Arial', 15)).pack(side=TOP)
Label(frm_L, text='博学', font=('Arial', 15)).pack(side=TOP)
frm_L.pack(side=LEFT)

#右
frm_R = Frame(frm)
Label(frm_R, text='敬业', font=('Arial', 15)).pack(side=TOP)
Label(frm_R, text='乐群', font=('Arial', 15)).pack(side=TOP)
frm_R.pack(side=RIGHT)

frm.pack()

root.mainloop()
```

执行效果如图 9-3 所示。

图 9-3 执行效果

实例 09-04：向窗口中添加按钮控件

库 tkinter 中有很多 GUI 控件，主要包括在图形化界面中常用的按钮、标签、文本框、菜单、单选框、复选框等，本节将首先介绍按钮控件的使用方法。在使用按钮控件 tkinter.Button() 时，通过向其传递属性参数的方式可以控制按钮的属性。例如，可以设置按钮文本的颜色、按钮的颜色、按钮的大小及按钮的状态等。例如，下面的实例文件 an.py 演示了使用 tkinter 向窗口中添加按钮控件的过程。

源码路径：daima\9\09-04\an.py

```
import tkinter                            #导入tkinter模块
root = tkinter.Tk()                       #生成一个主窗口对象
button1 = tkinter.Button(root,            #创建按钮1
                anchor = tkinter.E,       #设置文本的对齐方式
                text = 'Button1',         #设置按钮文本
                width = 30,               #设置按钮的宽度
                height = 7)               #设置按钮的高度
button1.pack()                            #将按钮添加到窗口
button2 = tkinter.Button(root,            #创建按钮2
                text = 'Button2',         #设置按钮文本
                bg = 'blue')              #设置按钮的背景色
button2.pack()                            #将按钮添加到窗口
button3 = tkinter.Button(root,            #创建按钮3
                text = 'Button3',         #设置按钮文本
                width = 12,               #设置按钮的宽度
                height = 1)               #设置按钮的高度
button3.pack()                            #将按钮添加到窗口
button4 = tkinter.Button(root,            #创建按钮4
                text = 'Button4',#设置按钮文本
                width = 40,               #设置按钮的宽度
                height = 7,               #设置按钮的高度
                state = tkinter.DISABLED) #设置按钮禁用
button4.pack()                            #将按钮添加到窗口
root.mainloop()                           #进入消息循环
```

在上述实例代码中，使用不同的属性参数实例化了 4 个按钮，并分别将这 4 个按钮添加到主窗口中。上述代码执行后会在主窗口中显示 4 种不同的按钮，执行效果如图 9-4 所示。

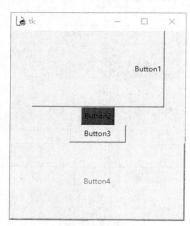

图 9-4　执行效果

实例 09-05：使用文本框控件

在库 tkinter 的控件中，文本框控件主要用于实现信息接收和用户的信息输入工作。在 Python 程序中，使用 tkinter.Entry() 和 tkinter.Text() 可以创建单行文本框和多行文本框。通过向其传递属性参数可以设置文本框的背景色、大小、状态等。例如，下面的实例文件 wen.py 演示了使用 tkinter 在窗口中使用文本框控件的过程。

源码路径：daima\9\09-05\wen.py

```
import tkinter                            #导入tkinter模块
root = tkinter.Tk()                       #生成一个主窗口对象
entry1 = tkinter.Entry(root,              #创建单行文本框1
                show = '*',)              #设置显示的文本是星号
```

```
entry1.pack()                                   #将文本框添加到窗口
entry2 = tkinter.Entry(root,                    #创建单行文本框2
             show = '#',                        #设置显示的文本是#号
             width = 50)                        #设置文本框的宽度
entry2.pack()                                   #将文本框添加到窗口
entry3 = tkinter.Entry(root,                    #创建单行文本框3
             bg = 'red',                        #设置文本框的背景色
             fg = 'blue')                       #设置文本框的前景色
entry3.pack()                                   #将文本框添加到窗口
entry4 = tkinter.Entry(root,                    #创建单行文本框4
             selectbackground = 'red',          #设置选中文本的背景色
             selectforeground = 'gray')         #设置选中文本的前景色
entry4.pack()                                   #将文本框添加到窗口
entry5 = tkinter.Entry(root,                    #创建单行文本框5
             state = tkinter.DISABLED)          #设置文本框禁用
entry5.pack()                                   #将文本框添加到窗口
edit1 = tkinter.Text(root,                      #创建多行文本框
             selectbackground = 'red',          #设置选中文本的背景色
             selectforeground = 'gray')
edit1.pack()                                    #将文本框添加到窗口
root.mainloop()                                 #进入消息循环
```

在上述实例代码中，使用不同的属性参数实例化了 6 种文本框，执行效果如图 9-5 所示。

图 9-5　执行效果

实例 09-06：实现会员注册界面效果

下面的实例文件 wen1.py 演示了使用 tkinter 在窗体中实现会员注册界面效果的过程。

源码路径：daima\9\09-06\wen1.py

```python
from tkinter import *
class MainWindow:
    def __init__(self):
        self.frame = Tk()

        self.label_name = Label(self.frame, text="name:")
        self.label_age = Label(self.frame, text="age:")
        self.label_sex = Label(self.frame, text="sex:")

        self.text_name = Text(self.frame, height="1", width=30)
        self.text_age = Text(self.frame, height="1", width=30)
        self.text_sex = Text(self.frame, height="1", width=30)

        self.label_name.grid(row=0, column=0)
        self.label_age.grid(row=1, column=0)
        self.label_sex.grid(row=2, column=0)

        self.button_ok = Button(self.frame, text="ok", width=10)
        self.button_cancel = Button(self.frame, text="cancel", width=10)
```

```
                self.text_name.grid(row=0, column=1)
                self.text_age.grid(row=1, column=1)
                self.text_sex.grid(row=2, column=1)

                self.button_ok.grid(row=3, column=0)
                self.button_cancel.grid(row=3, column=1)

                self.frame.mainloop()
frame = MainWindow()
```

执行效果如图 9-6 所示。

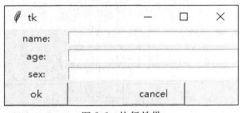

图 9-6　执行效果

实例 09-07：使用菜单控件

在库 tkinter 的控件中，使用菜单控件的方式与使用其他控件的方式有所不同。在创建菜单控件时，需要使用创建主窗口的方法 config()将菜单添加到窗口中。例如，下面的实例文件 cai.py 演示了使用 tkinter 在窗口中使用菜单控件的过程。

源码路径：daima\9\09-07\cai.py

```
import tkinter
root = tkinter.Tk()
menu = tkinter.Menu(root)
submenu = tkinter.Menu(menu, tearoff=0)
submenu.add_command(label="打开")
submenu.add_command(label="保存")
submenu.add_command(label="关闭")
menu.add_cascade(label="文件", menu=submenu)
submenu = tkinter.Menu(menu, tearoff=0)
submenu.add_command(label="复制")
submenu.add_command(label="粘贴")
submenu.add_separator()
submenu.add_command(label="剪切")
menu.add_cascade(label="编辑", menu=submenu)
submenu = tkinter.Menu(menu, tearoff=0)
submenu.add_command(label="关于")
menu.add_cascade(label="帮助", menu=submenu)
root.config(menu=menu)
root.mainloop()
```

在上述实例代码中，在主窗口中加入了 3 个主菜单，而在每个主菜单下面又创建了对应的子菜单。其中，主窗口中显示了 3 个主菜单（文件、编辑、帮助），而在"文件"主菜单下设置了 3 个子菜单(打开、保存、关闭)。在第 2 个主菜单"编辑"中，通过代码 submenu.add_separator()添加了一个分割线。执行效果如图 9-7 所示。

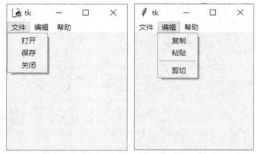

图 9-7　执行效果

实例 09-08：在窗口中创建标签

在 Python 程序中，标签控件的功能是在窗口中显示文本或图片。在库 tkinter 的控件中，使用 tkinter.Label()可以创建标签控件。例如，下面的实例文件 biao.py 演示了使用 tkinter 在窗口中创建标签的过程。

源码路径：daima\9\09-08\biao.py

```
import tkinter                              #导入tkinter模块
root = tkinter.Tk()                         #生成一个主窗口对象
label1 = tkinter.Label(root,                #创建标签1
            anchor = tkinter.E,             #设置标签文本的位置
            bg = 'red',                     #设置标签的背景色
            fg = 'blue',                    #设置标签的前景色
            text = 'Python',                #设置标签的显示文本
            width = 40,                     #设置标签的宽度
            height = 5)                     #设置标签的高度
label1.pack()                               #将标签添加到主窗口
label2 = tkinter.Label(root,                #创建标签2
            text = 'Python\ntkinter',       #设置标签的显示文本
            justify = tkinter.LEFT,         #设置多行文本左对齐
            width = 40,                     #设置标签的宽度
            height = 5)                     #设置标签的高度
label2.pack()                               #将标签添加到主窗口
label3 = tkinter.Label(root,                #创建标签3
            text = 'Python\ntkinter',       #设置标签的显示文本
            justify = tkinter.RIGHT,        #设置多行文本右对齐
            width = 40,                     #设置标签的宽度
            height = 5)                     #设置标签的高度
label3.pack()                               #将标签添加到主窗口
label4 = tkinter.Label(root,                #创建标签4
            text = 'Python\ntkinter',       #设置标签的显示文本
            justify = tkinter.CENTER,       #设置多行文本居中对齐
            width = 40,                     #设置标签的宽度
            height = 5)                     #设置标签的高度
label4.pack()                               #将标签添加到主窗口
root.mainloop()                             #进入消息循环
```

在上述实例代码中，在主窗口中创建了 4 个类型的标签，执行效果如图 9-8 所示。

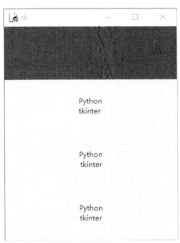

图 9-8　执行效果

实例 09-09：在 tkinter 窗口中创建单选按钮和复选框

在 Python 程序中，在一组单选按钮中只有一个选项可以被选中，而在复选框中可以同时选择多个选项。在库 tkinter 的控件中，使用 tkinter.Radiobutton()和 tkinter.Checkbutton()可以分别创建单选按钮和复选框。通过向其传递属性参数的方式，可以单独设置单选按钮和复选框的背景色、大小、状态等。例如，下面的实例文件 danfu.py 演示了在 tkinter 窗口中创建单选按钮和复选框的过程。

源码路径：daima\9\09-09\danfu.py

```
import tkinter                              #导入tkinter模块
root = tkinter.Tk()                         #生成一个主窗口对象
r = tkinter.StringVar()                     #生成字符串变量
r.set('1')                                  #初始化变量值
radio = tkinter.Radiobutton(root,           #创建单选按钮1
                variable = r,               #单选按钮关联的变量
                value = '1',                #设置选中单选按钮时的变量值
                text = '单选按钮1')          #设置单选按钮的显示文本
radio.pack()                                #将单选按钮1添加到窗口
radio = tkinter.Radiobutton(root,           #创建单选按钮2
                variable = r,               #单选按钮关联的变量
                value = '2',                #设置选中单选按钮时的变量值
                text = '单选按钮2' )         #设置单选按钮的显示文本
radio.pack()                                #将单选按钮2添加到窗口
radio = tkinter.Radiobutton(root,           #创建单选按钮3
                variable = r,               #单选按钮关联的变量
                value = '3',                #设置选中单选按钮时的变量值
                text = '单选按钮3' )         #设置单选按钮的显示文本
radio.pack()                                #将单选按钮3添加到窗口
radio = tkinter.Radiobutton(root,           #创建单选按钮4
                variable = r,               #单选按钮关联的变量
                value = '4',                #设置选中单选按钮时的变量值
                text = '单选按钮4' )         #设置单选按钮的显示文本
radio.pack()                                #将单选按钮4添加到窗口
c = tkinter.IntVar()                        #生成整型变量
c.set(1)                                    #变量初始化
check = tkinter.Checkbutton(root,           #创建复选框
                text = '复选按钮',           #设置复选框的显示文本
                variable = c,               #复选框关联的变量
                onvalue = 1,                #设置选中复选框时的变量值
                offvalue = 2)               #设置未选中复选框时的变量值
check.pack()                                #将复选框添加到窗口
root.mainloop()                             #进入消息循环
print(r.get())                              #调用函数get()输出r
print(c.get())                              #调用函数get()输出c
```

在上述实例代码中，在主窗口中分别创建了一个包含 4 个选项的单选按钮和包含 1 个选项的复选框。其中，使用函数 StringVar()生成字符串变量，将生成的字符串用于单选按钮组件；使用函数 IntVar()生成整型变量，并将生成的变量用于复选框。执行效果如图 9-9 所示。

图 9-9　执行效果

实例 09-10：在窗口中绘制图形

在 Python 程序中，可以使用 tkinter.Canvas()创建绘图控件，通过绘图控件提供的方法可以绘制直线、圆弧、矩形及图片。例如，下面的实例文件 tu.py 演示了使用 tkinter 在窗口中绘制图形的过程。

源码路径：daima\9\09-10\tu.py

```
import tkinter                              #导入tkinter模块
root = tkinter.Tk()                         #生成一个主窗口对象
canvas = tkinter.Canvas(root,               #创建绘图控件
                width = 600,                #设置绘图控件的宽度
                height = 480,               #设置绘图控件的高度
```

```
                            bg = 'white')                   #设置绘图控件的背景色
    #下一行代码用于打开一幅照片
    im = tkinter.PhotoImage(file='5-140FGZ249-51.gif')
    canvas.create_image(300,50,image = im)                  #将打开的照片添加到绘图控件
    canvas.create_text(302,77,
                       text = 'Use Canvas',fill = 'gray')   #设置绘制文字的内容
                       canvas.create_text(300,75,           #绘制文字
                       text = 'Use Canvas',                 #绘制的文字内容
                       fill = 'blue')                       #文字的颜色
    canvas.create_polygon(290,114,316,114,                  #绘制一个多（六）边形
                          330,130,310,146,284,146,270,130)
    canvas.create_oval(280,120,320,140,                     #绘制一个椭圆
                       fill = 'white')                      #设置椭圆颜色为白色
    canvas.create_line(250,130,350,130)                     #绘制直线
    canvas.create_line(300,100,300,160)                     #绘制直线
    canvas.create_rectangle(90,190,510,410,                 #绘制矩形
                            width=5)                        #设置矩形的线宽是5
    canvas.create_arc(100, 200, 500, 400,                   #绘制圆弧
                      start=0, extent=240,                  #设置圆弧的起始角度和终止角度
                      fill="pink")                          #设置圆弧的颜色为粉色
    canvas.create_arc(103,203,500,400,                      #绘制圆弧
                      start=241, extent=118,                #设置圆弧的起始角度和终止角度
                      fill="red")                           #设置圆弧的颜色为红色
    canvas.pack()                                           #将绘图控件添加到窗口
    root.mainloop()                                         #进入消息循环
```

在上述实例代码中，首先创建了一个绘图控件，然后在上面分别绘制了文字、图形、六边形、椭圆、直线、矩形和圆弧等图形，最后把绘图组件添加到主窗口中进行显示。执行效果如图 9-10 所示。

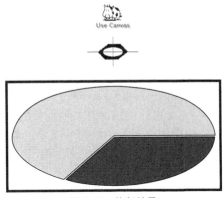

图 9-10　执行效果

实例 09-11：使用事件机制创建一个"英尺/米"转换器

在库 tkinter 中，事件是指在各个控件上发生的各种鼠标和键盘事件。对于按钮控件、菜单控件来说，可以在创建控件时通过参数 command 指定其事件的处理函数。除去控件所触发的事件外，在创建快捷菜单时还需处理右击事件，类似的事件还有鼠标事件、键盘事件和窗口事件。下面的实例文件 zhuan.py 演示了使用事件机制创建一个"英尺/米"转换器的过程。

源码路径：daima\9\09-11\zhuan.py

```
from tkinter import*
from tkinter import ttkZ
def calculate(*args):
    try:
        value = float(feet.get())
        meters.set((0.3048 * value * 10000.0 + 0.5) / 10000.0)
    except ValueError:
        pass


root = Tk()
```

```
root.title("英尺转换成米")
mainframe = ttk.Frame(root, padding="3 3 12 12")
mainframe.grid(column=0, row=0, sticky=(N, W, E, S))
mainframe.columnconfigure(0, weight=1)
mainframe.rowconfigure(0, weight=1)
feet = StringVar()
meters = StringVar()
feet_entry = ttk.Entry(mainframe, width=7, textvariable=feet)
feet_entry.grid(column=2, row=1, sticky=(W, E))
ttk.Label(mainframe, textvariable=meters).grid(column=2, row=2, sticky=(W, E))
ttk.Button(mainframe, text="计算", command=calculate).grid(column=3, row=3, sticky=W)
ttk.Label(mainframe, text="英尺").grid(column=3, row=1, sticky=W)
ttk.Label(mainframe, text="相当于").grid(column=1, row=2, sticky=E)
ttk.Label(mainframe, text="米").grid(column=3, row=2, sticky=W)
for child in mainframe.winfo_children(): child.grid_configure(padx=5, pady=5)
feet_entry.focus()
root.bind('<Return>', calculate)
root.mainloop()
```

上述代码的实现流程如下。

（1）导入 tkinter 所有的模块，这样可以直接使用 tkinter 的所有功能，这是 tkinter 的标准做法。然而在后面导入了 ttk 后，这意味着接下来要用到的控件前面都要加前缀。举一个例子，直接调用 Entry()会调用 tkinter 内部的模块，然而我们需要的是 ttk 中的 Entry()，所以要用 ttk.Entey()，其实许多函数在两者之中都有，如果同时用到这两个模块，则需要根据整体代码选择使用哪个模块，从而让 ttk 的调用更加清晰，本书也会采用这种风格。

（2）创建主窗口，设置窗口的标题为"英尺转换成米"，然后创建一个框架控件，用户界面上的所有东西都被包含在里面，并且被放在主窗口中。Columnconfigure()/rowconfigure()用于告诉 Tk，如果主窗口的大小被调整，则框架空间的大小也随之调整。

（3）创建 3 个主要的控件，第 1 个是用于输入英尺的文本框，第 2 个是用于输出转换成米单位结果的标签，第 3 个是用于执行计算的"计算"按钮。这 3 个控件都是窗口的"孩子""带主题"控件的类的实例。另外还为它们设置一些选项，如输入的宽度、按钮显示的文本等。输入框和标签都带了一个"神秘"的参数——textvariable。如果控件仅仅被创建了，则它是不会自动显示在屏幕上的，因为 Tk 并不知道这些控件和其他控件的位置关系，那是 grid()部分要做的事情。在程序的网格布局中，我们把每个控件放到对应行或者列中，sticky 选项指明控件在网格单元中的排列，用的是指南针方向。W 代表这个控件固定在左边的网格中。例如，WE 代表这个控件固定在左网格、右网格之间。

（4）创建 3 个静态标签，然后将其放在适合的网格位置。在最后 3 行代码中，第 1 行处理了框架中的所有控件，并且为每个空间四周添加了一些空隙，使其不会显得那么紧凑。我们可以在调用 grid()的时候做这件事，但上面这样做也是一个不错的选择。第 2 行告诉 Tk 让文本框获取到焦点。这样可以让光标一开始就在文本框中，用户就可以不用再去单击了。第 3 行告诉 Tk 如果用户在窗口中按了 Enter 键，就执行计算，等同于用户单击了"计算"按钮。

```
def calculate(*args):
    try:
        value = float(feet.get())
        meters.set((0.3048 * value * 10000.0 + 0.5)/10000.0)
    except ValueError:
        pass
```

上述代码定义了计算过程，无论是按 Enter 键还是单击"计算"按钮，都会从文本框中取得英尺，然后将其转换成米，并输出到标签中。执行效果如图 9-11 所示。

实例 09-12：实现一个动态绘图程序

下面的实例文件 huitu.py 演示了使用 tkinter

图 9-11　执行效果

事件实现一个动态绘图程序的过程。

源码路径：daima\9\09-12\huitu.py

(1) 先导入库 tkinter，然后在窗口中定义绘制不同图形的按钮，并且定义单击按钮时将要调用的操作函数。具体实现代码如下。

```python
import tkinter                                              #导入tkinter模块
class MyButton:                                             #定义一个按钮类MyButton
    def __init__(self,root,canvas,label,type):              #构造方法实现类的初始化
        self.root = root                                    #初始化属性root
        self.canvas = canvas                                #初始化属性canvas
        self.label = label                                  #初始化属性label
        if type == 0:                                       #type表示类型，如果type值为0
            button = tkinter.Button(root,text = '直线',     #创建一个绘制直线的按钮
                    command = self.DrawLine)                #通过command设置单击时要调用的操作函数
        elif type == 1:                                     #如果type值为1
            button = tkinter.Button(root,text = '弧形',     #创建一个绘制弧形的按钮
                    command = self.DrawArc)                 #通过command设置单击时要调用的操作函数
        elif type == 2:                                     #如果type值为2
            button = tkinter.Button(root,text = '矩形',     #创建一个绘制矩形的按钮
                    command = self.DrawRec)                 #通过command设置单击时要调用的操作函数
        else :                                              #如果type是其他值
            button = tkinter.Button(root,text = '椭圆',     #创建一个绘制椭圆的按钮
                    command = self.DrawOval)                #通过command设置单击时要调用的操作函数
        button.pack(side = 'left')                          #将按钮添加到主窗口
    def DrawLine(self):                                     #绘制直线函数
        self.label.text.set('直线')                         #显示的文本
        self.canvas.SetStatus(0)                            #设置绘图组件的状态
    def DrawArc(self):                                      #绘制弧形函数
        self.label.text.set('弧形')                         #显示的文本
        self.canvas.SetStatus(1)                            #设置绘图组件的状态
    def DrawRec(self):                                      #绘制矩形函数
        self.label.text.set('矩形')                         #显示的文本
        self.canvas.SetStatus(2)                            #设置绘图组件的状态
    def DrawOval(self):                                     #绘制椭圆函数
        self.label.text.set('椭圆')                         #显示的文本
        self.canvas.SetStatus(3)                            #设置绘图组件的状态
```

(2) 定义类 MyCanvas，根据用户单击的绘图按钮调用对应的操作函数，绑定不同的操作事件。具体实现代码如下。

```python
class MyCanvas:                                             #定义绘图类MyCanvas
    def __init__(self,root):                                #构造函数
        self.status = 0                                     #属性初始化
        self.draw = 0                                       #属性初始化
        self.root = root                                    #属性初始化
        self.canvas = tkinter.Canvas(root,bg = 'white',     #设置绘图组件的背景色为白色
                    width = 600,                            #设置绘图组件的宽度
                    height = 480)                           #设置绘图组件的高度
        self.canvas.pack()                                  #将绘图组件添加到主窗口
        #绑定鼠标释放事件，x=[1,2,3]，分别表示鼠标的左击、中击、右击操作
        self.canvas.bind('<ButtonRelease-1>',self.Draw)
        self.canvas.bind('<Button-2>',self.Exit)            #绑定鼠标中击事件
        self.daunas.bind('<Button-3>',self.Del)             #绑定鼠标右击事件
        self.canvas.bind_all('<Delete>',self.Del)           #绑定键盘中的Delete键
        self.canvas.bind_all('<KeyPress-d>',self.Del)       #绑定键盘中的D键
        self.canvas.bind_all('<KeyPress-e>',self.Exit)      #绑定键盘中的E键
    def Draw(self,event):                                   #定义绘图事件处理函数
        if self.draw == 0:                                  #开始绘图
            self.x = event.x                                #设置属性x
            self.y = event.y                                #设置属性y
            self.draw = 1                                   #设置属性draw
        else:
            if self.status == 0:                            #根据status的值绘制不同的图形
                self.canvas.create_line(self.x,self.y,      #绘制直线
                        event.x,event.y)
                self.draw = 0
            elif self.status == 1:
                self.canvas.create_arc(self.x,self.y,       #绘制弧形
                        event.x,event.y)
                self.draw = 0
            elif self.status == 2:
```

```
                    self.canvas.create_rectangle(self.x,self.y,    #绘制矩形
                                event.x,event.y)
                    self.draw = 0
            else:
                    self.canvas.create_oval(self.x,self.y,          #绘制椭圆
                                event.x,event.y)
                    self.draw = 0
    def Del(self,event):                        #鼠标右击或按D键时删除图形
        items = self.canvas.find_all()
        for item in items:
            self.canvas.delete(item)
    def Exit(self,event):                       #鼠标中击或按E键时则退出程序
        self.root.quit()
    def SetStatus(self,status):                 #使用status设置要绘制的图形
        self.status = status
```

（3）定义标签类 MyLabel，在绘制不同图形时显示对应的标签，具体实现代码如下。

```
class MyLabel:                                  #定义标签类MyLabel
    def __init__(self,root):                    #构造初始化
        self.root = root
        self.canvas = canvas
        self.text = tkinter.StringVar()         #生成标签引用变量
        self.text.set('Draw Line')              #设置标签文本
        self.label = tkinter.Label(root,textvariable = self.text,
                     fg = 'red',width = 50) #创建指定的标签
        self.label.pack(side = 'left')          #将标签添加到主窗口
```

（4）分别生成主窗口、绘图控件、标签和绘图按钮，具体实现代码如下。

```
root = tkinter.Tk()                             #生成主窗口
canvas = MyCanvas(root)                         #绘图组件
label = MyLabel(root)                           #生成标签
MyButton(root,canvas,label,0)                   #生成绘图按钮
MyButton(root,canvas,label,1)                   #生成绘图按钮
MyButton(root,canvas,label,2)                   #生成绘图按钮
MyButton(root,canvas,label,3)                   #生成绘图按钮
root.mainloop()                                 #进入消息循环
```

执行效果如图 9-12 所示。

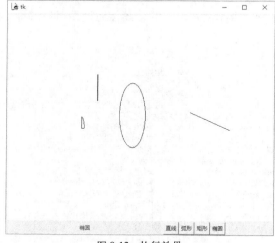

图 9-12　执行效果

实例 09-13：实现一个简单计算器程序

下面的实例文件 jsq.py 演示了使用 tkinter 事件实现一个简单计算器程序的过程。

源码路径：daima\9\09-13\jsq.py

```
from tkinter import *                           #导入tkinter库
root =Tk() #给窗体
root.title('calculator')                        #设置窗体名字
frm=Frame(root,bg='pink')                       #新建框架
frm.pack(expand = YES,fill = BOTH)              #放置框架
```

```
display=StringVar()
e=Entry(frm,textvariable=display)                    #添加文本框
e.grid(row=0,column=0,sticky=N,columnspan=4,rowspan=2) #放置文本框位置

def print_jia():
    e.insert(INSERT,'+')
def print_jian():
    e.insert(INSERT,'-')
def print_cheng():
    e.insert(INSERT,'*')
def print_chu():
    e.insert(INSERT,'/')
def print_dengyu():
    e.insert(INSERT,'=')

Button(frm,text='1',width=3,bg='yellow',command= lambda :e.insert(INSERT,'1')).grid(row=2,column=0,sticky=W)
#设置按钮，lambda为匿名函数
Button(frm,text='2',width=3,bg='yellow',command= lambda :e.insert(INSERT,'2')).grid(row=2,column=1)
Button(frm,text='3',width=3,bg='yellow',command= lambda :e.insert(INSERT,'3')).grid(row=2,column=2)
Button(frm,text='4',width=3,bg='yellow',command= lambda :e.insert(INSERT,'4')).grid(row=3,column=0,sticky=W)
Button(frm,text='5',width=3,bg='yellow',command= lambda :e.insert(INSERT,'5')).grid(row=3,column=1)
Button(frm,text='6',width=3,bg='yellow',command= lambda :e.insert(INSERT,'6')).grid(row=3,column=2)
Button(frm,text='7',width=3,bg='yellow',command= lambda :e.insert(INSERT,'7')).grid(row=4,column=0,sticky=W,rowspan=2)
Button(frm,text='8',width=3,bg='yellow',command= lambda :e.insert(INSERT,'8')).grid(row=4,column=1,rowspan=2)
Button(frm,text='9',width=3,bg='yellow',command= lambda :e.insert(INSERT,'9')).grid(row=4,column=2,rowspan=2)
Button(frm,text='/',width=4,bg='white',command=print_chu).grid(row=5,column=3,sticky=E)
Button(frm,text='*',width=4,bg='white',command=print_cheng).grid(row=4,column=3,sticky=E)
Button(frm,text='-',width=4,bg='white',command=print_jian).grid(row=3,column=3,sticky=E)
Button(frm,text='+',width=4,bg='white',command=print_jia).grid(row=2,column=3,sticky=E)
Button(frm,text='=',width=4,bg='white',command= lambda :cal(display)).grid(row=6,column=3,sticky=E)
Button(frm,text='clear',width=3,bg='red',command= lambda :display.set('')).grid(row=6,column=0,sticky=W)
Button(frm,text='0',width=3,bg='red',command= lambda :e.insert(INSERT,'0')).grid(row=6,column=2)
Button(frm,text='.',width=3,bg='red',command= lambda :e.insert(INSERT,'.')).grid(row=6,column=1)
def cal(display): #eval()函数将字符串转化为表达式
    display.set(eval(display.get()))

print('OK')
root.mainloop() #让程序一直循环
```

执行效果如图 9-13 所示。

图 9-13　执行效果

实例 09-14：创建消息对话框

tkinter.messagebox 模块为 Python 3 提供了几个内置的消息对话框模板。使用 tkinter.messagebox 模块中的函数 askokcancel()、askquestion()、askyesno()、showerror()、showinfo()和 showwarning()可以创建消息对话框，在使用这些方法时需要向其传递 title 和 message 参数。例如，下面的实例文件 duihua.py 演示了使用 tkinter.messagebox 模块创建消息对话框的过程。

源码路径：daima\9\09-14\duihua.py

```
import tkinter                    #导入tkinter模块
import tkinter.messagebox         #导入messagebox模块
def cmd():                        #定义处理按钮消息函数cmd()
    global n                      #定义全局变量n
    global buttontext             #定义全局变量buttontext
```

```
            n = n + 1              #设置n的值加1
            if n == 1:             #如果n的值是1
                tkinter.messagebox.askokcancel('Python tkinter','取消')   #调用askokcancel()函数
                buttontext.set('浪潮软件') #修改按钮上的文字
            elif n == 2:                                                  #如果n的值是2
                tkinter.messagebox.askquestion('Python tkinter','浪潮软件')  #调用askquestion()函数
                buttontext.set('AAAA')                                    #修改按钮上的文字
            elif n == 3:                                                  #如果n的值是3
                tkinter.messagebox.askyesno('Python tkinter','否')         #调用askyesno()函数
                buttontext.set('showerror')                               #修改按钮上的文字
            elif n == 4:                                                  #如果n的值是4
                tkinter.messagebox.showerror('Python tkinter','错误')      #调用showerror()函数
                buttontext.set('showinfo')                                #修改按钮上的文字
            elif n == 5:                                                  #如果n的值是5
                tkinter.messagebox.showinfo('Python tkinter','详情')       #调用showinfo()函数
                buttontext.set('显示警告')                                  #修改按钮上的文字
            else :                                                        #如果n是其他的值
                n = 0                                                     #将n赋值为0,并重新进行循环
                tkinter.messagebox.showwarning('Python tkinter','警告')    #调用showwarning()函数
                buttontext.set('AAAAl')                                   #修改按钮上的文字
        n = 0                                                             #设置n的初始值
        root = tkinter.Tk()                                               #生成一个主窗口对象
        buttontext = tkinter.StringVar()                                  #生成相关按钮的显示文字
        buttontext.set('AAAAl')                                           #设置buttontext显示的文字
        button = tkinter.Button(root,                                     #创建一个按钮
                    textvariable = buttontext,
                    command = cmd)
        button.pack()                                                     #将按钮添加到主窗口
        root.mainloop()                                                   #进入消息循环
```

在上述实例代码中创建了按钮消息处理函数，然后将其绑定到按钮上，单击按钮后会显示消息对话框效果，执行后会显示不同类型的消息对话框效果。执行效果如图9-14所示。

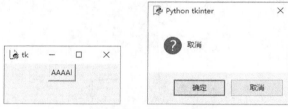

图 9-14　执行效果

实例 09-15：创建输入对话框

使用 tkinter.simpledialog 模块中的上述函数创建输入对话框后，可以返回输入对话框中文本框的值。例如，下面的实例文件 shuru.py 演示了使用 tkinter.simpledialog 模块创建输入对话框的过程。

源码路径：daima\9\09-15\shuru.py

```
import tkinter                            #导入tkinter模块
import tkinter.simpledialog               #导入simpledialog模块
def InStr():                              #定义处理按钮消息函数InStr()
    r = tkinter.simpledialog.askstring('Python',#输入字符串对话框
                '字符串',                   #提示字符
                initialvalue='字符串')#设置初始化文本
    print(r)                              #显示返回值
def InInt():                              #定义按钮事件处理函数
    r = tkinter.simpledialog.askinteger('Python','输入整数')   #输入整数对话框
    print(r)                              #显示返回值
def InFlo():                              #定义按钮事件处理函数
    r = tkinter.simpledialog.askfloat('Python','输入浮点数')   #输入浮点数对话框
    print(r)                              #显示返回值
root = tkinter.Tk()                       #生成一个主窗口对象
button1 = tkinter.Button(root,text = '输入字符串',   #创建按钮
            command = InStr)              #设置按钮事件处理函数
button1.pack(side='left')                 #将按钮添加到窗口
button2 = tkinter.Button(root,text = '输入整数',     #创建按钮
```

```
                command = InInt)                    #设置按钮事件处理函数
button2.pack(side='left')                            #将按钮添加到窗口
button2 = tkinter.Button(root,text = '输入浮点数',
                command = InFlo)                    #设置按钮事件处理函数
button2.pack(side='left')                            #将按钮添加到窗口
root.mainloop()                                      #进入消息循环
```

在上述实例代码中，定义了用于创建不同类型输入对话框的消息处理函数，然后分别将其绑定到相应的按钮上。单击3个按钮后将分别弹出3种不同类型的输入对话框。执行效果如图9-15所示。

图 9-15 执行效果

实例 09-16：创建打开/保存文件对话框

在 Python 程序中，可以使用 tkinter.filedialog 模块中的函数 askopenfilename()创建标准的打开文件对话框，使用函数 asksaveasfilename()可以创建标准的保存文件对话框。使用 filedialog 模块中的函数创建对话框后，可以返回选择的文件的完整路径。例如，下面的实例文件 da.py 演示了使用 tkinter 库创建打开/保存文件对话框的过程。

源码路径：daima\9\09-16\da.py

```
import tkinter                                      #导入tkinter模块
import tkinter.filedialog                           #导入filedialog模块
def FileOpen():                                     #创建打开文件事件处理函数
    r = tkinter.filedialog.askopenfilename(title = 'Python',filetypes=[('Python', '*.py *.pyw'),('All files', '*')])
    #创建打开文件对话框
    print(r)                                        #显示返回值
def FileSave():                                     #创建保存文件事件处理函数
    r = tkinter.filedialog.asksaveasfilename(title = 'Python',   #设置标题
                    initialdir=r'E:\Python\code',   #设置默认保存目录
                    initialfile = 'test.py')        #设置初始化文件
    print(r)                                        #显示返回值
root = tkinter.Tk()                                 #生成一个主窗口对象
button1 = tkinter.Button(root,text = '打开文件',    #创建按钮
            command = FileOpen)                     #设置按钮事件的处理函数
button1.pack(side='left')                            #将按钮添加到主窗口
button2 = tkinter.Button(root,text = '保存文件',    #创建按钮
            command = FileSave)                     #设置按钮事件的处理函数
button2.pack(side='left')                            #将按钮添加到主窗口
root.mainloop()                                      #进入消息循环
```

在上述实例代码中，首先定义了用于创建打开文件对话框和保存文件对话框的事件处理函数，然后将这两个函数绑定到相应的按钮上。执行程序后，单击"打开文件"按钮，将创建一个打开文件对话框；单击"保存文件"按钮，将创建一个文件保存对话框。执行效果如图9-16所示。

图 9-16 执行效果

实例 09-17：创建选择颜色对话框

在 Python 程序中，可以使用 tkinter.colorchooser 模块中的函数 askcolor()创建标准的选择颜

色对话框。使用 tkinter.colorchooser 模块创建选择颜色对话框后，会返回颜色的 RGB 值和可以在 tkinter 中使用的颜色字符值。例如，下面的实例文件 color.py 演示了使用 tkinter.colorchooser 模块创建选择颜色对话框的过程。

源码路径：daima\9\09-17\color.py

```
import tkinter                          #导入tkinter模块
import tkinter.colorchooser             #导入colorchooser模块
def ChooseColor():   #定义单击按钮后的事件处理函数
    #选择颜色对话框
    r = tkinter.colorchooser.askcolor(title = 'Python')
    print(r)                            #显示返回值
root = tkinter.Tk()                     #生成一个主窗口对象
button = tkinter.Button(root,text = '选择颜色',  #创建一个按钮
            command = ChooseColor)      #设置按钮单击事件处理函数
button.pack()                           #将按钮添加到窗口
root.mainloop()                         #进入消息循环
```

在上述实例代码中，定义了用于创建选择颜色对话框的事件处理函数，单击"选择颜色"按钮后弹出选择颜色对话框。执行效果如图 9-17 所示。

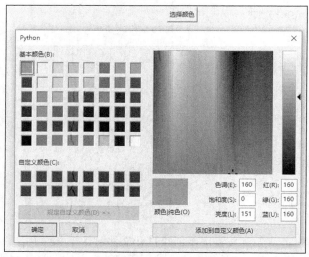

图 9-17 执行效果

实例 09-18：创建自定义对话框

在 Python 程序中，有时为了项目的特殊需求，需要创建自定义格式的对话框。可以使用库 tkinter 中的 Toplevel 控件来创建自定义对话框。开发者可以向 Toplevel 控件添加其他控件，并且可以定义事件处理函数或类。在使用库 tkinter 创建对话框的时候，如果在对话框中也需要进行事件处理，则建议大家以类的形式来定义对话框，否则只能大量使用全局变量来处理参数，这样会导致程序维护和调试困难。对于代码较多的 GUI 程序，建议大家使用类的方式来组织整个程序。例如，下面的实例文件 zidingyi.py 演示了使用库 tkinter 创建自定义对话框的过程。

源码路径：daima\9\09-18\zidingyi.py

```
import tkinter                          #导入tkinter模块
import tkinter.messagebox               #导入messagebox模块
class MyDialog:                         #定义对话框类MyDialog
    def __init__(self, root):           #构造函数初始化
        self.top = tkinter.Toplevel(root)   #创建Toplevel控件
        #创建Label控件
        label = tkinter.Label(self.top, text='请输入信息')
        label.pack()                    #将标签控件添加到主窗口
        self.entry = tkinter.Entry(self.top)   #文本框
```

```
                self.entry.pack()
                self.entry.focus()              #获得输入焦点
                button = tkinter.Button(self.top, text='好',
                        command=self.Ok)
                button.pack()                   #将按钮添加到主窗口
            def Ok(self):                       #定义单击按钮后的事件处理函数
                self.input = self.entry.get()   #获取文本框的内容
                self.top.destroy()              #销毁对话框
            def get(self):                      #定义单击按钮后的事件处理函数
                return self.input               #返回在文本框中输入的内容
        class MyButton():                       #定义类MyButton
            def __init__(self, root, type):     #构造函数
                self.root = root                #父窗口初始化
                if type == 0:                   #根据type的值创建不同的按钮，如果type值为0
                    self.button = tkinter.Button(root,
                            text='创建',         #设置按钮的显示文本
                            command = self.Create)  #设置单击按钮后的事件处理函数
                else:                           #如果type值不为0
                    self.button = tkinter.Button(root,   #创建"退出"按钮
                            text='退出',
                            command = self.Quit)   #设置单击按钮后的事件处理函数
                self.button.pack()
            def Create(self):                   #定义单击按钮后的事件处理函数
                d = MyDialog(self.root)         #创建一个对话框
                self.button.wait_window(d.top)  #等待对话框执行结果
                #显示在对话框中输入的内容
                tkinter.messagebox.showinfo('Python','你输入的是：\n' + d.get())
            def Quit(self):                     #定义单击按钮后的事件处理函数
                self.root.quit()                #用于退出主窗口
        root = tkinter.Tk()                     #生成一个主窗口对象
        MyButton(root,0)                        #生成"创建"按钮
        MyButton(root,1)                        #生成"退出"按钮
        root.mainloop()                         #进入消息循环
```

在上述实例代码中，使用 Toplevel 控件自定义创建了一个简单的对话框。首先自定义了两个类 MyDialog 和 MyButton，类实例化后被加入主窗口中。程序执行后会显示"创建"和"退出"两个按钮，单击"创建"按钮后将创建一个输入对话框，在里面的文本框中可以输入一些文字，单击"好"按钮后会弹出对话框，显示刚才输入的文字。单击"退出"按钮后则关闭窗口并退出程序。执行效果如图 9-18 所示。

图 9-18 执行效果

实例 09-19：开发一个记事本程序

例如，下面的实例文件 jishiben.py 演示了使用库 tkinter 开发一个记事本程序的过程。

源码路径：daima\9\09-19\jishiben.py

```
def author():
    showinfo('作者信息', '本软件由管管完成')

def about():
    showinfo('版权信息.Copyright', '本软件版权为管管所有')

def openfile():
    global filename
    filename = askopenfilename(defaultextension='.txt')
    if filename == '':
        filename = None
    else:
        root.title('FileName: ' + os.path.basename(filename))
        textPad.delete(1.0, END)
        f = open(filename,'r')
        textPad.insert(1.0, f.read())
        f.close()
```

```python
def new():
    global filename
    root.title('未命名文件')
    textPad.delete(1.0, END)

def save():
    global filename
    try:
        f = open(filename, 'w')
        msg = textPad.get(1.0, END)
        f.write(msg)
        f.close()
    except:
        save_as()

def save_as():
    f = asksaveasfilename(initialfile='未命名.txt', defaultextension='.txt')
    global filename
    filename = f
    fh = open(f, 'w')
    msg = textPad.get(1.0, END)
    fh.write(msg)
    fh.close()
    root.title('FileName: ' + os.path.basename(f))

def cut():
    textPad.event_generate('<<Cut>>')

def copy():
    textPad.event_generate('<<Copy>>')

def paste():
    textPad.event_generate('<<Paste>>')

def redo():       #重做
    textPad.event_generate('<<Redo>>')

def undo():       #撤销
    textPad.event_generate('<<Undo>>')

def select_all():
    textPad.tag_add('sel', '1.0', END)

def search():
    topsearch = Toplevel(root)
    topsearch.geometry('300x30+200+250')
    label1 = Label(topsearch, text='Find')
    label1.grid(row=0, column=0, padx=5)
    entry1 = Entry(topsearch, width=20)
    entry1.grid(row=0, column=1, padx=5)
    button1 = Button(topsearch, text='查找')
    button1.grid(row=0, column=2, padx=10)


root = Tk()
root.title('junxi note')
root.geometry("800x500+100+100")

#创建菜单
menubar = Menu(root)
root.config(menu = menubar)

# "文件" 菜单
filemenu = Menu(menubar)
filemenu.add_command(label='新建', accelerator='Ctrl + N', command=new)
filemenu.add_command(label='打开', accelerator='Ctrl + O', command=openfile)
filemenu.add_command(label='保存', accelerator='Ctrl + S', command=save)
filemenu.add_command(label='另存为', accelerator='Ctrl + Shift + S', command=save_as)
menubar.add_cascade(label= '文件', menu = filemenu)         #关联

# "编辑" 菜单
```

```
editmenu = Menu(menubar)
editmenu.add_command(label='撤销', accelerator='Ctrl + Z', command=undo)
editmenu.add_command(label='重做', accelerator='Ctrl + Y', command=redo)
editmenu.add_separator()
editmenu.add_command(label='剪切', accelerator='Ctrl + X', command=cut)
editmenu.add_command(label='复制', accelerator='Ctrl + C', command=copy)
editmenu.add_command(label='粘贴', accelerator='Ctrl + V', command=paste)
editmenu.add_separator()
editmenu.add_command(label='查找', accelerator='Ctrl + F', command=search)
editmenu.add_command(label='全选', accelerator='Ctrl + A', command=select_all)
menubar.add_cascade(label='编辑', menu = editmenu)

#"关于"菜单
aboutmenu = Menu(menubar)
aboutmenu.add_command(label='作者', command=author)
aboutmenu.add_command(label='版权', command=about)
menubar.add_cascade(label='关于', menu = aboutmenu)

#工具栏
toolbar = Frame(root, height=25, bg='Light sea green')

#按钮
shortButton = Button(toolbar, text='打开', command=openfile)
shortButton.pack(side=LEFT, padx=5, pady=5)
shortButton = Button(toolbar, text='保存', command=save)
shortButton.pack(side=LEFT)
toolbar.pack(expand=NO, fill=X)              #显示

#状态栏
status = Label(root, text='Ln20', bd=1, relief=SUNKEN, anchor=W)
status.pack(side=BOTTOM, fill=X)

#行号和文本编辑
linelabel = Label(root, width=2, bg='antique white')
linelabel.pack(side=LEFT, fill=Y)
textPad = Text(root, undo=True)
textPad.pack(expand=YES, fill=BOTH)

#右边的滚动下拉条
scroll = Scrollbar(textPad)
textPad.config(yscrollcommand=scroll.set)
scroll.config(command=textPad.yview)
scroll.pack(side=RIGHT, fill=Y)


root.mainloop()
```

执行效果如图 9-19 所示。

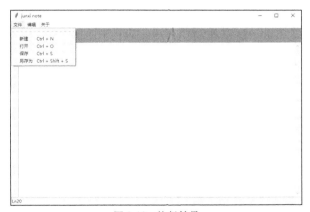

图 9-19 执行效果

实例 09-20：使用偏函数模拟实现交通标志

GUI 是偏函数的最佳发挥舞台之一，因为在很多时候需要 GUI 控件在外观上具有某种一

致性，而这种一致性来自使用相同参数创建的相似对象。我们现在要实现一个应用，这个应用中的很多按钮拥有相同的前景色和背景色。对于这种只有细微差别的按钮，每次都使用相同的参数创建相似的对象简直是一种浪费：前景色和背景色都是相同的，只有文本有一些差异。例如，下面的实例文件 pfaGUI3.py 演示了使用偏函数模拟实现交通标志的过程。在本实例文件中创建了文字版本的路标，并将其根据标志类型进行区分，如严重、警告、通知等（就像日志级别那样）。标志类型决定了创建时的颜色方案，例如，严重级别标志是白底红字，警告级别标志是黄底黑字，通知级别（即标准级别）标志是白底黑字。例如，"不准驶入"和"错误路线"标志属于严重级别，"汇入车道"和"十字路口"属于警告级别，而"限速路段"和"单行线"属于通知级别。当单击这些标志按钮时会弹出相应的 Tk 对话框：不准驶入，十字路口，60 限速路段等。

源码路径：daima\9\09-20\pfaGUI3.py

```
①from functools import partial as pto
  from tkinter import Tk, Button, X
  from tkinter.messagebox import showinfo, showwarning, showerror

  WARN = 'warn'
  CRIT = 'crit'
  REGU = 'regu'

  SIGNS = {
      '不准驶入': CRIT,
      '十字路口': WARN,
      '60\n限速路段': REGU,
      '误路线': CRIT,
      '汇入车道': WARN,
      '单行线': REGU,
②}

③critCB = lambda : showerror('Error', 'Error Button Pressed!')
  warnCB = lambda : showwarning('Warning', 'Warning Button Pressed!')
  infoCB = lambda : showinfo('Info', 'Info Button Pressed!')

  top = Tk()
  top.title('Road Signs')
  Button(top, text='QUIT', command=top.quit,
④      bg='red', fg='white').pack()

⑤MyButton = pto(Button, top)
  CritButton = pto(MyButton, command=critCB, bg='white', fg='red')
  WarnButton = pto(MyButton, command=warnCB, bg='goldenrod1')
⑥ReguButton = pto(MyButton, command=infoCB, bg='white')

⑦for eachSign in SIGNS:
      signType = SIGNS[eachSign]
      cmd = '%sButton(text=%r%s).pack(fill=X, expand=True)' % (
          signType.title(), eachSign,
          '.upper()' if signType == CRIT else '.title()')
      eval(cmd)

⑧top.mainloop()
```

①~②中先在应用程序中导入 functools.partial()、若干 tkinter 属性及若干 Tk 对话框，然后根据类别定义了一些标志。

③~④中使用 Tk 对话框实现按钮回调函数，在创建每个按钮时使用它们。然后启动 Tk 对话框并设置标题，并创建一个 QUIT 按钮。

⑤~⑥中使用了两阶偏函数。其中第一阶函数模板化了 Button 类和根窗口 top，这说明每次调用 MyButton 时就会调用 Button 类（tkinter.Button()会创建一个按钮），并将 top 作为它的第一个参数，我们将其冻结为 MyButton；第二阶偏函数会使用上面的第一阶偏函数，并对其进行模板化处理，会为每种标志类型创建单独的按钮类型。当用户创建一个严重类型的按钮

CritButton 时（如通过调用 CritButton()），就会调用包含适当的按钮回调函数、前景色和背景色的 MyButton，或者使用 top、回调函数和颜色这几个参数去调用 Button。我们可以看到它是如何一步步展开并最终调用到最底层的，如果没有偏函数，则这些调用本来应该是由用户自己执行的。WarnButton 和 ReguButton 也会执行同样的操作。

⑦~⑧中在设置好按钮后，会根据标志列表将其创建出来。使用一个 Python 可求值字符串，该字符串由正确的按钮名、传给按钮标签的文本参数及 pack()操作组成。如果这是一个严重级别的标志，那么所有字符大写；否则，按照标题格式进行输出。每个按钮会通过 eval()函数进行实例化处理，在最后进入主事件循环来启动 GUI 程序。

执行效果如图 9-20 所示。

图 9-20　执行效果

实例 09-21：创建桌面天气预报程序

例如，下面的实例文件 123.py 演示了创建桌面天气预报程序的过程。

源码路径：daima\9\09-21\123.py

```
class Weather:
    def __init__(self):
        self.root = Tk()
        self.root.geometry("500x300")
        self.root.wm_title("Weather Teller")
        self.label = Label(self.root, text="City: ")
        self.label.pack()
        self.entrytext = StringVar()
        Entry(self.root, textvariable=self.entrytext).pack()
        self.buttontext = StringVar()
        self.buttontext.set("Tell me")
        Button(self.root, textvariable=self.buttontext, command=self.click).pack()
        self.label = Label(self.root, text="")
        self.label.pack()
        self.root.mainloop()

    def click(self):
        input = str(self.entrytext.get())
        location = str(input)
        r = requests.get('http://api.openweathermap.org/data/2.5/weather?q=' + location +
                        '&APPID=c8b38acfc161132a0da1574bc472e74e')
        j = json.loads(r.text)
        temperature = str(j['main']['temp'] - 273) + "℃"
        result = temperature
        self.label.configure(text=result)


def main():
    Weather()
```

```
if __name__ == '__main__':
    main()
```

上述代码执行后会提示用户输入城市的名字，单击 Tell me 按钮后会显示这个城市的当前温度。本实例不支持中文，国内城市名字可以输入拼音全拼，国外城市名字是英文格式。例如，分别输入"jinan"和"London"后的效果如图 9-21 所示。

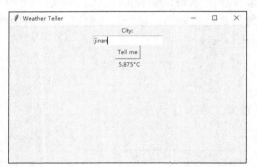 

图 9-21　获取济南和伦敦当前的温度

实例 09-22：创建精简版资源管理器

在下面的实例文件 ziyuan.py 创建了精简版资源管理器：目录遍历系统。其功能是显示当前文件所在目录的文件列表，双击列表中任意一个目录后会切换到这个新目录中，用新目录中的文件列表代替旧文件列表。实例文件 ziyuan.py 的具体实现流程如下。

源码路径：daima\9\09-22\ziyuan.py

（1）导入 os 模块、time.sleep()函数和 tkinter 模块的所有属性。

（2）定义类 DirList 的构造函数和一个待应用的对象，然后创建第一个标签控件，其中的文本是应用的主标题和版本号。对应实现代码如下。

```
class DirList(object):
    def __init__(self, initdir=None):
        self.top = Tk()
        self.label = Label(self.top,
            text='资源管理器（精简版）')
        self.label.pack()
```

（3）声明 Tk 的一个变量 cwd，用于保存当前所在的目录名。然后又创建了另一个标签控件，用于显示当前的目录名。对应实现代码如下。

```
self.cwd=StringVar(self.top)

self.dirl = Label(self.top, fg='blue',
    font=('Helvetica', 12, 'bold'))
self.dirl.pack()
```

（4）定义 GUI 程序的核心部分，Listbox 控件 dirs，该控件包含了要列出的目录的文件列表。Scrollbar 可以让用户在文件数超过 Listbox 的大小时能够移动列表。上述这两个控件都包含在框架控件中。通过使用 Listbox 的 bind()方法，Listbox 的列表项可以与回调函数（setdirandgo()）连接起来。绑定意味着将一个回调函数与键盘、鼠标操作或一些其他事件连接起来，当用户触发这类事件时，回调函数就会执行。当双击 Listbox 中的任意列表项时，就会调用 setdirandgo()函数。Scrollbar 通过调用 Scrollbar.config()方法与 Listbox 连接起来。对应实现代码如下。

```
self.dirfm = Frame(self.top)
self.dirsb = Scrollbar(self.dirfm)
self.dirsb.pack(side=RIGHT, fill=Y)
self.dirs = Listbox(self.dirfm, height=15,
    width=50, yscrollcommand=self.dirsb.set)
self.dirs.bind('<Double-1>', self.setdirandgo)
```

```
self.dirsb.config(command=self.dirs.yview)
self.dirs.pack(side=LEFT, fill=BOTH)
self.dirfm.pack()
```

（5）创建一个文本框，可以在其中输入想要遍历的目录名，从而可以在 Listbox 中看到该目录中的文件列表。在这个文本框中添加了一个 Enter 键的绑定，这样除了单击按钮，还可以通过按 Enter 键来更新文件列表。我们之前在 Listbox 中看到的鼠标绑定也是同样的应用。用鼠标右键双击 Listbox 中的列表项，其效果与在文本框中手动输入目录名然后单击 go 按钮有同样的效果。对应实现代码如下。

```
self.dirn = Entry(self.top, width=50,
        textvariable=self.cwd)
self.dirn.bind('<Return>', self.dols)
self.dirn.pack()
```

（6）定义按钮框架（bfm），用于放置 3 个按钮：一个"清除"按钮（clr）、一个"目标列表"按钮（ls）和一个"退出"按钮（quit）。每个按钮在单击时都有其自己的配置和回调函数。对应实现代码如下。

```
self.bfm = Frame(self.top)
self.clr = Button(self.bfm, text='清除',
        command=self.clrdir,
        activeforeground='white',
        activebackground='blue')
self.ls = Button(self.bfm,
        text='目录列表',
        command=self.dols,
        activeforeground='white',
        activebackground='green')
self.quit = Button(self.bfm, text='退出',
        command=self.top.quit,
        activeforeground='white',
        activebackground='red')
self.clr.pack(side=LEFT)
self.ls.pack(side=LEFT)
self.quit.pack(side=LEFT)
self.bfm.pack()
```

（7）实现构造函数的最后一部分来初始化 GUI 程序，以当前工作目录作为起始点。对应实现代码如下。

```
if initdir:
    self.cwd.set(os.curdir)
    self.dols()
```

（8）函数 clrdir()会清空 Tk 字符串变量 cwd（包含当前活动目录），该变量不但会跟踪当前所处的目录，而且当发生错误时可以帮助我们回到之前的目录。回调函数中变量 ev 的默认值是 None，任何像这样的值都是由窗口系统传入的，它们在回调函数中可能会被用到，也可能用不到。对应实现代码如下。

```
def clrdir(self, ev=None):
    self.cwd.set('')
```

（9）函数 setdirandgo()用于设置要遍历的目录，并通过调用 dols()实现遍历目录的行为。对应实现代码如下。

```
def setdirandgo(self, ev=None):
    self.last = self.cwd.get()
    self.dirs.config(selectbackground='red')
    check = self.dirs.get(self.dirs.curselection())
    if not check:
        check = os.curdir
    self.cwd.set(check)
    self.dols()
```

（10）函数 dols()是上述整个 GUI 程序的最核心部分，功能是进行所有安全检查（例如，目标是否是一个目录？它是否存在？）。如果发生错误，则之前的目录就会被重设为当前目录。如果一切正常，就会调用 os.listdir()获取实际文件列表并在 Listbox 中进行替换。当后台忙于拉取新目录中的信息时，突出显示的蓝色条就会变成红色，直到新目录设置完毕后，它又会变回蓝色。

(11) 最后一段代码是这段代码的主要部分。只有当直接调用脚本时，main()函数才会执行。当 main()函数执行时，会先创建 GUI 程序，然后调用 mainloop()来启动 GUI 程序，之后由其控制程序的执行。对应实现代码如下。

```
def main():
    d = DirList(os.curdir)
    mainloop()
if __name__ == '__main__':
    main()
```

执行效果如图 9-22 所示。

图 9-22 执行效果

9.2 使用 tkinter 的扩展小部件 tkinter.tix

在 Python 程序中，模块 tkinter.tix 是 tkinter 的接口扩展，提供一些额外的丰富的小部件。虽然标准的 tkinter 库中有许多有用的小部件，但是它们远远不够完整。tkinter.tix 模块提供了标准 tkinter 中缺少的大多数常用窗口小部件（如 HList、ComboBox、ControlSpinBox）和各种可滚动小部件。另外，模块 tkinter.tix 还提供了许多通用于广泛应用的小部件：NoteBook、FileEntry 和 PanedWindow。

实例 09-23：使用 Balloon 组件

在使用 tkinter.tix 模块时必然会用到 TK()方法，其语法格式如下。

```
class tkinter.tix.Tk(screenName=None, baseName=None, className='Tix')
```

TK()是 tkinter.tix 的顶层小部件，其大部分参数代表应用程序的主窗口属性。模块 tkinter.tix 中的类对 tkinter 中的类进行子类化，因为是前者导入后者，所以在使用时需要导入 tkinter.tix 与 tkinter。其中，组件是 tkinter.tix 模块的核心功能，例如，下面的实例文件 zujian01.py 演示了使用 Balloon 组件的过程。

源码路径：daima\9\09-23\zujian01.py

```
import tkinter.tix as Tix

TCL_ALL_EVENTS              = 0

def RunSample (root):
    balloon = DemoBalloon(root)
    balloon.mainloop()
    balloon.destroy()
```

```python
class DemoBalloon:
    def __init__(self, w):
        self.root = w
        self.exit = -1

        z = w.winfo_toplevel()
        z.wm_protocol("WM_DELETE_WINDOW", lambda self=self: self.quitcmd())

        status = Tix.Label(w, width=40, relief=Tix.SUNKEN, bd=1)
        status.pack(side=Tix.BOTTOM, fill=Tix.Y, padx=2, pady=1)

        button1 = Tix.Button(w, text='Something Unexpected',
                             command=self.quitcmd)
        button2 = Tix.Button(w, text='Something Else Unexpected')
        button2['command'] = lambda w=button2: w.destroy()
        button1.pack(side=Tix.TOP, expand=1)
        button2.pack(side=Tix.TOP, expand=1)

        b = Tix.Balloon(w, statusbar=status)

        b.bind_widget(button1, balloonmsg='Close Window',
                      statusmsg='Press this button to close this window')
        b.bind_widget(button2, balloonmsg='Self-destruct button',
                      statusmsg='Press this button and it will destroy itself')

    def quitcmd (self):
        self.exit = 0

    def mainloop(self):
        foundEvent = 1
        while self.exit < 0 and foundEvent > 0:
            foundEvent = self.root.tk.dooneevent(TCL_ALL_EVENTS)

    def destroy (self):
        self.root.destroy()

if __name__ == '__main__':
    root = Tix.Tk()
    RunSample(root)
```

执行效果如图 9-23 所示。

图 9-23　执行效果

实例 09-24：使用 DirList 组件

组件 tkinter.tix.DirList 用于显示目录，例如，下面的实例文件 zujian02.py 演示了使用 DirList 组件的过程。

源码路径：daima\9\09-24\zujian02.py

```python
TCL_ALL_EVENTS = 0

def RunSample(root):
    dirlist = DemoDirList(root)
    dirlist.mainloop()
    dirlist.destroy()

class DemoDirList:
```

```
def __init__(self, w):
    self.root = w
    self.exit = -1

    z = w.winfo_toplevel()
    z.wm_protocol("WM_DELETE_WINDOW", lambda self=self: self.quitcmd())

    top = tkinter.tix.Frame(w, relief=RAISED, bd=1)

    top.dir = tkinter.tix.DirList(top)
    top.dir.hlist['width'] = 40
    top.btn = tkinter.tix.Button(top, text="  >>  ", pady=0)
    top.ent = tkinter.tix.LabelEntry(top, label="Installation Directory:",
                                    labelside='top',
                                    options='''
                                    entry.width 40
                                    label.anchor w
                                    ''')

    font = self.root.tk.eval('tix option get fixed_font')
    top.ent.entry['font'] = font

    self.dlist_dir = copy.copy(os.curdir)
    top.ent.entry['textvariable'] = self.dlist_dir
    top.btn['command'] = lambda dir=top.dir, ent=top.ent, self=self: \
        self.copy_name(dir, ent)

    top.ent.entry.bind('<Return>', lambda self=self: self.okcmd())

    top.pack(expand='yes', fill='both', side=TOP)
    top.dir.pack(expand=1, fill=BOTH, padx=4, pady=4, side=LEFT)
    top.btn.pack(anchor='s', padx=4, pady=4, side=LEFT)
    top.ent.pack(expand=1, fill=X, anchor='s', padx=4, pady=4, side=LEFT)
    box = tkinter.tix.ButtonBox(w, orientation='horizontal')
    box.add('ok', text='Ok', underline=0, width=6,
            command=lambda self=self: self.okcmd())
    box.add('cancel', text='Cancel', underline=0, width=6,
            command=lambda self=self: self.quitcmd())

    box.pack(anchor='s', fill='x', side=BOTTOM)
```

执行效果如图 9-24 所示。

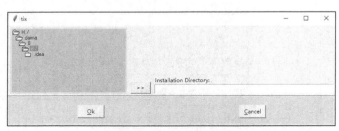

图 9-24　执行效果

实例 09-25：使用分组列表组件

- tkinter.tix.HList：用于显示具有层次结构的任何数据。例如，对于文件系统目录树，列表项缩进，并通过分支线根据它们在层次结构中的位置连接。
- tkinter.tix.CheckList：显示用户要选择的项目列表，其行为类似于 Tk 检查按钮或单选按钮，但它能够处理比检查按钮或单选按钮更多的项目。
- tkinter.tix.Tree：用于以树的形式显示分层数据，可以通过打开或关闭部分树来调整树的视图。

例如，下面的实例文件 zujian03.py 演示了使用分组列表组件的过程。

源码路径:daima\9\09-25\zujian03.py

```python
class DemoSHList:
    def __init__(self, w):
        self.root = w
        self.exit = -1

        z = w.winfo_toplevel()
        z.wm_protocol("WM_DELETE_WINDOW", lambda self=self: self.quitcmd())
        top = tkinter.tix.Frame( w, relief=tkinter.tix.RAISED, bd=1)

        top.a = tkinter.tix.ScrolledHList(top)
        top.a.pack( expand=1, fill=tkinter.tix.BOTH, padx=10, pady=10, side=tkinter.tix.TOP)
        bosses = [
            ('jeff',  'Jeff Waxman'),
            ('john',  'John Lee'),
            ('peter', 'Peter Kenson')
        ]

        employees = [
            ('alex',  'john',  'Alex Kellman'),
            ('alan',  'john',  'Alan Adams'),
            ('andy',  'peter', 'Andreas Crawford'),
            ('doug',  'jeff',  'Douglas Bloom'),
            ('jon',   'peter', 'Jon Baraki'),
            ('chris', 'jeff',  'Chris Geoffrey'),
            ('chuck', 'jeff',  'Chuck McLean')
        ]

        hlist=top.a.hlist

        hlist.config( separator='.', width=25, drawbranch=0, indent=10)

        count=0
        for boss,name in bosses :
            if count :
                f=tkinter.tix.Frame(hlist, name='sep%d' % count, height=2, width=150,
                    bd=2, relief=tkinter.tix.SUNKEN )

                hlist.add_child( itemtype=tkinter.tix.WINDOW,
                    window=f, state=tkinter.tix.DISABLED )

            hlist.add(boss, itemtype=tkinter.tix.TEXT, text=name)
            count = count+1


        for person,boss,name in employees :

            hlist.add( key, text=name )

        box= tkinter.tix.ButtonBox(top, orientation=tkinter.tix.HORIZONTAL )
        box.add( 'ok',     text='Ok',     underline=0,  width=6,
            command = self.okcmd)

        box.add( 'cancel', text='Cancel', underline=0, width=6,
            command = self.quitcmd)

        box.pack( side=tkinter.tix.BOTTOM, fill=tkinter.tix.X)
        top.pack( side=tkinter.tix.TOP,    fill=tkinter.tix.BOTH, expand=1 )
    def okcmd (self):
        self.quitcmd()

    def quitcmd (self):
        self.exit = 0

    def mainloop(self):
        while self.exit < 0:
            self.root.tk.dooneevent(TCL_ALL_EVENTS)

    def destroy (self):
        self.root.destroy()
```

执行效果如图 9-25 所示。

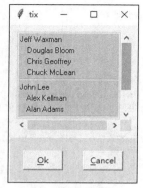

图 9-25 执行效果

实例 09-26：使用管理组件

- tkinter.tix.PanedWindow：允许用户交互式操作窗格的大小，窗格可以垂直或水平布局。用户通过在两个窗格之间拖动的方式来更改窗格的大小。
- tkinter.tix.ListNoteBook：与 TixNoteBook 组件非常相似，可以在窗口中显示多个页面。ListNoteBook 被分成一堆页面，一次只能显示其中一个页面。用户可以通过在 Hlist 子窗口中选择所需页面的名称来浏览这些页面。
- tkinter.tix.NoteBook：用于在窗口中使用 NoteBook 显示多个页面，NoteBook 被分为一堆页面，一次只能显示其中一个页面。可以通过选择 NoteBook 窗口顶部的可视标签来浏览这些页面。

例如，下面的实例文件 zujian04.py 演示了使用管理组件的过程。

源码路径：daima\9\09-26\zujian04.py

```
def RunSample(w):
    global root
    root = w
    prefix = tkinter.tix.OptionName(w)
    if prefix:
        prefix = '*'+prefix
    else:
        prefix = ''
    w.option_add(prefix+'*TixControl*entry.width', 10)
    w.option_add(prefix+'*TixControl*label.width', 18)
    w.option_add(prefix+'*TixControl*label.anchor', tkinter.tix.E)
    w.option_add(prefix+'*TixNoteBook*tagPadX', 8)
    nb = tkinter.tix.NoteBook(w, name='nb', ipadx=6, ipady=6)
    nb['bg'] = 'gray'
    nb.nbframe['backpagecolor'] = 'gray'

    nb.add('hard_disk', label="Hard Disk", underline=0)
    nb.add('network', label="Network", underline=0)

    nb.pack(expand=1, fill=tkinter.tix.BOTH, padx=5, pady=5 ,side=tkinter.tix.TOP)
    tab=nb.hard_disk
    f = tkinter.tix.Frame(tab)
    common = tkinter.tix.Frame(tab)

    f.pack(side=tkinter.tix.LEFT, padx=2, pady=2, fill=tkinter.tix.BOTH, expand=1)
    common.pack(side=tkinter.tix.RIGHT, padx=2, fill=tkinter.tix.Y)

    a = tkinter.tix.Control(f, value=12,    label='Access time: ')
    w = tkinter.tix.Control(f, value=400,   label='Write Throughput: ')
    r = tkinter.tix.Control(f, value=400,   label='Read Throughput: ')
    c = tkinter.tix.Control(f, value=1021, label='Capacity: ')

    a.pack(side=tkinter.tix.TOP, padx=20, pady=2)
    w.pack(side=tkinter.tix.TOP, padx=20, pady=2)
```

```
r.pack(side=tkinter.tix.TOP, padx=20, pady=2)
c.pack(side=tkinter.tix.TOP, padx=20, pady=2)

# Create the common buttons
createCommonButtons(common)

#---------------------------------------
#创建第二页
#---------------------------------------

tab = nb.network

f = tkinter.tix.Frame(tab)
common = tkinter.tix.Frame(tab)

f.pack(side=tkinter.tix.LEFT, padx=2, pady=2, fill=tkinter.tix.BOTH, expand=1)
common.pack(side=tkinter.tix.RIGHT, padx=2, fill=tkinter.tix.Y)

a = tkinter.tix.Control(f, value=12,   label='Access time: ')
w = tkinter.tix.Control(f, value=400,  label='Write Throughput: ')
r = tkinter.tix.Control(f, value=400,  label='Read Throughput: ')
c = tkinter.tix.Control(f, value=1021, label='Capacity: ')
u = tkinter.tix.Control(f, value=10,   label='Users: ')

a.pack(side=tkinter.tix.TOP, padx=20, pady=2)
w.pack(side=tkinter.tix.TOP, padx=20, pady=2)
r.pack(side=tkinter.tix.TOP, padx=20, pady=2)
c.pack(side=tkinter.tix.TOP, padx=20, pady=2)
u.pack(side=tkinter.tix.TOP, padx=20, pady=2)

createCommonButtons(common)
```

执行效果如图 9-26 所示。

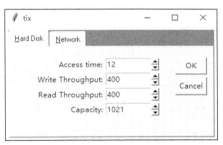

图 9-26　执行效果

实例 09-27：实现一个日历程序

下面的实例文件 fff.py 演示了使用库 tkinter.tix 实现一个日历程序的过程。

源码路径：daima\9\09-27\fff.py

```
class Calendar(TKx.Frame):
    datetime = calendar.datetime.date
    timedelta = calendar.datetime.timedelta
    today = datetime.today()

    MONTH_COLORS = ['grey92', 'thistle2', 'white', 'lightblue']

    def __init__(self, master=None, **kw):

        #在初始化ttk之前，从kw中删除自定义选项
        today = self.datetime.today()
        self.year = kw.pop('year', today.year)
        self.month = kw.pop('month', today.month)
        self.day = kw.pop('day', None)
        self.settoday = kw.pop('settoday', False)
        self.sel_bg = kw.pop('selectbackground', 'gold')
        self.preweeks = kw.pop('preweeks', 0)
        self.postweeks = kw.pop('postweeks', 0)
```

```python
        self.userange = kw.pop('selectrange', False)
        self.months_on_calendar = kw.pop('monthsoncalendar', True)

        self.range = []

        # StringVar parameter for returning a date selection.
        self.strvar = kw.pop('textvariable', TKx.StringVar())

        TKx.Frame.__init__(self, master, **kw)

        self._cal = calendar.Calendar(calendar.SUNDAY)

        #在当前空日历中插入日期
        self._build_calendar()

        if self.settoday:
            self.date_set(today)
        elif self.day:
            self.date_set(self.year, self.month, self.day)

        self._build_dategrid()

    def _build_calendar(self):
        hframe = TKx.Frame(self)
        hframe.pack(fill='x', expand=1)
        self.month_str = TKx.StringVar()
        lbtn = TKx.Button(hframe, text=u'\u25b2', command=self._prev_month)
        rbtn = TKx.Button(hframe, text=u'\u25bc', command=self._next_month)
        lbtn.pack(side='left')
        rbtn.pack(side='right')
        tl = TKx.Label(hframe, textvariable=self.month_str)
        tl.pack(side='top')
        self.top_label = tl
        self._set_month_str()
        self.days_frame = TKx.Frame(self)
        self.days_frame.pack(fill='x', expand=1)

    def _set_month_str(self):
        if self.date_obj:
            if self.userange:
                self.month_str.set(u' \u25c0\u25b6 '.join([unicode(d) for d in self.date_obj]))
            else:
                text = u'{}-{}-{}'.format(self.date_obj.year, calendar.month_name[self.date_obj.month], self.date_obj.day)
                self.month_str.set(text)
        else:
            #恢复中文系统上的月名编码错误
            text = u'{}-{}'.format(self.year, calendar.month_name[self.month])
            self.month_str.set(text)

        if self.userange:
            try:
                self._color_date_range()
            except AttributeError:
                pass
```

执行效果如图 9-27 所示。

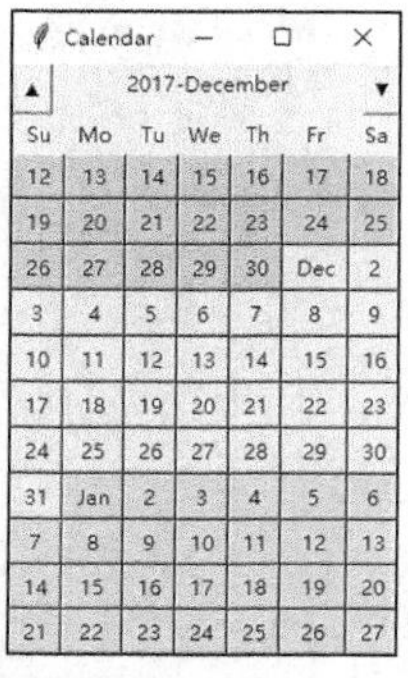

图 9-27 执行效果

9.3　Pmw 库开发实战

在 Python 程序中，可以使用库 Pmw 实现更加高级的桌面程序。Pmw 组件是在 Python 中利用 tkinter 模块构建的高级 GUI 组件，每个 Pmw 组件都合并了一个或多个 tkinter 组件，以实现更有用和更复杂的功能。Pmw 是一个第三方库，Python 没有内置 Pmw。

实例 09-28：下载并安装 Pmw

因为 Pmw 不是 Python 的内置库，所以需要读者自行下载并安装。下面介绍 3 种安装方式，读者可以根据自己的喜好进行选择，作者建议使用第 3 种安装方式。

1．第 1 种安装方式
（1）登录下载页面，单击 Download 按钮下载，如图 9-28 所示。

图 9-28　单击 Download 按钮下载

（2）下载完成后对获取的压缩包进行解压缩处理，一般会包含两个版本的 Pmw，如图 9-29 所示。

__pycache__	2017/12/13 10:07	文件夹
Pmw_1_3_3	2014/3/30 1:52	文件夹
Pmw_2_0_0	2017/12/13 10:07	文件夹
__init__.py	2012/8/4 8:56	Python File

图 9-29　两个版本的 Pmw

（3）我们选择最新（作者写本书时）版本文件夹 Pmw_2_0_0，将其重命名为"Pmw"。然后将重命名后的文件夹 Pmw 复制到本地 Python 安装目录的"site-packages"路径下，此时便完成了在本地计算机安装 Pmw 库的所有操作。在此需要注意，不同计算机的安装路径会有所不同，例如，作者计算机的安装路径是：

C:\Users\apple\AppData\Local\Programs\Python\Python36\Lib\site-packages

2．第 2 种安装方式
第 2 种安装方式是使用命令行命令进行安装，例如，使用如下两个命令都可以成功安装 Pmw。
pip install Pmw
easy_install Pmw

3．第 3 种安装方式
第 3 种安装方式是通过第三方网站下载 Pmw，下载完成后会得.whl 文件。假设这个文件保存在本地 H 盘的根目录下，接下来需要打开 CMD 控制台，并将目录切换到该文件的所在目录 H:，再使用如下的 pip 命令来安装 Pmw。

```
python -m pip install --user Pmw-2.0.1-py3-none-any.whl
```

以作者的计算机为例，完整安装过程的 CMD 控制台界面效果如图 9-30 所示。

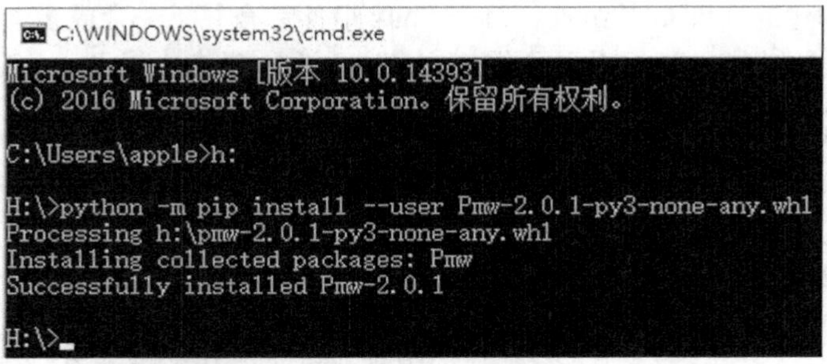

图 9-30　完整安装过程的 CMD 控制台界面效果

实例 09-29：使用 ButtonBox 组件

在库 Pmw 中，ButtonBox 组件的功能是实现按钮效果。下面的实例文件 Pmw01.py 演示了使用 ButtonBox 组件的过程。

源码路径：daima\9\09-29\Pmw01.py

```python
class Demo:
    def __init__(self, parent):
        self.balloon = Pmw.Balloon(parent)

        frame = tkinter.Frame(parent)
        frame.pack(padx = 10, pady = 5)
        field = Pmw.EntryField(frame,
                labelpos = 'nw',
                label_text = 'Command:')
        field.setentry('mycommand -name foo')
        field.pack(side = 'left', padx = 10)
        self.balloon.bind(field, 'Command to\nstart/stop',
                'Enter the shell command to control')

        start = tkinter.Button(frame, text='Start')
        start.pack(side='left', padx = 10)
        self.balloon.bind(start, 'Start the command')

        stop = tkinter.Button(frame, text='Stop')
        stop.pack(side='left', padx = 10)
        self.balloon.bind(stop, 'Stop the command')

        self.suicide = tkinter.Button(frame, text='help',
                command = self.killButton)
        self.suicide.pack(side='left', padx = 10)
        self.balloon.bind(self.suicide, 'Watch this button disappear!')

        scrolledCanvas = Pmw.ScrolledCanvas(parent,
                canvas_width = 300,
                canvas_height = 115,
        )
        scrolledCanvas.pack()
        canvas = scrolledCanvas.component('canvas')
        self.canvas = canvas

        item = canvas.create_arc(5, 5, 35, 35, fill = 'red', extent = 315)
        self.balloon.tagbind(canvas, item, 'This is help for\nan arc item')
        item = canvas.create_bitmap(20, 150, bitmap = 'question')
        self.balloon.tagbind(canvas, item, 'This is help for\na bitmap')
        item = canvas.create_line(50, 60, 70, 80, 85, 20, width = 5)
        self.balloon.tagbind(canvas, item, 'This is help for\na line item')
        item = canvas.create_text(10, 90, text = 'Canvas items with balloons',
```

```
                anchor = 'nw', font = field.cget('entry_font'))
        self.balloon.tagbind(canvas, item, 'This is help for\na text item')

        canvas.create_rectangle(100, 10, 170, 50, fill = 'aliceblue',
                tags = 'TAG1')
        self.bluecircle = canvas.create_oval(110, 30, 160, 80, fill = 'blue',
                tags = 'TAG1')
        self.balloon.tagbind(canvas, 'TAG1',
                'This is help for the two blue items' + '\n' * 10 +
                    'It is very, very big.',
                'This is help for the two blue items')
        item = canvas.create_text(180, 10, text = 'Delete',
                anchor = 'nw', font = field.cget('entry_font'))
        self.balloon.tagbind(canvas, item,
                'After 2 seconds,\ndelete the blue circle')
        canvas.tag_bind(item, '<ButtonPress>', self._canvasButtonpress)
        scrolledCanvas.resizescrollregion()

        scrolledText = Pmw.ScrolledText(parent,
                text_width = 32,
                text_height = 4,
                text_wrap = 'none',
        )
        scrolledText.pack(pady = 5)
        text = scrolledText.component('text')
        self.text = text

        text.insert('end',
                'This is a text widget with ', '',
                ' balloon', 'TAG1',
                '\nhelp. Find the ', '',
                ' text ', 'TAG1',
                ' tagged with', '',
                ' help.', 'TAG2',
                '\n', '',
                'Remove tag 1.', 'TAG3',
                '\nAnother line.\nAnd another', '',
        )
        text.tag_configure('TAG1', borderwidth = 2, relief = 'sunken')
if __name__ == '__main__':
    root = tkinter.Tk()
    Pmw.initialise(root, 12, fontScheme = 'default')
    root.title(title)

    exitButton = tkinter.Button(root, text = 'Exit', command = root.destroy)
    exitButton.pack(side = 'bottom')
    widget = Demo(root)
    root.mainloop()
```

执行效果如图 9-31 所示。

(a) 界面效果   (b) 单击按钮后的提示

图 9-31 执行效果

实例 09-30：使用 ComboBox 组件

在库 Pmw 中，ComboBox 组件的功能是实现组合框效果。下面的实例文件 Pmw02.py 演示了使用 ComboBox 组件的过程。

源码路径：daima\9\09-30\Pmw02.py

```python
class Demo:
    def __init__(self, parent):
        parent.configure(background = 'white')

        # Create and pack the widget to be configured.
        self.target = tkinter.Label(parent,
                relief = 'sunken',
                padx = 20,
                pady = 20,
        )
        self.target.pack(fill = 'x', padx = 8, pady = 8)

        words = ('Monti', 'Python', 'ik', 'den', 'Holie', 'Grailen', '(Bok)')
        simple = Pmw.ComboBox(parent,
                label_text = 'Simple ComboBox:',
                labelpos = 'nw',
                selectioncommand = self.changeText,
                scrolledlist_items = words,
                dropdown = 0,
        )
        simple.pack(side = 'left', fill = 'both',
                expand = 1, padx = 8, pady = 8)

        #显示第一个文本
        first = words[0]
        simple.selectitem(first)
        self.changeText(first)

        colours = ('cornsilk1', 'snow1', 'seashell1', 'antiquewhite1',
                'bisque1', 'peachpuff1', 'navajowhite1', 'lemonchiffon1',
                'ivory1', 'honeydew1', 'lavenderblush1', 'mistyrose1')
        dropdown = Pmw.ComboBox(parent,
                label_text = 'Dropdown ComboBox:',
                labelpos = 'nw',
                selectioncommand = self.changeColour,
                scrolledlist_items = colours,
        )
        dropdown.pack(side = 'left', anchor = 'n',
                fill = 'x', expand = 1, padx = 8, pady = 8)

        first = colours[0]
        dropdown.selectitem(first)
        self.changeColour(first)

    def changeColour(self, colour):
        print(('Colour: ' + colour))
        self.target.configure(background = colour)

    def changeText(self, text):
        print(('Text: ' + text))
        self.target.configure(text = text)
```

执行效果如图 9-32 所示。

(a) 界面效果   (b) 选择列表项后的提示

图 9-32 执行效果

实例 09-31：使用 Counter 组件

在库 Pmw 中，Counter 组件的功能是使用上/下箭头按钮自动实现数据输入功能。下面的实例文件 Pmw03.py 演示了使用 Counter 组件的过程。

源码路径：daima\9\09-31\Pmw03.py

```python
class Demo:
    def __init__(self, parent):
        now = (int(time.time()) / 300) * 300

        self._date = Pmw.Counter(parent,
                labelpos = 'w',
                label_text = 'Date (4-digit year):',
                entryfield_value =
                        time.strftime('%d/%m/%Y', time.localtime(now)),
                entryfield_command = self.execute,
                entryfield_validate = {'validator' : 'date', 'format' : 'dmy'},
                datatype = {'counter' : 'date', 'format' : 'dmy', 'yyyy' : 1})

        self._isodate = Pmw.Counter(parent,
                labelpos = 'w',
                label_text = 'ISO-Date (4-digit year):',
                entryfield_value =
                        time.strftime('%Y-%m-%d', time.localtime(now)),
                entryfield_command = self.execute,
                entryfield_validate = {'validator' : 'date', 'format' : 'ymd',
                        'separator' : '-' },
                datatype = {'counter' : 'date', 'format' : 'ymd', 'yyyy' : 1,
                        'separator' : '-' })

        self._time = Pmw.Counter(parent,
                labelpos = 'w',
                label_text = 'Time:',
                entryfield_value =
                        time.strftime('%H:%M:%S', time.localtime(now)),
                entryfield_validate = {'validator' : 'time',
                        'min' : '00:00:00', 'max' : '23:59:59',
                        'minstrict' : 0, 'maxstrict' : 0},
                datatype = {'counter' : 'time', 'time24' : 1},
                increment=5*60)
        self._real = Pmw.Counter(parent,
                labelpos = 'w',
                label_text = 'Real (with comma)\nand extra\nlabel lines:',
                label_justify = 'left',
                entryfield_value = '1,5',
                datatype = {'counter' : 'real', 'separator' : ','},
                entryfield_validate = {'validator' : 'real',
                        'min' : '-2,0', 'max' : '5,0',
                        'separator' : ','},
                increment = 0.1)
        self._custom = Pmw.Counter(parent,
                labelpos = 'w',
                label_text = 'Custom:',
                entryfield_value = specialword[:4],
                datatype = _custom_counter,
                entryfield_validate = _custom_validate)
        self._int = Pmw.Counter(parent,
                labelpos = 'w',
                label_text = 'Vertical integer:',
                orient = 'vertical',
                entry_width = 2,
                entryfield_value = 50,
                entryfield_validate = {'validator' : 'integer',
                        'min' : 0, 'max' : 99}
        )
```

执行效果如图 9-33 所示。

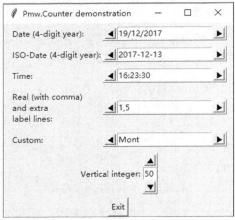

图 9-33 执行效果

实例 09-32：使用 Group 组件

在库 Pmw 中，Group 组件的功能是使用标签和边框实现分组功能。下面的实例文件 Pmw04.py 演示了使用 Group 组件的过程。

源码路径：daima\9\09-32\Pmw04.py

```
class Demo:
    def __init__(self, parent):

        w = Pmw.Group(parent, tag_text='label')
        w.pack(fill = 'both', expand = 1, padx = 6, pady = 6)
        cw = tkinter.Label(w.interior(),
                text = 'A group with the\ndefault Label tag')
        cw.pack(padx = 2, pady = 2, expand='yes', fill='both')

        w = Pmw.Group(parent, tag_pyclass = None)
        w.pack(fill = 'both', expand = 1, padx = 6, pady = 6)
        cw = tkinter.Label(w.interior(), text = 'A group\nwithout a tag')
        cw.pack(padx = 2, pady = 2, expand='yes', fill='both')

        radiogroups = []
        self.var = tkinter.IntVar()
        self.var.set(0)
        radioframe = tkinter.Frame(parent)
        w = Pmw.Group(radioframe,
                tag_pyclass = tkinter.Radiobutton,
                tag_text='radiobutton 1',
                tag_value = 0,
                tag_variable = self.var)
        w.pack(fill = 'both', expand = 1, side='left')
        cw = tkinter.Frame(w.interior(),width=200,height=20)
        cw.pack(padx = 2, pady = 2, expand='yes', fill='both')
        radiogroups.append(w)

        w = Pmw.Group(radioframe,
                tag_pyclass = tkinter.Radiobutton,
                tag_text='radiobutton 2',
                tag_font = Pmw.logicalfont('Helvetica', 4),
                tag_value = 1,
                tag_variable = self.var)
        w.pack(fill = 'both', expand = 1, side='left')
        cw = tkinter.Frame(w.interior(),width=200,height=20)
        cw.pack(padx = 2, pady = 2, expand='yes', fill='both')
        radiogroups.append(w)
        radioframe.pack(padx = 6, pady = 6, expand='yes', fill='both')
        Pmw.aligngrouptags(radiogroups)

        w = Pmw.Group(parent,
                tag_pyclass = tkinter.Checkbutton,
```

```
                tag_text='checkbutton',
                tag_foreground='blue')
w.pack(fill = 'both', expand = 1, padx = 6, pady = 6)
cw = tkinter.Frame(w.interior(),width=150,height=20)
cw.pack(padx = 2, pady = 2, expand='yes', fill='both')

w = Pmw.Group(parent,
            tag_pyclass = tkinter.Button,
            tag_text='Tkinter.Button')
w.configure(tag_command = w.toggle)
w.pack(fill = 'both', expand = 1, padx = 6, pady = 6)
cw = tkinter.Label(w.interior(),
            background = 'aliceblue',
            text = 'A group with\na Button tag!?'
)
cw.pack(padx = 2, pady = 2, expand='yes', fill='both')

w = Pmw.Group(parent,
            tag_pyclass = tkinter.Button,
            tag_text='Show/Hide')
w.configure(tag_command = w.toggle)
w.pack(fill = 'both', expand = 1, padx = 6, pady = 6)
cw = tkinter.Label(w.interior(),
            background = 'aliceblue',
            text = 'Now you see me.\nNow you don\'t.'
)
cw.pack(padx = 2, pady = 2, expand='yes', fill='both')
```

执行效果如图 9-34 所示。

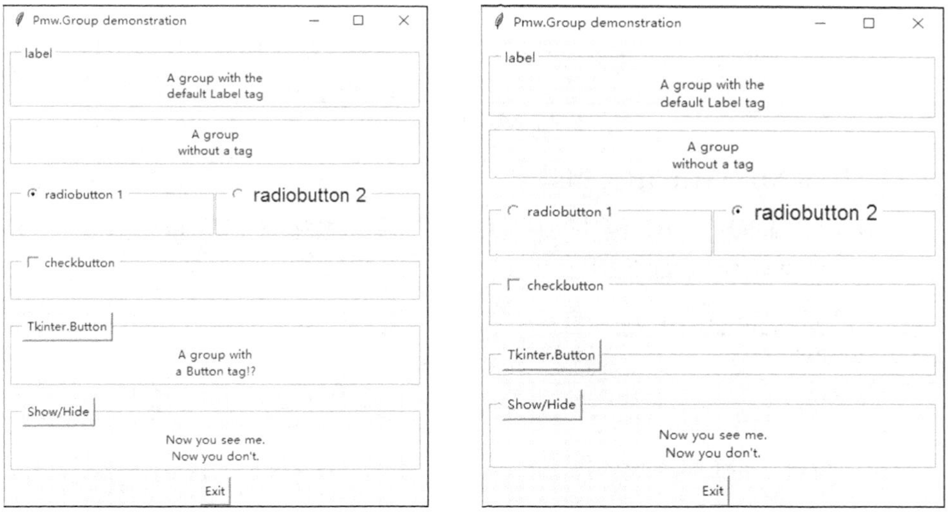

(a) 初始执行效果　　　　　　(b) 单击 tkinter.Button 组件后的效果

图 9-34　执行效果

实例 09-33：使用 LabeledWidget 组件

在库 Pmw 中，LabeledWidget 组件的功能是实现内外嵌套组合的文本标签效果。下面的实例文件 Pmw05.py 演示了使用 LabeledWidget 组件的过程。

源码路径：daima\9\09-33\Pmw05.py

```
class Demo:
    def __init__(self, parent):

        frame = tkinter.Frame(parent, background = 'grey90')
        frame.pack(fill = 'both', expand = 1)

        column = 0
        row = 0
```

```
                    for pos in ('n', 'nw', 'wn', 'w'):
                        lw = Pmw.LabeledWidget(frame,
                                labelpos = pos,
                                label_text = pos + ' label')
                        lw.component('hull').configure(relief='sunken', borderwidth=2)
                        lw.grid(column=column, row=row, padx=10, pady=10)
                        cw = tkinter.Button(lw.interior(), text='child\nsite')
                        cw.pack(padx=10, pady=10, expand='yes', fill='both')

                        column = column + 1
                        if column == 2:
                            column = 0
                            row = row + 1
```

执行效果如图 9-35 所示。

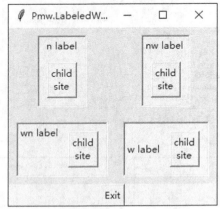

图 9-35　执行效果

实例 09-34：使用 MainMenuBar 组件

在库 Pmw 中，MainMenuBar 组件的功能是实现主菜单栏效果。下面的实例文件 Pmw06.py 演示了使用 MainMenuBar 组件的过程。

源码路径：daima\9\09-34\Pmw06.py

```
class MainMenuBarToplevel:
    def __init__(self, parent):
        megaToplevel = Pmw.MegaToplevel(parent, title = title)
        toplevel = megaToplevel.interior()

        self.balloon = Pmw.Balloon(toplevel)

        menuBar = Pmw.MainMenuBar(toplevel,
                balloon = self.balloon)
        toplevel.configure(menu = menuBar)
        self.menuBar = menuBar

        menuBar.addmenu('File', 'Close this window or exit')
        menuBar.addmenuitem('File', 'command', 'Close this window',
                command = PrintOne('Action: close'),
                label = 'Close')
        menuBar.addmenuitem('File', 'separator')
        menuBar.addmenuitem('File', 'command', 'Exit the application',
                command = PrintOne('Action: exit'),
                label = 'Exit')

        menuBar.addmenu('Edit', 'Cut, copy or paste')
        menuBar.addmenuitem('Edit', 'command', 'Delete the current selection',
                command = PrintOne('Action: delete'),
                label = 'Delete')

        menuBar.addmenu('Options', 'Set user preferences')
        menuBar.addmenuitem('Options', 'command', 'Set general preferences',
```

```
                command = PrintOne('Action: general options'),
                label = 'General...')
self.toggleVar = tkinter.IntVar()
self.toggleVar.set(1)
menuBar.addmenuitem('Options', 'checkbutton', 'Toggle me on/off',
        label = 'Toggle',
        command = self._toggleMe,
        variable = self.toggleVar)
self._toggleMe()

menuBar.addcascademenu('Options', 'Size',
        'Set some other preferences', traverseSpec = 'z', tearoff = 1)
for size in ('tiny', 'small', 'average', 'big', 'huge'):
    menuBar.addmenuitem('Size', 'command', 'Set size to ' + size,
            command = PrintOne('Action: size ' + size),
            label = size)

menuBar.addmenu('Help', 'User manuals', name = 'help')
menuBar.addmenuitem('Help', 'command', 'About this application',
        command = PrintOne('Action: about'),
        label = 'About...')

self.mainPart = tkinter.Label(toplevel,
        text = 'This is the\nmain part of\nthe window',
        background = 'black',
        foreground = 'white',
        padx = 30,
        pady = 30)
self.mainPart.pack(fill = 'both', expand = 1)

self.messageBar = Pmw.MessageBar(toplevel,
        entry_width = 40,
        entry_relief='groove',
        labelpos = 'w',
        label_text = 'Status:')
self.messageBar.pack(fill = 'x', padx = 10, pady = 10)
self.messageBar.message('state',
    'Balloon/status help not working properly - Tk menubar bug')

buttonBox = Pmw.ButtonBox(toplevel)
buttonBox.pack(fill = 'x')
buttonBox.add('Disable\nall', command = menuBar.disableall)
buttonBox.add('Enable\nall', command = menuBar.enableall)
buttonBox.add('Create\nmenu', command = self.add)
buttonBox.add('Delete\nmenu', command = self.delete)
buttonBox.add('Create\nitem', command = self.additem)
buttonBox.add('Delete\nitem', command = self.deleteitem)

# message bar.
self.balloon.configure(statuscommand = self.messageBar.helpmessage)

self.testMenuList = []
```

执行效果如图 9-36 所示。

（a）初始执行效果

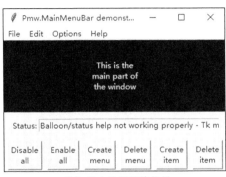

（b）单击按钮后的效果

图 9-36　执行效果

实例 09-35：使用 MessageBar 组件

在库 Pmw 中，MessageBar 组件的功能是实现消息框提示效果。下面的实例文件 Pmw07.py 演示了使用 MessageBar 组件的过程。

源码路径：daima\9\09-35\Pmw07.py

```python
class Demo:
    def __init__(self, parent):
        self._messagebar = Pmw.MessageBar(parent,
                entry_width = 40,
                entry_relief='groove',
                labelpos = 'w',
                label_text = 'Status:')
        self._messagebar.pack(side = 'bottom', fill = 'x',
                expand = 1, padx = 10, pady = 10)

        self.box = Pmw.ScrolledListBox(parent,
                listbox_selectmode='single',
                items=('state', 'help', 'userevent', 'systemevent',
                        'usererror', 'systemerror', 'busy',),
                label_text='Message type',
                labelpos='n',
                selectioncommand=self.selectionCommand)
        self.box.pack(fill = 'both', expand = 'yes', padx = 10, pady = 10)

        self._index = 0
        self._stateCounter = 0

    def selectionCommand(self):
        sels = self.box.getcurselection()
        if len(sels) > 0:
            self._index = self._index + 1
            messagetype = sels[0]
            if messagetype == 'state':
                self._stateCounter = (self._stateCounter + 1) % 3
                text = stateMessages[self._stateCounter]
                if text != '':
                    text = text + ' (' + messagetype + ')'
                self._messagebar.message('state', text)
            else:
                text = messages[messagetype]
                text = text + ' (' + messagetype + ')'
                self._messagebar.message(messagetype, text)
                if messagetype == 'busy':
                    Pmw.showbusycursor()
                    self.box.after(2000)
                    Pmw.hidebusycursor()
                    self._messagebar.resetmessages('busy')
                    text = 'All files successfully removed'
                    text = text + ' (userevent)'
                    self._messagebar.message('userevent', text)
```

执行效果如图 9-37 所示。

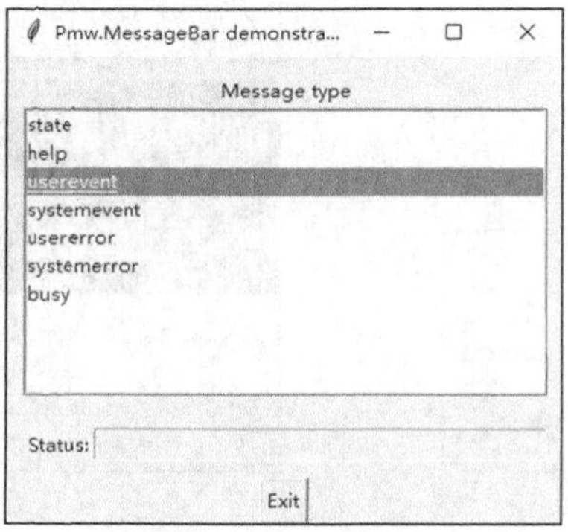

图 9-37　执行效果

实例 09-36：使用 OptionMenu 组件

在库 Pmw 中，OptionMenu 组件的功能是实现选项菜单效果。下面的实例文件 Pmw08.py 演示了使用 OptionMenu 组件的过程。

源码路径：daima\9\09-36\Pmw08.py

```
class Demo:
    def __init__(self, parent):
        self.var = tkinter.StringVar()
        self.var.set('steamed')
        self.method_menu = Pmw.OptionMenu(parent,
                labelpos = 'w',
                label_text = 'Choose method:',
                menubutton_textvariable = self.var,
                items = ['baked', 'steamed', 'stir fried', 'boiled', 'raw'],
                menubutton_width = 10,
        )
        self.method_menu.pack(anchor = 'w', padx = 10, pady = 10)

        self.vege_menu = Pmw.OptionMenu (parent,
                labelpos = 'w',
                label_text = 'Choose vegetable:',
                items = ('broccoli', 'peas', 'carrots', 'pumpkin'),
                menubutton_width = 10,
                command = self._printOrder,
        )
        self.vege_menu.pack(anchor = 'w', padx = 10, pady = 10)

        self.direction_menu = Pmw.OptionMenu (parent,
                labelpos = 'w',
                label_text = 'Menu direction:',
                items = ('flush', 'above', 'below', 'left', 'right'),
                menubutton_width = 10,
                command = self._changeDirection,
        )
        self.direction_menu.pack(anchor = 'w', padx = 10, pady = 10)

        menus = (self.method_menu, self.vege_menu, self.direction_menu)
        Pmw.alignlabels(menus)
```

执行效果如图 9-38 所示。

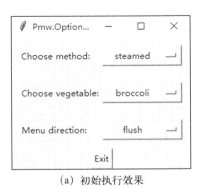

(a) 初始执行效果　　　　　　(b) 单击按钮后弹出选项菜单

图 9-38　执行效果

实例 09-37：使用 RadioSelect 组件

在库 Pmw 中，RadioSelect 组件的功能是实现单选按钮菜单效果。下面的实例文件 Pmw09.py 演示了使用 RadioSelect 组件的过程。

源码路径：daima\9\09-37\Pmw09.py

```python
class Demo:
    def __init__(self, parent):
        horiz = Pmw.RadioSelect(parent,
                labelpos = 'w',
                command = self.callback,
                label_text = 'Horizontal',
                frame_borderwidth = 2,
                frame_relief = 'ridge'
        )
        horiz.pack(fill = 'x', padx = 10, pady = 10)

        for text in ('Fruit', 'Vegetables', 'Cereals', 'Legumes'):
            horiz.add(text)
        horiz.invoke('Cereals')

        self.multiple = Pmw.RadioSelect(parent,
                labelpos = 'w',
                command = self.multcallback,
                label_text = 'Multiple\nselection',
                frame_borderwidth = 2,
                frame_relief = 'ridge',
                selectmode = 'multiple',
        )
        self.multiple.pack(fill = 'x', padx = 10)
        for text in ('Apricots', 'Eggplant', 'Rice', 'Lentils'):
            self.multiple.add(text)
        self.multiple.invoke('Rice')
        self.checkbuttons = Pmw.RadioSelect(parent,
                buttontype = 'checkbutton',
                orient = 'vertical',
                labelpos = 'w',
                command = self.checkbuttoncallback,
                label_text = 'Vertical,\nusing\ncheckbuttons',
                hull_borderwidth = 2,
                hull_relief = 'ridge',
        )
        self.checkbuttons.pack(side = 'left', expand = 1, padx = 10, pady = 10)

        for text in ('Male', 'Female'):
            self.checkbuttons.add(text)
        self.checkbuttons.invoke('Male')
        self.checkbuttons.invoke('Female')

        radiobuttons = Pmw.RadioSelect(parent,
                buttontype = 'radiobutton',
                orient = 'vertical',
                labelpos = 'w',
                command = self.callback,
                label_text = 'Vertical,\nusing\nradiobuttons',
                hull_borderwidth = 2,
                hull_relief = 'ridge',
        )
        radiobuttons.pack(side = 'left', expand = 1, padx = 10, pady = 10)

        for text in ('Male', 'Female', 'Both', 'Neither'):
            radiobuttons.add(text)
        radiobuttons.invoke('Both')
```

执行效果如图 9-39 所示。

9.3 Pmw 库开发实战

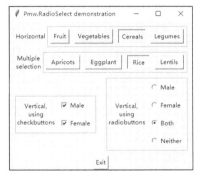

(a) 初始执行效果　　　　　　　　　　　(b) 命令行界面显示选中的选项

图 9-39　执行效果

实例 09-38：使用 ScrolledCanvas 组件

在库 Pmw 中，ScrolledCanvas 组件的功能是实现滚动画布效果。滚动条可以是动态的，这意味着一个滚动条只会在必要的时候才会被显示出来，例如当显示的内容超出画布时。下面的实例文件 Pmw10.py 演示了使用 ScrolledCanvas 组件的过程。

　　源码路径：daima\9\09-38\Pmw10.py

```
w = Pmw.Group(parent, tag_text='Scroll mode')
w.pack(side = 'bottom', padx = 5, pady = 5)

hmode = Pmw.OptionMenu(w.interior(),
        labelpos = 'w',
        label_text = 'Horizontal:',
        items = ['none', 'static', 'dynamic'],
        command = self.sethscrollmode,
        menubutton_width = 8,
)
hmode.pack(side = 'left', padx = 5, pady = 5)
hmode.invoke('dynamic')

vmode = Pmw.OptionMenu(w.interior(),
        labelpos = 'w',
        label_text = 'Vertical:',
        items = ['none', 'static', 'dynamic'],
        command = self.setvscrollmode,
        menubutton_width = 8,
)
vmode.pack(side = 'left', padx = 5, pady = 5)
vmode.invoke('dynamic')

buttonBox = Pmw.ButtonBox(parent)
buttonBox.pack(side = 'bottom')
buttonBox.add('yview', text = 'Show\nyview', command = self.showYView)
buttonBox.add('scroll', text = 'Page\ndown', command = self.pageDown)
buttonBox.add('center', text = 'Center', command = self.centerPage)

self.sc.pack(padx = 5, pady = 5, fill = 'both', expand = 1)

self.sc.component('canvas').bind('<1>', self.addcircle)

testEntry = tkinter.Entry(parent)
self.sc.create_line(20, 20, 100, 100)
self.sc.create_oval(100, 100, 200, 200, fill = 'green')
self.sc.create_text(100, 20, anchor = 'nw',
        text = 'Click in the canvas\nto draw ovals',
        font = testEntry.cget('font'))
button = tkinter.Button(self.sc.interior(),
        text = 'Hello,\nWorld!\nThis\nis\na\nbutton.')
self.sc.create_window(200, 200,
        anchor='nw',
        window = button)
```

```
            self.sc.resizescrollregion()

        self.colours = ('red', 'green', 'blue', 'yellow', 'cyan', 'magenta',
                'black', 'white')
```

执行效果如图 9-40 所示。

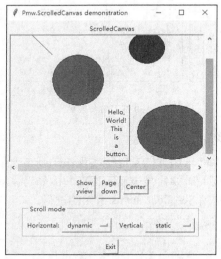

图 9-40　执行效果

实例 09-39：使用 AboutDialog 组件

在库 Pmw 中，AboutDialog 组件主要用于显示软件的版本和联系人信息。下面的实例文件 Pmw11.py 演示了使用 AboutDialog 组件的过程。

源码路径：daima\9\09-39\Pmw11.py

```
class Demo:
    def __init__(self, parent):
        Pmw.aboutversion('9.9')
        Pmw.aboutcopyright('Copyright My Company 1999\nAll rights reserved')
        Pmw.aboutcontact(
                'For information about this application contact:\n' +
                '  My Help Desk\n' +
                '  Phone: +61 2 9876 5432\n' +
                '  email: help@my.company.com.au'
        )
        self.about = Pmw.AboutDialog(parent, applicationname = 'My Application')
        self.about.withdraw()

        w = tkinter.Button(parent, text = 'Show about dialog',
                command = self.execute)
        w.pack(padx = 8, pady = 8)

    def execute(self):
        self.about.show()
if __name__ == '__main__':
    root = tkinter.Tk()
    Pmw.initialise(root)
    root.title(title)

    exitButton = tkinter.Button(root, text = 'Exit', command = root.destroy)
    exitButton.pack(side = 'bottom')
    widget = Demo(root)
    root.mainloop()
```

执行效果如图 9-41 所示。

9.3 Pmw 库开发实战

（a）初始效果

（b）单击按钮后的效果

图 9-41 执行效果

实例 09-40：使用 Balloon 组件

在库 Pmw 中，Balloon 组件的功能是实现悬浮提示信息效果，当用户将鼠标指针放在按钮或其他控件上的时候会弹出这些悬浮提示信息，它通常用于显示画布或文本类型的帮助信息。下面的实例文件 Pmw12.py 演示了使用 Balloon 组件的过程。

源码路径：daima\9\09-40\Pmw12.py

```python
class Demo:
    def __init__(self, parent):
        # Create the Balloon.
        self.balloon = Pmw.Balloon(parent)

        # Create some widgets and megawidgets with balloon help.
        frame = tkinter.Frame(parent)
        frame.pack(padx = 10, pady = 5)
        field = Pmw.EntryField(frame,
                labelpos = 'nw',
                label_text = 'Command:')
        field.setentry('mycommand -name foo')
        field.pack(side = 'left', padx = 10)
        self.balloon.bind(field, 'Command to\nstart/stop',
                'Enter the shell command to control')

        start = tkinter.Button(frame, text='Start')
        start.pack(side='left', padx = 10)
        self.balloon.bind(start, 'Start the command')

        stop = tkinter.Button(frame, text='Stop')
        stop.pack(side='left', padx = 10)
        self.balloon.bind(stop, 'Stop the command')

        self.suicide = tkinter.Button(frame, text='关闭',
                command = self.killButton)
        self.suicide.pack(side='left', padx = 10)
        self.balloon.bind(self.suicide, 'Watch this button disappear!')

        scrolledCanvas = Pmw.ScrolledCanvas(parent,
                canvas_width = 300,
                canvas_height = 115,
        )
        scrolledCanvas.pack()
        canvas = scrolledCanvas.component('canvas')
        self.canvas = canvas

        #创建一些画布项目和帮助信息
        item = canvas.create_arc(5, 5, 35, 35, fill = 'red', extent = 315)
        self.balloon.tagbind(canvas, item, 'This is help for\nan arc item')
```

```python
        item = canvas.create_bitmap(20, 150, bitmap = 'question')
        self.balloon.tagbind(canvas, item, 'This is help for\na bitmap')
        item = canvas.create_line(50, 60, 70, 80, 85, 20, width = 5)
        self.balloon.tagbind(canvas, item, 'This is help for\na line item')
        item = canvas.create_text(10, 90, text = 'Canvas items with balloons',
                anchor = 'nw', font = field.cget('entry_font'))
        self.balloon.tagbind(canvas, item, 'This is help for\na text item')

        #创建两个具有相同标记且使用相同帮助的画布项
        canvas.create_rectangle(100, 10, 170, 50, fill = 'aliceblue',
                tags = 'TAG1')
        self.bluecircle = canvas.create_oval(110, 30, 160, 80, fill = 'blue',
                tags = 'TAG1')
        self.balloon.tagbind(canvas, 'TAG1',
                'This is help for the two blue items' + '\n' * 10 +
                    'It is very, very big.',
                'This is help for the two blue items')
        item = canvas.create_text(180, 10, text = 'Delete',
                anchor = 'nw', font = field.cget('entry_font'))
        self.balloon.tagbind(canvas, item,
                'After 2 seconds,\ndelete the blue circle')
        canvas.tag_bind(item, '<ButtonPress>', self._canvasButtonpress)
        scrolledCanvas.resizescrollregion()

        scrolledText = Pmw.ScrolledText(parent,
                text_width = 32,
                text_height = 4,
                text_wrap = 'none',
        )
        scrolledText.pack(pady = 5)
        text = scrolledText.component('text')
        self.text = text

        text.insert('end',
                'This is a text widget with ', '',
                ' balloon', 'TAG1',
                '\nhelp. Find the ', '',
                ' text ', 'TAG1',
                ' tagged with', '',
                ' help.', 'TAG2',
                '\n', '',
                'Remove tag 1.', 'TAG3',
                '\nAnother line.\nAnd another', '',
        )
        text.tag_configure('TAG1', borderwidth = 2, relief = 'sunken')
        text.tag_configure('TAG3', borderwidth = 2, relief = 'raised')

        self.balloon.tagbind(text, 'TAG1',
                'There is one secret\nballoon help.\nCan you find it?')
        self.balloon.tagbind(text, 'TAG2',
                'Well done!\nYou found it!')
        self.balloon.tagbind(text, 'TAG3',
                'After 2 seconds\ndelete the tag')
        text.tag_bind('TAG3', '<ButtonPress>', self._textButtonpress)

        frame = tkinter.Frame(parent)
        frame.pack(padx = 10)
        self.toggleBalloonVar = tkinter.IntVar()
        self.toggleBalloonVar.set(1)
        toggle = tkinter.Checkbutton(frame,
                variable = self.toggleBalloonVar,
                text = 'Balloon help', command = self.toggle)
        toggle.pack(side = 'left', padx = 10)
        self.balloon.bind(toggle, 'Toggle balloon help\non and off')

        self.toggleStatusVar = tkinter.IntVar()
        self.toggleStatusVar.set(1)
        toggle = tkinter.Checkbutton(frame,
                variable = self.toggleStatusVar,
                text = 'Status help', command = self.toggle)
        toggle.pack(side = 'left', padx = 10)
        self.balloon.bind(toggle,
                'Toggle status help on and off, on and off' + '\n' * 10 +
```

```
                        'It is very, very big, too.',
                        'Toggle status help on and off')
            messageBar = Pmw.MessageBar(parent,
                        entry_width = 40,
                        entry_relief='groove',
                        labelpos = 'w',
                        label_text = 'Status:')
            messageBar.pack(fill = 'x', expand = 1, padx = 10, pady = 5)

            self.balloon.configure(statuscommand = messageBar.helpmessage)
```

执行效果如图 9-42 所示。

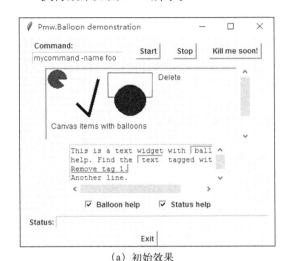

(a) 初始效果　　　　　　　　　　(b) 鼠标指针悬浮在控件上后显示提示

图 9-42　执行效果

实例 09-41：使用 PyQt 创建第一个 GUI 程序

在 Python 程序中，可以使用库 PyQt 来创建美观的用户界面。在作者编写本书时，PyQt 已经发展到 PyQt5。所以，这里使用如下命令安装 PyQt5 库。

```
pip install PyQt5
```

安装成功 PyQt 后，就可以开始我们的 PyQt 开发之旅了。例如，下面的实例文件 PyQt01.py 演示了使用 PyQt 创建第一个 GUI 程序的过程。

源码路径：daima\9\09-41\PyQt01.py

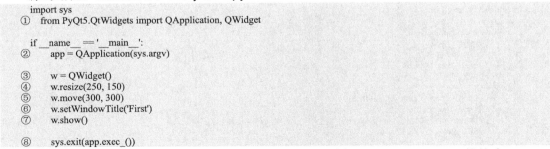

①中导入一些必要模块，PyQt 中的基础 QWidget 组件位于 PyQt5.QtWidget 模块中。

②中所有的 PyQt5 程序必须创建一个应用（QApplication）对象。参数 sys.argv 是一个来自命令行的参数列表。Python 脚本可以在 Python Shell 中执行，这是开发者用来控制应用启动的一种方法。

③中 QWidget 组件是 PyQt5 中所有用户界面类的基础类，本行使用了 QWidget 提供的默认的构造方法。默认构造方法没有父类，没有父类的 QWidget 组件将被作为窗口来使用。

④中使用 resize()方法调整 QWidget 组件的大小，设置为 250px 宽、150px 高。

⑤中使用方法 move()将组件 QWidget 移动到一个位置，这个位置是屏幕上横坐标值为 300、纵坐标值为 300 处。

⑥中设置显示的标题，这个标题显示在标题栏中。

⑦中使用方法 show()在屏幕上显示 QWidget 组件，Qwidget 组件第一次在内存中被创建，并且之后在屏幕上显示。

⑧中整个程序进入主循环，从本行代码开始执行事件处理。主循环用于接收来自窗口的事件，并且转发它们到 QWidget 上处理。如果调用 exit()方法或主 QWidget 组件被销毁，则主循环将退出。sys.exit()方法确保不留垃圾地退出程序。系统环境将会被通知应用是怎样被结束的。值得注意的是，方法 exec_()有一个下划线，这是因为"exec"是 Python 的保留关键字，所以使用 exec_()来代替。

上述代码执行后将显示一个最简单的 GUI 程序，如图 9-43 所示。

图 9-43　执行效果

实例 09-42：在 PyQt 窗体中创建一个图标

在下面的实例文件 PyQt02.py 中，演示了在 PyQt 窗体中创建一个图标的过程。

源码路径：daima\9\09-42PyQt02.py

```
import sys
from PyQt5.QtWidgets import QApplication, QWidget
from PyQt5.QtGui import QIcon                    #从PyQt5.QtGui中引入类QIcon，为了便于图标的设定

class Ico(QWidget):

    def __init__(self):
        super().__init__()
        self.initUI()                             #用函数initUI()创建GUI

    def initUI(self):

①       self.setGeometry(300, 300, 300, 220)
        self.setWindowTitle('书创文化')
②       self.setWindowIcon(QIcon('123.ico'))
        self.show()
if __name__ == '__main__':

    app = QApplication(sys.argv)
    ex = Ico()
    sys.exit(app.exec_())
```

①和②之间的这 3 行代码中的方法都是从 QWidget 类中继承的，方法 setGeometry()用于在屏幕上定位窗口并设置它的大小，其前 2 个参数是窗口的横坐标值和纵坐标值，第 3 个参数是

窗口的宽度,第 4 个参数是窗口的高度。实际上,方法 setGeometry()实现了 resize()和 move()两个方法的功能。方法 setWindowIcon()用于设置应用程序图标,首先创建了一个 QIcon 对象,QIcon 能够接收要显示的图标的路径(和当前程序在同一个目录下)。执行效果如图 9-44 所示。

图 9-44　执行效果

实例 09-43:在 PyQt 窗体中实现一个提示信息

下面的实例文件 PyQt03.py 演示了在 PyQt 窗体中实现一个提示信息的过程。

源码路径:daima\9\09-43\PyQt03.py

```
import sys
from PyQt5.QtWidgets import (QWidget, QToolTip, QPushButton, QApplication)
from PyQt5.QtGui import QFont

class Example(QWidget):

    def __init__(self):
        super().__init__()

        self.initUI()

    def initUI(self):
        QToolTip.setFont(QFont('SansSerif', 10))

        self.setToolTip('这是一个<b>QWidget</b>组件')

        btn = QPushButton('按钮', self)
        btn.setToolTip('这是一个<b>QPushButton</b>组件')
        btn.resize(btn.sizeHint())
        btn.move(50, 50)

        self.setGeometry(300, 300, 300, 200)
        self.setWindowTitle('工具栏')
        self.show()

if __name__ == '__main__':
    app = QApplication(sys.argv)
    ex = Example()
    sys.exit(app.exec_())
```

通过上述代码为两个 PyQt5 组件设置了提示信息,具体流程如下。

❑ 使用静态方法 setFont(QFont('SansSerif', 10))设置提示信息的字体,此处使用 10px 的 SansSerif 字体。

❑ 调用方法 setTooltip()创建提示框,在提示框中可以使用更加丰富的文本格式。

❑ 使用 QPushButton()创建一个按钮组件,使用 setToolTip()为按钮设置一个提示框。

上述代码执行后,当将鼠标指针放到按钮上面后会显示提示信息,如图 9-45 所示。

图 9-45 执行效果

实例 09-44：在 PyQt 窗体中创建状态栏信息

在 GUI 程序中，类 QMainWindow 用于实现主窗口效果，会默认创建一个拥有状态栏、工具栏和菜单栏的经典应用窗口框架。其中，状态栏是用于显示状态信息的组件，在 PyQt 中由 QMainWindow（依赖于 QMainWindow 组件）组件创建。例如，下面的实例文件 PyQt04.py 演示了在 PyQt 窗体中创建状态栏信息的过程。

源码路径：daima\9\09-44\PyQt04.py

```
import sys
from PyQt5.QtWidgets import QMainWindow, QApplication
class Example(QMainWindow):

    def __init__(self):
        super().__init__()

        self.initUI()

    def initUI(self):
        self.statusBar().showMessage('这是状态信息')#调用了QMainWindow类的statusBar()方法

        self.setGeometry(300, 300, 250, 150)
        self.setWindowTitle('状态栏演示')
        self.show()
```

在上述代码中，通过调用 QMainWindow 类的 statusBar() 方法来得到状态栏。第一次调用该方法时 GUI 程序会创建一个状态栏，随后该方法返回状态栏对象。然后调用 showMessage() 方法在状态栏上显示一些信息。上述代码执行后会在状态栏中显示信息"这是状态信息"，如图 9-46 所示。

图 9-46 状态栏信息

实例 09-45：在 PyQt 窗体中同时创建菜单栏和状态栏信息

菜单栏是由各种菜单中的一组操作命令组成的组件，是 GUI 程序的重要组成部分之一。因为 macOS 操作系统对待菜单栏的方式与其他操作系统不同，为了获得全平台一致的效果，所以开发者通常在代码中加入如下代码。

```
menubar.setNativeMenuBar(False)
```
例如，下面的实例文件 PyQt05.py 演示了在 PyQt 窗体中同时创建菜单栏和状态栏信息的过程。

源码路径：daima\9\09-45\PyQt05.py
```
class Example(QMainWindow):
    def __init__(self):
        super().__init__()
        self.initUI()

    def initUI(self):
        exitAction = QAction(QIcon('123.ico'), '&退出', self)
        exitAction.setShortcut('Ctrl+Q')
        exitAction.setStatusTip('退出程序')
        exitAction.triggered.connect(qApp.quit)
        self.statusBar()
        menubar = self.menuBar()
        fileMenu = menubar.addMenu('&文件')
        fileMenu.addAction(exitAction)

        self.setGeometry(300, 300, 300, 200)
        self.setWindowTitle('菜单栏练习')
        self.show()

if __name__ == '__main__':
    app = QApplication(sys.argv)
    ex = Example()
    sys.exit(app.exec_())
```
对上述代码的具体说明如下。

❑ QAction 是一个用于菜单栏、工具栏或自定义快捷键的抽象动作行为。通过如下 3 行代码创建了一个有指定图标和文本为"退出"的命令，并且为这个命令定义了一个快捷键。如下第 3 行代码创建了一个当鼠标指针被放在命令之上就会显示的一个状态提示信息。

```
exitAction = QAction(QIcon('123.ico'), '&退出', self)
exitAction.setShortcut('Ctrl+Q')
exitAction.setStatusTip('退出程序')
```

❑ 代码 exitAction.triggered.connect(qApp.quit)的功能是，当选中特定命令后会发送一个触发信号，将信号连接到 QApplication 组件的 quit()方法，这样就会中断当前整个应用程序。

❑ 使用方法 menuBar()创建一个菜单栏，首先创建一个"文件"菜单，然后将退出命令添加到"文件"菜单中。

通过上述代码创建了含有一个菜单"文件"的菜单栏，执行后会发现在这个菜单中包含一个"退出"命令，选择"退出"命令后会退出当前程序，执行效果如图 9-47 所示。

图 9-47 执行效果

第 9 章 图形化界面开发实战

实例 09-46：在 PyQt 窗体中创建工具栏

前面介绍的菜单栏可以集成所有的操作命令，这样我们可以方便地在程序中使用这些被集成的命令。而接下来讲解的工具栏则是由按钮和一些常规操作命令组成的组件，提供了一个快速访问常用命令的方式。例如，下面的实例文件 PyQt06.py 演示了在 PyQt 窗体中创建工具栏的过程。

源码路径：daima\9\09-46\PyQt06.py

```
class Example(QMainWindow):

    def __init__(self):
        super().__init__()
        self.initUI()

    def initUI(self):
        exitAction = QAction(QIcon('123.ico'), '退出', self)
        exitAction.setShortcut('Ctrl+Q')
        exitAction.triggered.connect(qApp.quit)
        self.toolbar = self.addToolBar('退出')
        self.toolbar.addAction(exitAction)
        self.setGeometry(300, 300, 300, 200)
        self.setWindowTitle('工具栏练习')
        self.show()
```

在上述代码中创建了一个 QAction 对象，整个过程和前面介绍的菜单栏中的部分代码相似。这个 QAction 对象包含一个标签、图标和快捷键，将 QMainWindow 的 quit() 方法连接到了触发信号上。执行后会创建了一个简单的工具栏图标，单击这个工具栏图标后会退出当前程序。执行效果如图 9-48 所示。

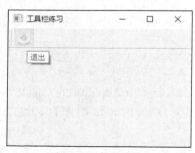

图 9-48　执行效果

实例 09-47：在 PyQt 窗体中使用绝对定位方式

绝对定位是指在程序设置每一个组件的位置坐标，并且以像素为单位设置每个组件的大小。在 PyQt 窗体中使用绝对定位方式时需要清楚如下 4 点限制：
- 如果改变了窗口大小，则组件的位置和大小并不会发生改变；
- 在不同操作系统上，应用程序的外观可能会不同；
- 如果改变应用程序中的字体，则可能会打乱整个应用程序的布局；
- 如果想修改之前的布局，则必须完全重写布局，这样会非常浪费时间。

例如，下面的实例文件 PyQt07.py 演示了在 PyQt 窗体中使用绝对定位方式的过程。

源码路径：daima\9\09-47\PyQt07.py

```
class Example(QWidget):
    def __init__(self):
        super().__init__()
        self.initUI()

    def initUI(self):
```

```
lbl1 = QLabel('第1行文本', self)
lbl1.move(15, 10)
lbl2 = QLabel('第2行文本', self)
lbl2.move(35, 40)
lbl3 = QLabel('第3行文本', self)
lbl3.move(55, 70)
self.setGeometry(300, 300, 250, 150)
self.setWindowTitle('Absolute')
self.show()
```

在上述代码中,使用 move()方法定位了窗体中的 3 行文本。在使用 move()方法时,给 move()方法提供了横坐标值和纵坐标值作为参数。方法 move()使用的坐标系统是从左上角开始计算的。横坐标值从左到右增长,纵坐标值从上到下增长。这种使用详细坐标定位的方式就是绝对定位。执行效果如图 9-49 所示。

图 9-49　执行效果

实例 09-48:使用箱布局

库 PyQt 专门提供了内置的界面布局管理类,这是将组件定位在窗口上的首选方式。其中,QHBoxLayout 和 QVBoxLayout 是两个基础布局管理类,分别实现水平和垂直的箱布局功能。假如需要在窗口右下角设置两个按钮,可以使用一个水平箱布局和垂直箱布局来实现。下面的实例文件 PyQt08.py 使用箱布局。

源码路径:daima\9\09-48\PyQt08.py

```
class Example(QWidget):
    def __init__(self):
        super().__init__()
        self.initUI()

    def initUI(self):
        okButton = QPushButton("确定")
        cancelButton = QPushButton("取消")
        hbox = QHBoxLayout()
        hbox.addStretch(1)
        hbox.addWidget(okButton)
        hbox.addWidget(cancelButton)
        vbox = QVBoxLayout()
        vbox.addStretch(1)
        vbox.addLayout(hbox)
        self.setLayout(vbox)
        self.setGeometry(300, 300, 300, 150)
        self.setWindowTitle('按钮布局')
        self.show()
```

在上述代码中使用了两个布局管理类,即 QHBoxLayout 和 QVBoxLayout,具体流程如下。

❑ 使用 QPushButton()创建两个按钮;
❑ 使用如下 4 行代码创建一个水平箱布局,并且增加一个拉伸因子和两个按钮。通过拉伸因子在两个按钮的前面增加了一个可伸缩空间,这样可在窗口的靠右部分显示按钮。

```
hbox = QHBoxLayout()
hbox.addStretch(1)
hbox.addWidget(okButton)
hbox.addWidget(cancelButton)
```

❑ 通过如下 3 行代码把水平箱布局放置在垂直布局内，使用拉伸因子把包含两个按钮的水平箱布局挪到窗口的底部。

```
vbox = QVBoxLayout()
vbox.addStretch(1)
vbox.addLayout(hbox)
```

❑ 使用 setLayout(vbox)设置窗口的主布局。

上述代码执行后会在窗口右下角设置两个按钮，当改变应用窗口大小时，这两个按钮的位置相对于窗口不改变，这也正是箱布局与绝对布局的最大区别。执行效果如图 9-50 所示。

图 9-50　执行效果

实例 09-49：使用网格布局模拟实现一个计算器界面

网格布局是指使用单元格（行和列）来分割空间。在 PyQt 程序中可以使用类 QGridLayout 实现网格布局功能。例如，下面的实例文件 PyQt09.py 演示了使用网格布局模拟实现一个计算器界面的过程。

源码路径：daima\9\09-49\PyQt09.py

```
class Example(QWidget):
    def __init__(self):
        super().__init__()
        self.initUI()

    def initUI(self):
        grid = QGridLayout()
        self.setLayout(grid)
        names = ['删除', '后退', '', '关闭',
                 '7', '8', '9', '除',
                 '4', '5', '6', '乘',
                 '1', '2', '3', '减',
                 '0', '.', '=', '加']

        positions = [(i, j) for i in range(5) for j in range(4)]
        for position, name in zip(positions, names):
            if name == '':
                continue
            button = QPushButton(name)
            grid.addWidget(button, *position)
        self.move(300, 150)
        self.setWindowTitle('模拟计算器')
        self.show()
```

❑ 通过 QGridLayout()创建一个网格布局对象。
❑ 使用 positions 设置单元格位置设置组件跨行和跨列的参数，在此设置 reviewEdit 组件跨 5 行显示。

上述代码执行后会创建多个数字按钮和四则运算按钮，并使用网格布局排列上述元素。执行效果如图 9-51 所示。

图 9-51 执行效果

实例 09-50：使用表单布局实现一个留言板界面

在 PyQt 程序中，可以使用类 QFormLayout 来布局输入型控件和关联标签组成的表单（Form）。QFormLayout 是一个使用方便的布局类，其中的控件以两列的形式被布局在表单中。左列包括标签，右列包括输入控件，如 QLineEdit、QSpinBox 和 QTextEdit 等。例如，下面的实例文件 PyQt10.py 演示了使用表单布局实现一个留言板界面的过程。

源码路径：daima\9\09-50\PyQt10.py

```
class Example(QWidget):
    def __init__(self):
        super().__init__()
        self.Init_UI()

    def Init_UI(self):
        self.setGeometry(300,300,300,200)
        self.setWindowTitle('书创文化留言板系统')
        formlayout = QFormLayout()
        nameLabel = QLabel("标题")
        nameLineEdit = QLineEdit("")
        introductionLabel = QLabel("内容")
        introductionLineEdit = QTextEdit("")
        formlayout.addRow(nameLabel,nameLineEdit)
        formlayout.addRow(introductionLabel,introductionLineEdit)
        self.setLayout(formlayout)
        self.show()
```

- 使用 QFormLayout() 创建一个表单布局对象。
- 使用 formlayout.addRow() 向表单中增加一行，内容是定义的组件。

执行效果如图 9-52 所示。

图 9-52 执行效果

实例 09-51：使用单击按钮事件处理程序

PyQt5 提供了一个独一无二的信号和槽机制来处理事件，信号和槽用于实现对象之间的通信功能。当发生指定事件时，一个事件信号会被发送。槽可以被任何 Python 脚本调用，当和槽连接的信号被发送时会调用槽。例如，下面的实例文件 PyQt11.py 演示了使用单击按钮事件处

源码路径：daima\9\09-51\PyQt11.py

```
class Ico(QWidget):
    def __init__(self):
        super().__init__()
        self.initUI()

    def initUI(self):
        self.setGeometry(300, 300, 300, 220)
        self.setWindowTitle('书创文化')
        self.setWindowIcon(QIcon('123.ico'))
        qbtn = QPushButton('退出', self)
        qbtn.clicked.connect(QCoreApplication.instance().quit)
        qbtn.resize(70,30)
        qbtn.move(50, 50)
        self.show()
```

在上述代码中，使用 QPushButton 类创建了一个按钮，该按钮是 QPushButton 类的一个实例。构造函数的第一个参数"退出"是按钮的标签，第二个参数 self 是父窗口小部件。父窗口小部件是示例窗口小部件，它是通过 QWidget 继承的。后面的 qbtn.clicked.connect()用于处理单击按钮事件，PyQt5 中的事件处理系统采用信号和槽机制构建。如果单击按钮，则单击动作的信号被发送出去。槽可以是 Qt 槽函数或任何 Python 可调用的函数。QCoreApplication 包含主事件循环，它用于处理和调度所有事件。上述代码执行会显示拥有一个按钮的窗体，执行效果如图 9-53 所示。单击"退出"按钮后，当前窗体关闭。

图 9-53 执行效果

实例 09-52：在 PyQt5 中使用信号和槽

例如，下面的实例文件 PyQt12.py 演示了在 PyQt5 中使用信号和槽的过程。

源码路径：daima\9\09-52\PyQt12.py

```
class Example(QWidget):
    def __init__(self):
        super().__init__()
        self.initUI()

    def initUI(self):
        lcd = QLCDNumber(self)
        sld = QSlider(Qt.Horizontal, self)
        vbox = QVBoxLayout()
        vbox.addWidget(lcd)
        vbox.addWidget(sld)
        self.setLayout(vbox)
        sld.valueChanged.connect(lcd.display)
        self.setGeometry(300, 300, 250, 150)
        self.setWindowTitle('信号和槽')
        self.show()
```

在上述代码中用到了 QLCDNumber 类和 QSlider 类,当拖动滑块时,上面的 lcd 数字会随之发生变化。滑块的 valueChanged 信号和 lcd 数字显示的 display 槽连接在一起。发送者是一个发送了信号的对象,接收者是一个接收了信号的对象,槽是对信号做出反应的方法。执行效果如图 9-54 所示。

图 9-54　执行效果

实例 09-53:重新实现按键盘按键后的操作功能

在 PyQt 程序中,通常通过重写事件处理函数实现事件处理功能。例如,下面的实例文件 PyQt13.py 演示了重新实现按键盘按键后的操作功能的过程。

源码路径:daima\9\09-53\PyQt13.py

```
class Example(QWidget):
    def __init__(self):
        super().__init__()
        self.initUI()

    def initUI(self):
        self.setGeometry(300, 300, 250, 150)
        self.setWindowTitle('事件处理')
        self.show()

    def keyPressEvent(self, e):
        if e.key() == Qt.Key_Escape:
            self.close()
```

在上述代码中,通过自定义编写的函数重写了 keyPressEvent()事件处理函数。执行程序后,如果按键盘中的 Esc 键,则当前程序窗体关闭。

实例 09-54:重新实现按住方向键后的操作功能

下面的实例文件 PyQt14.py 演示了重新实现按住方向键后的操作功能的过程。

源码路径:daima\9\09-54\PyQt14.py

```
class Example(QWidget):
    def __init__(self):
        super().__init__()
        self.initUi()

    def initUi(self):
        self.setGeometry(300, 300, 350, 250)
        self.setWindowTitle('书创文化')
        self.lab = QLabel('移动方向',self)
        self.lab.setGeometry(150,100,50,50)
        self.show()

    def keyPressEvent(self, e):
        if e.key() == Qt.Key_Up:
            self.lab.setText('↑')
        elif e.key() == Qt.Key_Down:
            self.lab.setText('↓')
```

```
        elif e.key() == Qt.Key_Left:
            self.lab.setText('←')
        else:
            self.lab.setText('→')
```

在上述代码中重新实现了 keyPressEvent()事件处理函数,当按住键盘中的上、下、左、右方向键的时候,窗口中会显示对应的方位标记。例如,按住上方向键时的效果如图 9-55 所示。

图 9-55 按住上方向键时的效果

实例 09-55:实现人机对战"石头、剪刀、布"小游戏

在现实程序开发过程中,有时需要知道哪个窗口组件是信号的发送者,这一功能可以通过 PyQt5 中的 sender()方法实现。例如,下面的实例文件 PyQt15.py 演示了实现人机对战"石头、剪刀、布"小游戏的过程。

源码路径:daima\9\09-55\PyQt15.py

```
class Example(QWidget):

    def __init__(self):
        super().__init__()
        self.initUI()
    def initUI(self):
        self.setGeometry(200, 200, 300, 300)
        self.setWindowTitle('书创文化传播')
        bt1 = QPushButton('剪刀',self)
        bt1.setGeometry(30,180,50,50)
        bt2 = QPushButton('石头',self)
        bt2.setGeometry(100,180,50,50)
        bt3 = QPushButton('布',self)
        bt3.setGeometry(170,180,50,50)
        bt1.clicked.connect(self.buttonclicked)
        bt2.clicked.connect(self.buttonclicked)
        bt3.clicked.connect(self.buttonclicked)
        self.show()

    def buttonclicked(self):
        computer = randint(1,3)
        player = 0
        sender = self.sender()
        if sender.text() == '剪刀':
            player = 1
        elif sender.text() == '石头':
            player = 2
        else:
            player = 3
        if player == computer:
            QMessageBox.about(self, '结果', '平手')
        elif player == 1 and computer == 2:
            QMessageBox.about(self, '结果', '智者:石头,我赢了!')
        elif player == 2 and computer == 3:
            QMessageBox.about(self, '结果', '智者:布,我赢了!')
```

```
        elif player == 3 and computer == 1:
            QMessageBox.about(self,'结果','智者：剪刀，我赢了！')
        elif computer == 1 and player == 2:
            QMessageBox.about(self,'结果','智者：剪刀，人类赢了！')
        elif computer == 2 and player == 3:
            QMessageBox.about(self,'结果','智者：石头，人类赢了！')
        elif computer == 3 and player == 1:
            QMessageBox.about(self,'结果','智者：布，人类赢了！')
```

在上述代码中设置了 3 个按钮，分别代表石头、剪刀、布。在 buttonclicked()方法中，通过调用 sender()方法来确定用户单击了哪个按钮。bt1、bt2 和 bt3 这 3 个按钮的 clicked 信号都被连接到同一个槽 buttonclicked。通过调用 sender()方法来确定信号源，然后根据信号源确定人类究竟选择了石头、剪刀、布中的哪一个，从而与智者随机给出的结果进行比较，最终进行判断。执行效果如图 9-56 所示。

图 9-56　执行效果

实例 09-56：发送自定义信号

在 PyQt 程序中，可以从 QObejct 对象中发送自定义信号。例如，下面的实例文件 PyQt16.py 演示了发送自定义信号的过程。

源码路径：daima\9\09-56\PyQt16.py

```
class Signal(QObject):
    showmouse = pyqtSignal()

class Example(QWidget):

    def __init__(self):
        super().__init__()
        self.initUI()
    def initUI(self):
        self.setGeometry(200, 200, 300, 300)
        self.setWindowTitle('书创文化')
        self.s = Signal()
        self.s.showmouse.connect(self.about)

        self.show()
    def about(self):
        QMessageBox.about(self,'警告','从实招来！你是不是单击鼠标了？')

    def mousePressEvent(self, e):
        self.s.showmouse.emit()
```

❑ 创建一个名为 showmouse 的新信号，该信号在鼠标单击事件期间会被发出，将该信号连接到 QWidget 的 about 槽。

- 使用方法 pyqtSignal()作为外部 Signal 类的类属性，以创建一个信号。
- 在代码 self.s.showmouse.connect(self.about)中，将自定义的 showmouse 信号连接到 QWidget 的 about 槽。
- 定义函数 mousePressEvent()，当单击时，会发出 showmouse 信号，并调用相应的槽函数。

执行上述代码，当我们单击鼠标的时候，就会弹出对话框提醒我们单击了鼠标。执行效果如图 9-57 所示。

图 9-57　执行效果

实例 09-57：使用对话框获取用户名信息

在库 PyQt 中，类 QInputDialog 提供了实现一个简单的输入对话框功能，用于从用户处获得一个值。输入值可以是字符串、数字，或者一个列表中的列表项。例如，下面的实例文件 PyQt17.py 演示了使用对话框获取用户名信息的过程。

源码路径：daima\9\09-57\PyQt17.py

```python
class Example(QWidget):

    def __init__(self):
        super().__init__()
        self.initUI()

    def initUI(self):
        self.btn = QPushButton('Dialog', self)
        self.btn.move(20, 20)
        self.btn.clicked.connect(self.showDialog)
        self.le = QLineEdit(self)
        self.le.move(130, 22)
        self.setGeometry(300, 300, 290, 150)
        self.setWindowTitle('输入对话框')
        self.show()

    def showDialog(self):
        text, ok = QInputDialog.getText(self, '输入对话框', '用户名:')
        if ok:
            self.le.setText(str(text))
```

在上述代码中，通过 QInputDialog.getText()方法实现一个输入对话框效果，其第 1 个字符串参数 self 是对话框的标题，第 2 个字符串参数"输入对话框"是对话框内的消息文本。对话框返回输入的文本内容和一个布尔值。如果单击了 OK 按钮，则布尔值就是 True，反之布尔值是 False。只有在单击 OK 按钮时，返回的文本内容才会有值。

上述代码执行后会，界面显示一个按钮和一个单行文本框组件，单击 Dialog 按钮后会弹出输入对话框，用于获得用户的输入字符串，在对话框中输入的字符串会在单行文本框组件中显示。执行效果如图 9-58 所示。

（a）初始效果　　　　　　（b）在对话框中输入"aaa"　　　　　（c）文本框中显示"aaa"

图 9-58　执行效果

实例 09-58：使用颜色选择对话框设置背景颜色

在库 PyQt 中，类 QColorDialog 提供了一个实现颜色选择对话框效果的组件。例如，下面的实例文件 PyQt18.py 演示了使用颜色选择对话框设置背景颜色的过程。

源码路径：daima\9\09-58\PyQt18.py

```
class Example(QWidget):

    def __init__(self):
        super().__init__()
        self.initUI()

    def initUI(self):
        col = QColor(0, 0, 0)
        self.btn = QPushButton('选择颜色', self)
        self.btn.move(20, 20)

        self.btn.clicked.connect(self.showDialog)
        self.frm = QFrame(self)
        self.frm.setStyleSheet("QWidget { background-color: %s }"
                                % col.name())
        self.frm.setGeometry(130, 22, 100, 100)
        self.setGeometry(300, 300, 250, 180)
        self.setWindowTitle('颜色选择对话框')
        self.show()

    def showDialog(self):
        col = QColorDialog.getColor()
        if col.isValid():
            self.frm.setStyleSheet("QWidget { background-color: %s }" % col.name())
```

- ❏ 使用 QColor(0,0,0)初始化 QWidget 组件的背景颜色为黑色。
- ❏ 通过 QColorDialog.getColor()弹出一个颜色选择对话框。
- ❏ 通过 if col.isValid():语句进行判断，如果用户选中一个颜色并且单击 OK 按钮，则会返回一个有效的颜色值。如果用户单击 Cancel 按钮，则不会返回选中的颜色值。另外，还可以使用样式表来定义背景颜色。

上述代码执行后，窗体中显示一个按钮和一个 Qframe 组件，QFrame 组件的背景颜色被设置为黑色。可以使用颜色选择对话框来改变 Qframe 组件的背景颜色。执行效果如图 9-59 所示。

第 9 章 图形化界面开发实战

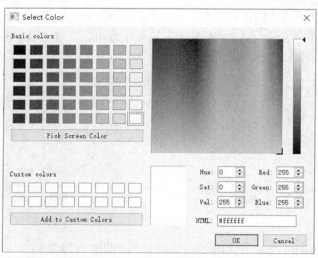

(a) 初始效果　　　　　　　　　(b) 单击"选择颜色"按钮后弹出颜色选择（Select Color）对话框

图 9-59　执行效果

实例 09-59：使用字体选择对话框设置字体

在库 PyQt 中，类 QFontDialog 是一个实现字体选择对话框效果的组件。例如，下面的实例文件 PyQt19.py 演示了使用字体选择框设置字体的过程。

源码路径：daima\9\09-59\PyQt19.py

```python
class Example(QWidget):

    def __init__(self):
        super().__init__()
        self.initUI()

    def initUI(self):
        vbox = QVBoxLayout()
        btn = QPushButton('选择字体', self)
        btn.setSizePolicy(QSizePolicy.Fixed,
                          QSizePolicy.Fixed)
        btn.move(20, 20)
        vbox.addWidget(btn)
        btn.clicked.connect(self.showDialog)
        self.lbl = QLabel('请看我的显示字体', self)
        self.lbl.move(130, 20)
        vbox.addWidget(self.lbl)
        self.setLayout(vbox)
        self.setGeometry(300, 300, 250, 180)
        self.setWindowTitle('字体对话框效果')
        self.show()

    def showDialog(self):
        font, ok = QFontDialog.getFont()
        if ok:
            self.lbl.setFont(font)
```

- 通过 QFontDialog.getFont() 弹出一个字体选择对话框，方法 getFont() 能够返回字体名字和布尔值。如果用户单击 OK 按钮，则布尔值为 True，否则为 False。
- 通过代码 if ok 进行判断，如果用户单击 OK 按钮，则文本的字体发生改变。

上述代码执行后，窗体中显示一个按钮和一行文本。单击"选择字体"按钮，弹出字体选择对话框，在此可以改变文本的字体。执行效果如图 9-60 所示。

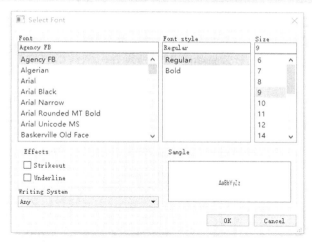

(a)初始效果　　　　　　(b)单击"选择字体"按钮后弹出字体选择对话框

图 9-60　执行效果

实例 09-60：使用文件选择对话框选择一个文件

在库 PyQt 中，类 QFileDialog 实现了文件选择对话框效果。通过文件选择对话框，用户可以选择某个文件或目录，可以进行文件的打开和保存操作。例如，下面的实例文件 PyQt20.py 演示了使用文件选择对话框选择一个文件的过程。

源码路径：daima\9\09-60\PyQt20.py

```python
class Example(QMainWindow):
    def __init__(self):
        super().__init__()
        self.initUI()

    def initUI(self):
        self.textEdit = QTextEdit()
        self.setCentralWidget(self.textEdit)
        self.statusBar()
        openFile = QAction(QIcon('123.ico'), '打开', self)
        openFile.setShortcut('Ctrl+O')
        openFile.setStatusTip('打开一个文件')
        openFile.triggered.connect(self.showDialog)
        menubar = self.menuBar()
        fileMenu = menubar.addMenu('&文件')
        fileMenu.addAction(openFile)
        self.setGeometry(300, 300, 350, 300)
        self.setWindowTitle('文件对话框')
        self.show()

    def showDialog(self):
        fname = QFileDialog.getOpenFileName(self, '打开文件', '/home')
        if fname[0]:
            f = open(fname[0], 'r')
            with f:
                data = f.read()
                self.textEdit.setText(data)
```

❑ 因为要设置一个文本编辑框组件，所以定义一个基于 QMainWindow 组件的类 Example (QMainWindow)。

❑ 通过 QFileDialog.getOpenFileName(self, '打开文件', '/home')实现弹出文件选择对话框的功能。参数"打开文件"是 getOpenFileName()方法的标题，后面的字符串参数"/home"指定了对话框的工作目录。默认将文件过滤器设置成 All files(*), 表示所有格式的文件。

❑ 选中文件后通过 data = f.read()和 self.textEdit.setText(data)读取文件的内容，并设置成文本编辑框组件的显示文本。

上述代码执行后，窗体中显示一个菜单栏，其中设置了一个文本编辑框组件和一个状态栏。选择菜单项会显示文件选择对话框，用于选择一个文件。文件的内容会被读取并显示在文本编辑框组件中。执行效果如图 9-61 所示。

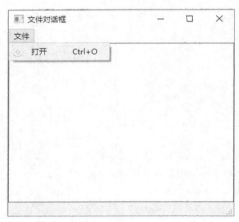
（a）初始效果

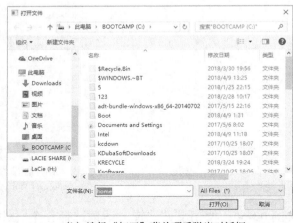
（b）选择"打开"菜单项后弹出对话框

图 9-61　执行效果

实例 09-61：使用 QCheckBox 实现复选框功能

PyQt 中的复选框组件有选中和未选中两种状态，它是由一个选择框和一个标签组成的。QCheckBox（复选框）和 QRadioButton（单选按钮）都是选项按钮，这是因为它们都可以在开（选中）或者关（未选中）之间切换。两者区别是对用户选择的限制：单选按钮提供的是"多选一"的选择，而复选框提供的是"多选多"的选择。只要复选框被选中或清除，都会发送一个 stateChanged 信号。如果想在复选框状态改变的时候触发一个行为，则需要连接这个信号，我们可以使用 isChecked() 来查询复选框是否被选中。

除了常用的选中和未选中两个状态之外，QCheckBox 还提供了可选状态（半选）。如果需要用到可选状态，则可以通过 setTristate() 使它生效，并使用 checkState() 来查询当前的切换状态。

例如，下面的实例文件 PyQt21.py 演示了使用 QCheckBox 实现复选框功能的过程。

源码路径：daima\9\09-61\PyQt21.py

```
class Example(QWidget):

    def __init__(self):
        super().__init__()
        self.initUI()

    def initUI(self):

        cb = QCheckBox('显示标题', self)
        cb.move(20, 20)
        cb.toggle()
        cb.stateChanged.connect(self.changeTitle)
        self.setGeometry(300, 300, 250, 150)
        self.setWindowTitle('QCheckBox')
        self.show()

    def changeTitle(self, state):
        if state == Qt.Checked:
            self.setWindowTitle('QCheckBox')
        else:
            self.setWindowTitle('')
```

❑ 首先用到了类 QCheckBox 的构造方法 QCheckBox('显示标题', self)。

- 通过 cb.toggle()设置窗口标题。在默认情况下，复选框不会被选中，窗口标题也不会被设置。
- 通过 cb.stateChanged.connect(self.changeTitle)将自定义的槽方法 changeTitle()和信号 stateChanged 进行连接，方法 changeTitle()能够切换窗口的显示标题。
- 定义函数 changeTitle(self, state)，复选框组件的状态会传入 changeTitle()方法的 state 参数。如果复选框被选中，则我们需要设置窗口标题；否则，我们需要把窗口标题设置成一个空字符串。

上述代码执行后，窗体中会创建一个复选框，通过复选框能够切换窗体的标题。执行效果如图 9-62 所示。

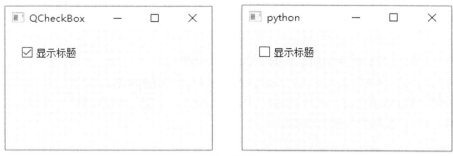

图 9-62 执行效果

实例 09-62：使用 QRadioButton 实现单选按钮功能

QRadioButton 是一个选项按钮，可以打开（选中）或关闭（取消选中），通常为用户提供"多选一"操作。在一组单选按钮中，一次只能选中一个单选按钮；如果用户选中另一个按钮，则先前选中的按钮被取消选中。例如，下面的实例文件 PyQt22.py 演示了使用 QRadioButton 实现单选按钮功能的过程。

源码路径：daima\9\09-62\PyQt22.py

```
class RadioButton(QtWidgets.QWidget):
    def __init__(self, parent=None):
        QtWidgets.QWidget.__init__(self)

        self.setGeometry(300, 300, 250, 150)
        self.setWindowTitle('Check')

        self.rb = QRadioButton('显示标题', self)
        self.rb.setFocusPolicy(Qt.NoFocus)

        self.rb.move(10, 10)
        self.rb.toggle()
        self.rb.toggled.connect(self.changeTitle)

    def changeTitle(self, value):
        if self.rb.isChecked():
            self.setWindowTitle('已经选择')
        else:
            self.setWindowTitle('没有选择')
```

- 使用 QRadioButton('显示标题', self)语句创建一个标签为"显示标题"的单选按钮。
- 通过 self.rb.toggled.connect(self.changeTitle)将用户定义的 changeTitle()函数与单选按钮的 toggled 信号连接起来，自定义函数 changeTitle()将重置窗口的标题。
- 使用 self.rb.setFocusPolicy(Qt.NoFocus)设置无聚焦样式。

- 因为在初始化状态下设置窗口的标题，因此需要使用代码 self.rb.toggle()将单选按钮选中。在默认情况下，单选按钮处于未被选中状态。

上述代码执行后，窗体中会创建一个用来改变窗口标题的单选按钮。执行效果如图 9-63 所示。

图 9-63　执行效果

实例 09-63：使用 QPushButton 实现切换按钮功能

在库 PyQt 中，切换按钮是 QPushButton 的特殊模式。切换按钮有两种状态：按下和没有按下。用户可以通过单击它来在两种状态之间切换。例如，下面的实例文件 PyQt23.py 演示了使用 QPushButton 实现切换按钮功能的过程。

源码路径：daima\9\09-63\PyQt23.py

```
class Example(QWidget):

    def __init__(self):
        super().__init__()

        self.initUI()

    def initUI(self):

        self.col = QColor(0, 0, 0)

        redb = QPushButton('红', self)
        redb.setCheckable(True)
        redb.move(10, 10)

        redb.clicked[bool].connect(self.setColor)

        redb = QPushButton('绿', self)
        redb.setCheckable(True)
        redb.move(10, 60)

        redb.clicked[bool].connect(self.setColor)

        blueb = QPushButton('蓝', self)
        blueb.setCheckable(True)
        blueb.move(10, 110)

        blueb.clicked[bool].connect(self.setColor)

        self.square = QFrame(self)
        self.square.setGeometry(150, 20, 100, 100)
        self.square.setStyleSheet("QWidget { background-color: %s }" %
                                    self.col.name())

        self.setGeometry(300, 300, 280, 170)
        self.setWindowTitle('切换按钮')
        self.show()

    def setColor(self, pressed):

        source = self.sender()
```

```
if pressed:
    val = 255
else:
    val = 0

if source.text() == "红":
    self.col.setRed(val)
elif source.text() == "绿":
    self.col.setGreen(val)
else:
    self.col.setBlue(val)

self.square.setStyleSheet("QFrame { background-color: %s }" % self.col.name())
```

- 通过 QColor(0, 0, 0)实现颜色初始化功能，设置颜色为黑色。
- 通过 QPushButton('红', self)创建切换按钮，并且通过 setCheckable(True)调用 setCheckable()方法让它可被按下。
- 通过代码 redb.clicked[bool].connect(self.setColor)把 clicked 信号连接到定义的方法上，使用 clicked 信号来操作布尔值。
- 使用 self.sender()获得发生状态切换的按钮。
- 使用判断语句 if source.text() == "红"，如果发生切换的按钮是"红"按钮，则更新 RGB 值中的红色部分的颜色值。
- 通过代码 self.square.setStyleSheet("QWidget { background-color: %s }" % self.col.name())使用样式表来改变背景颜色。

上述代码执行后，窗体中会创建 3 个切换按钮和 1 个 QWidget 组件，QWidget 组件的背景颜色初始化为黑色。单击切换按钮后，QWidget 组件的背景颜色在红色、绿色和蓝色之间进行切换。QWidget 组件的背景颜色取决于哪一个切换按钮被按下，执行效果如图 9-64 所示。

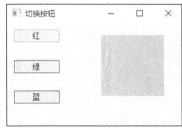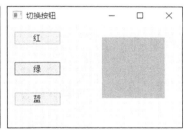

图 9-64　执行效果

实例 09-64：使用 QSlider 实现一个音量控制器

在库 PyQt 中，可以使用滑块（Qslider）组件实现滑动条效果。滑动条中有一个简单的可调节手柄，通过前后拖动这个手柄，可以选择一个具体的数值。有时使用滑动条比直接输入数字或使用数值选择框更方便。例如，下面的实例文件 PyQt24.py 演示了使用 QSlider 实现一个音量控制器的过程。

源码路径：daima\9\09-64\PyQt24.py

```
class Example(QWidget):

    def __init__(self):
        super().__init__()

        self.initUI()

    def initUI(self):

        sld = QSlider(Qt.Horizontal, self)
        sld.setFocusPolicy(Qt.NoFocus)
```

```
                    sld.setGeometry(30, 40, 100, 30)
                    sld.valueChanged[int].connect(self.changeValue)

                    self.label = QLabel(self)
                    self.label.setPixmap(QPixmap('123.ico'))
                    self.label.setGeometry(160, 40, 80, 30)

                    self.setGeometry(300, 300, 280, 170)
                    self.setWindowTitle('QSlider')
                    self.show()

            def changeValue(self, value):

                    if value == 0:
                            self.label.setPixmap(QPixmap('mute.png'))
                    elif value > 0 and value <= 30:
                            self.label.setPixmap(QPixmap('min.png'))
                    elif value > 30 and value < 80:
                            self.label.setPixmap(QPixmap('med.png'))
                    else:
                            self.label.setPixmap(QPixmap('max.png'))
```

- 使用 QSlider(Qt.Horizontal, self)创建一个横向的滑动条。
- 使用 valueChanged[int].connect(self.changeValue)把 valueChanged 信号连接到自定义的 changeValue()方法上。
- 使用 QLabel(self)创建一个标签组件，使用 setPixmap(QPixmap('mute.png'))设置一个初始的无声图片。
- 通过 if 语句根据滑动条的值设置不同的标签图片，例如，代码 self.label.setPixmap (QPixmap ('mute.png'))表示如果滑块条的值等于 0，则标签显示 mute.png 图片。

上述代码执行后，窗体中会显示一个滑动条和一个标签，标签中会显示一个图像，每个图像代表一种音量大小级别。通过拖动滑动条的调节手柄，可以显示不同的标签图像。执行效果如图 9-65 所示。

  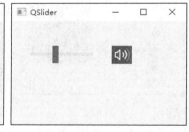

图 9-65　执行效果

实例 09-65：使用 QProgressBar 实现一个进度条效果

在库 PyQt5 中，可以使用进度条组件 QProgressBar 实现横向和纵向的进度条功能。我们可以设置进度条的最大值和最小值，进度条的默认值是 0～99。例如，下面的实例文件 PyQt25.py 演示了使用 QProgressBar 实现一个进度条效果的过程。

源码路径：daima\9\09-65\PyQt25.py
```
class Example(QWidget):
    def __init__(self):
        super().__init__()
        self.initUI()

    def initUI(self):
        self.pbar = QProgressBar(self)
        self.pbar.setGeometry(30, 40, 200, 25)
        self.btn = QPushButton('开始', self)
        self.btn.move(40, 80)
```

```
        self.btn.clicked.connect(self.doAction)
        self.timer = QBasicTimer()
        self.step = 0
        self.setGeometry(300, 300, 280, 170)
        self.setWindowTitle('进度条练习')
        self.show()

    def timerEvent(self, e):
        if self.step >= 100:
            self.timer.stop()
            self.btn.setText('完成')
            return
        self.step = self.step + 1
        self.pbar.setValue(self.step)

    def doAction(self):
        if self.timer.isActive():
            self.timer.stop()
            self.btn.setText('开始')
        else:
            self.timer.start(100, self)
            self.btn.setText('停止')
```

- 使用进度条类的构造方法 QProgressBar(self) 创建一个进度条对象；
- 使用定时器对象 QtCore.QBasicTimer() 激活进度条。
- 使用方法 self.timer.start(100,self) 开启定时器事件，参数 "100" 表示定时时间，参数 self 表示接收定时器事件的对象。
- 因为每个 QObject 类和的子类都有用于处理定时事件的事件处理函数 timerEvent()，为了对定时器事件做出反馈，重新实现了这个事件处理函数 timerEvent(self, e)，
- 上述代码执行后，窗体中会显示一个横向进度条和一个按钮，通过这个按钮可以控制进度条的开始和停止。
- 编写函数 doAction(self)，用于开始和停止定时器。

执行效果如图 9-66 所示。

图 9-66 执行效果

实例 09-66：使用 QCalendarWidget 实现一个日历

在库 PyQt5 中，类 QCalendarWidget 提供了一个基于月的日历组件，允许用户通过简单、直观的方式选择日期。例如，下面的实例文件 PyQt26.py 演示了使用 QCalendarWidget 实现一个日历的过程。

源码路径：daima\9\09-66\PyQt26.py

```
class Example(QWidget):
    def __init__(self):
        super().__init__()
        self.initUI()

    def initUI(self):
        cal = QCalendarWidget(self
```

```
            cal.setGridVisible(True)
            cal.move(20, 20)
            cal.clicked[QDate].connect(self.showDate)

            self.lbl = QLabel(self)
            date = cal.selectedDate()
            self.lbl.setText(date.toString())
            self.lbl.move(130, 260)
            self.setGeometry(300, 300, 350, 300)
            self.setWindowTitle('日历')
            self.show()

        def showDate(self, date):
            self.lbl.setText(date.toString())
```

- 使用 QCalendarWidget(self)创建 QCalendarWidget 对象。
- 编写代码 cal.clicked[QDate].connect(self.showDate)，如果在组件上选择了一个日期，则会发送 clicked[QDate]信号，把这个信号和自定义的 showDate()方法连接。
- 定义方法 showDate(self, date)，通过方法 selectedDate()检索选中的日期，然后把选中的日期转化成字符串显示在标签组件上。

上述代码执行后，窗体中会显示一个日历组件和标签组件，在日历中选择的日期会显示在标签组件中。执行效果如图 9-67 所示。

图 9-67　执行效果

实例 09-67：在窗口中显示一个图片

在库 PyQt5 中，像素图组件是在屏幕上显示图片的最佳选择。例如，下面的实例文件 PyQt27.py 演示了使用像素图组件在窗口中显示一个图片的过程。

源码路径：daima\9\09-67\PyQt27.py

```
class Example(QWidget):

    def __init__(self):
        super().__init__()
        self.initUI()

    def initUI(self):
        hbox = QHBoxLayout(self)
        pixmap = QPixmap("111.jpg")
        lbl = QLabel(self)
        lbl.setPixmap(pixmap)
        hbox.addWidget(lbl)
        self.setLayout(hbox)
        self.move(300, 200)
        self.setWindowTitle('显示图片')
        self.show()
```

- 使用代码 QPixmap("111.jpg ")创建 QPixmap 对象，该对象的构造方法将一个文件的名字作为参数。
- 通过代码 QLabel(self)把像素图对象设置给标签，使用 lbl.setPixmap(pixmap)通过标签来显示像素图。

上述代码执行后，指定图片 111.jpg 会显示在窗体中，执行效果如图 9-68 所示。

图 9-68 执行效果

实例 09-68：创建一个单行文本编辑框

在库 PyQt5 中，允许在单行文本编辑框组件中输入单行的纯文本数据，此组件支持撤销、重做、剪切、粘贴和拖动等方法。例如，下面的实例文件 PyQt28.py 演示了使用组件 QLineEdit 创建一个单行文本编辑框的过程。

源码路径：daima\9\09-68\PyQt28.py

```
class Example(QWidget):

    def __init__(self):
        super().__init__()
        self.initUI()

    def initUI(self):
        self.lbl = QLabel(self)
        qle = QLineEdit(self)
        qle.move(60, 100)
        self.lbl.move(60, 40)
        qle.textChanged[str].connect(self.onChanged)
        self.setGeometry(300, 300, 280, 170)
        self.setWindowTitle('单行文本框')
        self.show()

    def onChanged(self, text):
        self.lbl.setText(text)
        self.lbl.adjustSize()
```

- 通过 QLineEdit(self)创建单行文本编辑框（QLineEdit）组件。
- 编写代码 qle.textChanged[str].connect(self.onChanged)，如果单行文本编辑框中的文本内容发生改变，则调用 onChanged()方法进行处理。

❑ 方法 onChanged(self, text)设置了标签的显示文本，通过调用 adjustSize()方法来调整标签相对于显示文本的长度。

上述代码执行后，窗体中会显示一个单行编辑文本框和一个标签。在单行文本编辑框中输入文本时，文本会在标签中同步显示。执行效果如图 9-69 所示。

图 9-69　执行效果

实例 09-69：创建两个分割框组件

在库 PyQt5 中，使用分割框组件 QSplitter 可以通过拖拽分割线的方式来控制子组件的大小。例如，下面的实例文件 PyQt29.py 演示了创建两个分割框组件来控制 3 个 QFrame 组件范围的过程。

源码路径：daima\9\09-69\PyQt29.py

```
class Example(QWidget):

    def __init__(self):
        super().__init__()
        self.initUI()

    def initUI(self):
        hbox = QHBoxLayout(self)
        topleft = QFrame(self)
        topleft.setFrameShape(QFrame.StyledPanel)
        topright = QFrame(self)
        topright.setFrameShape(QFrame.StyledPanel)
        bottom = QFrame(self)
        bottom.setFrameShape(QFrame.StyledPanel)
        splitter1 = QSplitter(Qt.Horizontal)
        splitter1.addWidget(topleft)
        splitter1.addWidget(topright)
        splitter2 = QSplitter(Qt.Vertical)
        splitter2.addWidget(splitter1)
        splitter2.addWidget(bottom)
        hbox.addWidget(splitter2)
        self.setLayout(hbox)
        self.setGeometry(300, 300, 300, 200)
        self.setWindowTitle('分割框组件（QSplitter）')
        self.show()

    def onChanged(self, text):
        self.lbl.setText(text)
        self.lbl.adjustSize()
```

❑ 创建一个 QFrame 对象，通过代码 topleft.setFrameShape(QFrame.StyledPanel)使用一个样式框架，这样做是为了让 QFrame 组件之间的分割线显示得更加明显。

❑ 通过代码 QSplitter(Qt.Horizontal)创建一个分割框组件，并分别通过 addWidget(topleft) 和 addWidget(topright)在这个分割框组件中添加两个 QFrame 组件。

❑ 编写代码 splitter2 = QSplitter(Qt.Vertical)和 splitter2.addWidget(splitter1)，把第一个分割框组件添加进另一个分割框组件中。

上述代码执行后，窗体中会显示 3 个 QFrame 组件和 2 个分割框组件，如图 9-70 所示。需要注意的是，在某些样式主题下，分割框组件可能不会显示。

图 9-70　执行效果

实例 09-70：使用 Eric6 提高开发效率

Eric6 是 PyQt 编程的最佳集成开发环境（Integrated Development Environment，IDE），我们可以在其官网进行下载。下载后会得到一个压缩文件，对其解压缩后进入解压的目录，然后执行如下命令安装 Eric6。

```
python install.py
```

注意：如果读者的计算机中没有安装 QScintilla，则需要先通过如下命令安装 QScintilla，然后才能执行上面的命令安装 Eric6。

```
pip install QScintilla
```

成功安装 Eric6 后会显示如下提示信息。

```
Checking dependencies
Python Version: 3.6.0
Found PyQt5
Found pyuic5
Found QScintilla2
Found QtGui
Found QtNetwork
Found QtPrintSupport
Found QtSql
Found QtSvg
Found QtWidgets
Found QtWebEngineWidgets
Qt Version: 5.10.1
sip Version: 4.19.8
PyQt Version: 5.10.1
QScintilla Version: 2.10.3
All dependencies ok.

Cleaning up old installation ...

Creating configuration file ...

Compiling user interface files ...

Compiling source files ...

Installing eric6 ...

Installation complete.
Press enter to continue...
```

Eric6 的使用方法和传统语言 IDE 的类似，只需拖拽对应的组件即可实现窗体界面的规划和布局工作，并且可以自动生成 Python 代码。Eric6 的相关用法不是本书的重点，具体用法请读者参阅其他相关资料。

9.4 使用 pyglet 库

在 Python 程序中，可以使用库 pyglet 开发跨平台窗口及多媒体程序。本节将通过具体实例来讲解库 pyglet 的使用方法。

实例 09-71：创建第一个 pyglet 程序

可以使用如下命令安装 pyglet 库。

```
pip install pyglet
```

也可以使用如下命令安装最新版本的 pyglet 库。

```
pip install --upgrade https://bitbucket.org/pyglet/pyglet/get/tip.zip
```

例如，下面的实例文件 pyglet01.py 演示了创建第一个 pyglet 程序的过程。

源码路径：daima\9\09-71\pyglet01.py

```python
import pyglet
window = pyglet.window.Window()

label = pyglet.text.Label('Hello, world',
                          font_name='Times New Roman',
                          font_size=36,
                          x=window.width//2, y=window.height//2,
                          anchor_x='center', anchor_y='center')
@window.event
def on_draw():
    window.clear()
    label.draw()

pyglet.app.run()
```

在上述代码中，首先使用 import 语句引入库 pyglet，然后通过 pyglet.text.Label()在窗体中定义了一个标签，分别设置了标签中显示的文本内容、字体、字体大小、位置和对齐方式。执行效果如图 9-71 所示。

图 9-71 执行效果

实例 09-72：在窗体中显示指定图片

下面的实例文件 pyglet02.py 演示了使用库 pyglet 在窗体中显示指定图片的过程。

源码路径：daima\9\09-72\pyglet02.py

```python
window = pyglet.window.Window()
image = pyglet.resource.image('111.jpg')
```

```
@window.event
def on_draw():
    window.clear()
    image.blit(0, 0)

pyglet.app.run()
```

在上述代码中，使用 pyglet.resource 中的 image()函数来加载图像，此函数会自动查找图像文件与源文件（而不是工作目录）。在使用 blit()方法绘制图像时，会告知 pyglet 在坐标(0, 0)的位置绘制图像，这个坐标在窗体的左下角。上述代码执行后，窗体中会显示指定的图片 111.jpg，如图 9-72 所示。

图 9-72　执行效果

实例 09-73：使用库 pyglet 处理键盘事件程序

和其他的 GUI 库类似，库 pyglet 也能够处理鼠标和键盘等的事件程序。例如，下面的实例文件 pyglet03.py 演示了使用库 pyglet 处理键盘事件程序的过程。

源码路径：daima\9\09-73\pyglet03.py

```
def on_key_press(symbol, modifiers):
    if symbol == key.A:
        print('The "A" key was pressed.')
    elif symbol == key.LEFT:
        print('The left arrow key was pressed.')
    elif symbol == key.ENTER:
        print('The enter key was pressed.')

@window.event
def on_draw():
    window.clear()

pyglet.app.run()
```

上述代码执行后将会根据用户按的按键输出对应的提示信息。

```
The left arrow key was pressed.
The left arrow key was pressed.
The left arrow key was pressed.
The enter key was pressed.
The enter key was pressed.
The enter key was pressed.
```

实例 09-74：在屏幕上绘制一个三角形

库 pyglet 提供一个接口，便于开发者开发 OpenGL 和 GLU 程序，该接口可以供所有的 pyglet 高级 API 使用，可以快速完成所有的图形渲染功能，而不是操作系统实现这些功能。在实现 OpenGL 操作之前需要先用如下代码导入 pyglet.gl。

```
from pyglet.gl import *
```
pyglet.gl 中的所有函数名和常量都与 C 接口对应。例如，下面的实例文件 pyglet04.py 演示了使用库 pyglet 在屏幕上绘制一个三角形的过程。

源码路径：daima\9\09-74\pyglet04.py

```
def on_draw():
    glClear(GL_COLOR_BUFFER_BIT)
    glLoadIdentity()
    glBegin(GL_TRIANGLES)
    glVertex2f(0, 0)
    glVertex2f(window.width, 0)
    glVertex2f(window.width, window.height)
    glEnd()

pyglet.app.run()
```

执行效果如图 9-73 所示。

图 9-73　执行效果

实例 09-75：使用顶点数组绘制三角形

在使用某些 OpenGL 函数实现绘图功能时需要用到一组数据，这组数据通常是由不同类型的数组构成的阵列。下面的实例文件 pyglet05.py 演示了绘制与前面实例相同的三角形的过程，但是本实例是使用顶点数组实现的，而不是像前面实例文件 pyglet04.py 那样直接使用模式函数实现。

源码路径：daima\9\09-75\pyglet05.py

```
vertices = [0, 0,
            window.width, 0,
            window.width, window.height]
vertices_gl_array = (GLfloat * len(vertices))(*vertices)

glEnableClientState(GL_VERTEX_ARRAY)
glVertexPointer(2, GL_FLOAT, 0, vertices_gl_array)

@window.event
def on_draw():
    glClear(GL_COLOR_BUFFER_BIT)
    glLoadIdentity()
    glDrawArrays(GL_TRIANGLES, 0, len(vertices) // 2)

pyglet.app.run()
```

执行效果如图 9-74 所示。

另外，还可以使用上述类似的阵列结构实现顶点缓冲对象、创建数据结构、多边形点数据，以及地图功能。

图 9-74 执行效果

实例 09-76：开发一个 Minecraft 游戏

下面的实例文件 pyglet06.py 演示了使用库 pyglet 开发一个 Minecraft 游戏的过程。Minecraft 是一个几乎无所不能的沙盒游戏，中文非官方译名为"我的世界"。本实例使用 Python 和 pyglet 实现一个简单的 Minecraft 游戏，通过这个项目来学习 pyglet 和 Python 游戏编程。

在前期规划本项目时，设置通过如下按键控制精灵的移动。

- W：向前。
- S：向后。
- A：向左。
- D：向右。
- Mouse：环顾四周。
- Space：跳跃。
- Tab：切换飞行模式。

在游戏中创建了 4 种方块类型：石块、砖块、草地和沙块。按下鼠标左键会消除方块，按下鼠标右键会创造方块，按 Esc 键会释放鼠标，然后关闭窗口。

实例文件 pyglet06.py 的具体实现流程如下。

（1）设置在项目中需要的几个公共常量值，具体实现代码如下。

源码路径：daima\9\09-76\pyglet06.py

```
TICKS_PER_SEC = 60  #每秒刷新60次

# 用于减小块负荷的扇区的大小
SECTOR_SIZE = 16

WALKING_SPEED = 5        #移动速度
FLYING_SPEED = 15        #飞行速度

GRAVITY = 20.0
MAX_JUMP_HEIGHT = 1.0    #一个块的高度
# 跳跃速度公式
JUMP_SPEED = math.sqrt(2 * GRAVITY * MAX_JUMP_HEIGHT)
TERMINAL_VELOCITY = 50   #自由下落终端速度

PLAYER_HEIGHT = 2        #玩家高度

if sys.version_info[0] >= 3:
    xrange = range
```

（2）定义函数 cube_vertices()，返回以 (x,y,z) 为中心、边长为 2n 的正方体 6 个面的顶点坐标。

（3）定义纹理坐标函数 tex_coord()，给出纹理图左下角坐标，返回一个正方形纹理的 4 个顶点坐标。因为可以将纹理图看成 4×4 的纹理 patch（小块），所以此处的 n=4（图中实际有 6 个 patch，其他 patch 是空白）。例如，欲返回左下角的那个正方形纹理 patch，如果输入左下角的整数坐标(0,0)，则输出是 0、0、1/4、0、1/4、1/4、0、1/4。

（4）编写函数 tex_coord() 计算一个正方体 6 个面的纹理贴图坐标，将结果放入一个列表中，这 6 个面分别是顶面（top）、底面（bottom）和 4 个侧面（side）。

（5）设置使用的纹理图片是 texture.png，然后计算草块、沙块、砖块、石块 6 个面的纹理贴图坐标（用一个列表保存）。具体实现代码如下。

```
TEXTURE_PATH = 'texture.png'
# 由此可见，除了草块，其他的正方体6个面的贴图都一样
GRASS = tex_coords((1, 0), (0, 1), (0, 0))
SAND = tex_coords((1, 1), (1, 1), (1, 1))
BRICK = tex_coords((2, 0), (2, 0), (2, 0))
STONE = tex_coords((2, 1), (2, 1), (2, 1))
```

（6）设置当前位置向 6 个方向移动 1 个单位要用到的增量坐标。

```
FACES = [
    ( 0, 1, 0),
    ( 0,-1, 0),
    (-1, 0, 0),
    ( 1, 0, 0),
    ( 0, 0, 1),
    ( 0, 0,-1),
]
```

（7）编写函数 normalize() 对位置 x、y 和 z 取整，具体实现代码如下。

```
def normalize(position):
    x, y, z = position
    x, y, z = (int(round(x)), int(round(y)), int(round(z)))
    return (x, y, z)
```

（8）编写函数 sectorize() 计算位置，首先对位置坐标（x,y,z）取整，然后将结果各除以 SECTOR_SIZE，返回（x,0,z）。这样会将许多不同的位置映射到同一个(x,0,z)，一个(x,0,z)对应一个 xzy=16×16×y 的区域内的所有立方体中心位置。

（9）编写函数 _initialize() 绘制地图，大小是 80×80。具体实现代码如下。

```
def _initialize(self):
    n = 80    #地图大小
    s = 1     #步长
    y = 0     #初始化y
    for x in xrange(-n, n + 1, s):
        for z in xrange(-n, n + 1, s):
            #在地下画一层石块，上面是一层草块
            #地面从y=-2开始
            self.add_block((x, y - 2, z), GRASS, immediate=False)
            self.add_block((x, y - 3, z), STONE, immediate=False)
            #地图的四周用墙围起来
            if x in (-n, n) or z in (-n, n):
                #创建外墙
                for dy in xrange(-2, 3):
                    self.add_block((x, y + dy, z), STONE, immediate=False)

    #为了避免建到墙上，o取n-10
    o = n - 10
    #在地面上随机建造一些草块、沙块、砖块
    for _ in xrange(120):         #只想迭代120次，不需要迭代变量i，直接用 _
        a = random.randint(-o, o) #在[-o,o]内随机取一个整数
        b = random.randint(-o, o) #山丘的z位置
        c = -1                    #山丘的底部
        h = random.randint(1, 6)  #山丘的高度
        s = random.randint(4, 8)  #s是山丘的边长
        d = 1                     #如何迅速地从山丘上逐渐消失
        t = random.choice([GRASS, SAND, BRICK])
        for y in xrange(c, c + h):
            for x in xrange(a - s, a + s + 1):
                for z in xrange(b - s, b + s + 1):
                    if (x - a) ** 2 + (z - b) ** 2 > (s + 1) ** 2:
                        continue
                    if (x - 0) ** 2 + (z - 0) ** 2 < 5 ** 2:
```

```
            continue
        self.add_block((x, y, z), t, immediate=False)
    s -= d                    #递减边的长度, 山丘逐渐减少
```

(10) 编写函数 hit_test(), 检测使用鼠标是否能对一个立方体进行操作。此函数返回 key、previous, 其中 key 是鼠标可操作的块的中心坐标, 根据人所在位置和方向向量求出。而 previous 是与 key 处立方体相邻的空位置的中心坐标。如果返回非空, 则按下鼠标左键删除 key 处立方体, 按下鼠标右键在 previous 处添加砖块。

(11) 编写函数 exposed(), 设置只要指定位置周围 6 个面有一个没有被立方体包围, 返回真值, 表示要绘制指定位置处的立方体。如果 6 个面都被立方体包围, 则可以不绘制指定位置处的立方体, 因为即使绘制了也看不到。

(12) 编写函数 add_block() 添加立方体。

(13) 编写函数 remove_block() 删除立方体, 具体实现代码如下。

```
def remove_block(self, position, immediate=True):
    del self.world[position]                        #把world中的位置、纹理删除
    self.sectors[sectorize(position)].remove(position)  #把区域中相应的位置删除
    if immediate:                                    #如果同步
        if position in self.shown:                  #如果位置在显示列表中
            self.hide_block(position)               #立即删除它
        self.check_neighbors(position)
```

(14) 编写检查函数 check_neighbors(), 在删除一个立方体后检查它周围 6 个邻接的位置是否有因此暴露出来的立方体, 如果有则要把它绘制出来。具体实现代码如下。

```
def check_neighbors(self, position):
    x, y, z = position
    for dx, dy, dz in FACES:                        #检查周围6个位置
        key = (x + dx, y + dy, z + dz)
        if key not in self.world:                   #如果该位置没有立方体则跳过
            continue
        if self.exposed(key):                       #如果该位置有立方体且暴露在外
            if key not in self.shown:               #且没有在显示列表中
                self.show_block(key)                #则立即绘制出来
        else:   #如果没有暴露在外, 而又在显示列表中, 则立即隐藏(删除)它
            if key in self.shown:
                self.hide_block(key)
```

(15) 编写函数 show_block(), 功能是将 world 中还没显示且显露在外的立方体绘制出来。

(16) 编写函数 _show_block(), 将顶点列表 (VertexList) 添加到渲染对象 (on_draw() 函数会负责渲染), 并将"position:VertexList"对存入 _shown。

(17) 编写函数 hide_block() 隐藏立方体。

(18) 编写函数 _hide_block() 立即移除立方体, 将位置的顶点列表弹出并删除, 相应的立方体立即被移除, 其实整个操作是在 update() 之后进行的。

(19) 编写函数 show_sector() 绘制一个区域内的立方体, 如果区域内的立方体位置不在显示列表中, 且该立方体是显露在外的则显示立方体。

(20) 编写函数 hide_sector() 设置隐藏区域, 如果一个立方体在显示列表中则隐藏它。

(21) 编写函数 change_sectors() 移动立方体, 移动区域是一个连续的 x、y 子区域。

(22) 编写函数 _enqueue() 添加事件到队列 (queue) 中, 具体实现代码如下。

```
def _enqueue(self, func, *args):
    self.queue.append((func, args))
```

(23) 编写函数 _dequeue() 处理队头事件, 具体实现代码如下。

```
def _dequeue(self):
    func, args = self.queue.popleft()
    func(*args)
```

(24) 编写函数 process_queue() 用 1/60s 的时间来处理队列中的事件, 但是不一定要处理完。具体实现代码如下。

```
def process_queue(self):
    start = time.clock()
```

```
while self.queue and time.clock() - start < 1.0 / TICKS_PER_SEC:
    self._dequeue()
```

（25）编写函数 process_entire_queue() 处理事件队列中的所有事件，具体实现代码如下。

```
def process_entire_queue(self):
    while self.queue:
        self._dequeue()
```

（26）定义类 Window 来处理窗体界面，通过函数 __init__() 实现界面初始化处理。

（27）编写函数 set_exclusive_mouse() 设置鼠标事件是否绑定到游戏窗口，具体实现代码如下。

```
def set_exclusive_mouse(self, exclusive):
    super(Window, self).set_exclusive_mouse(exclusive)
    self.exclusive = exclusive
```

（28）编写函数 get_sight_vector()，功能是根据前进方向 rotation 来决定移动 1 单位距离时各轴的移动分量。

（29）编写函数 get_motion_vector()，功能是在运动时计算 3 个轴的位移增量。

（30）编写函数 update()，设置每 1/60s 调用一次来进行更新。

（31）编写函数 update() 更新 self.dy 和 self.position。

（32）编写函数 collide() 实现碰撞检测，返回值 p 表示碰撞检测后应该移动到的位置。如果没有遇到障碍物，则 p 仍然是当前位置；否则，p 是新的值（会使其沿着墙移动）。

（33）编写函数 on_mouse_press() 处理鼠标按下事件，具体实现代码如下。

```
def on_mouse_press(self, x, y, button, modifiers):
    if self.exclusive:#当鼠标事件已经绑定了此窗口
        vector = self.get_sight_vector()
        block, previous = self.model.hit_test(self.position, vector)
        #如果按下左键且该处有block
        if (button == mouse.RIGHT) or \
                ((button == mouse.LEFT) and (modifiers & key.MOD_CTRL)):
            if previous:
                self.model.add_block(previous, self.block)
        elif button == pyglet.window.mouse.LEFT and block:
            #如果按下右键，且有previous位置，则在previous处增加方块
            texture = self.model.world[block]
            if texture != STONE:# 如果block不是石块，则移除它
                self.model.remove_block(block)
    else: #否则隐藏鼠标，并绑定鼠标事件到该窗口
        self.set_exclusive_mouse(True)
```

（34）编写函数 on_mouse_motion() 处理鼠标移动事件实现视角的变化，参数 dx 和 dy 分别表示鼠标从上一位置移动到当前位置 x 轴、y 轴上的位移量。

（35）编写函数 on_key_press() 处理按下键盘事件，长按 W、S、A、D 键后会不断地改变坐标。

（36）编写函数 on_key_release() 处理释放按键事件。

（37）编写函数 on_resize() 处理窗口大小变化的响应事件。

（38）编写函数 set_2d() 设置在 OpenGL 中绘制三维图形。

（39）编写函数 set_3d() 设置在 OpenGL 中绘制三维图形，具体实现代码如下。

```
def set_3d(self):
    width, height = self.get_size()
    glEnable(GL_DEPTH_TEST)
    glViewport(0, 0, width, height)
    glMatrixMode(GL_PROJECTION)
    glLoadIdentity()
    gluPerspective(65.0, width / float(height), 0.1, 60.0)
    glMatrixMode(GL_MODELVIEW)
    glLoadIdentity()
    x, y = self.rotation
    glRotatef(x, 0, 1, 0)
    glRotatef(-y, math.cos(math.radians(x)), 0, math.sin(math.radians(x)))
    x, y, z = self.position
    glTranslatef(-x, -y, -z)
```

（40）编写函数 on_draw()，功能是重写 Window 的 on_draw() 函数。当需要重绘窗口时，事

件循环（EventLoop）就会调度该事件。

（41）编写函数 draw_focused_block()绘制鼠标指针聚焦的立方体，在它的外层绘制一个立方体线框。

（42）编写函数 draw_label()显示帧率，和在当前位置坐标显示的方块数及总共的方块数。具体实现代码如下。

```
def draw_label(self):
    x, y, z = self.position
    self.label.text = '%02d (%.2f, %.2f, %.2f) %d / %d' % (
        pyglet.clock.get_fps(), x, y, z,
        len(self.model._shown), len(self.model.world))
    self.label.draw()# 绘制label的text
```

（43）编写函数 draw_reticle()绘制游戏窗口中间的十字图标，也就是一条横线加一条竖线。具体实现代码如下。

```
def draw_reticle(self):
    glColor3d(0, 0, 0)
    self.reticle.draw(GL_LINES)
```

（44）编写函数 setup_fog()和 setup()设置雾效果。

（45）在主窗体函数 main()中调用前面的函数实现界面显示。

本实例的执行效果如图 9-75 所示。

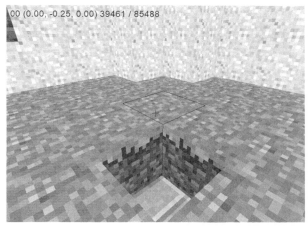

图 9-75　执行效果

9.5　使用 Toga 库

库 Toga 是一个 Python 原生、操作系统原生的 GUI 工具包。Toga 是一个跨平台工具包，可以在 macOS、Windows、Linux（GTK）、Android 和 iOS 操作系统中使用。本节将详细讲解使用库 Toga 开发 GUI 程序的方法。

实例 09-77：使用 Toga 创建第一个 GUI 程序

可以使用如下命令安装 Toga 库。

```
pip install toga
```

例如在下面的实例文件 toga01.py 中，演示了使用 Toga 创建第一个 GUI 程序的过程。

源码路径：daima\9\09-77\toga01.py

```
import toga

def button_handler(widget):
```

```
        print("hello")
def build(app):
    box = toga.Box()

    button = toga.Button('第一个GUI程序', on_press=button_handler)
    button.style.padding = 50
    button.style.flex = 1
    box.add(button)

    return box

def main():
    return toga.App('First App', '书创文化传播', startup=build)

if __name__ == '__main__':
    main().main_loop()
```

在上述代码中，使用一个按钮创建了一个 GUI 程序，当单击"第一个 GUI 程序"按钮时输出信息"hello"到控制台。其中，函数 button_handler()是一个处理程序，用于处理用户单击按钮时的行为。函数 button_handler()函数将激活的 widget 作为第 1 个参数，根据正在处理的事件的类型还可以提供其他参数。在只有一个简单的按钮被单击的情况下，该函数没有额外的参数。

再来看函数 build(app)，通过创建一个 app 程序创建一个主窗口，其中含有主菜单。然而 Toga 不知道我们想要在主窗口中包含什么内容。所以接下来定义一个设置在应用程序包含指定 UI 的方法，此方法是可接受的应用程序实例的可调用方法。在此我们设置了一个按钮，设置了在按钮中显示的文本，设置了按钮四周的空白都是 50 像素。

最后通过函数 main()实例化程序本身。程序有一个名称和唯一标识符，例如，上述程序设置的是"书创文化传播"，还设置了在窗体中显示的标题是"First App"。

上述代码执行后，窗体中会显示一个"第一个 GUI 程序"按钮，单击该按钮后，控制台中显示设置的提示信息。执行效果如图 9-76 所示。

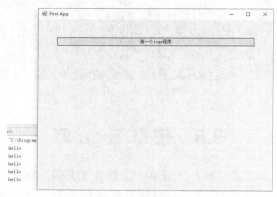

图 9-76　执行效果

实例 09-78：创建一个温度转换器

和前面介绍的库一样，库 Toga 也包含了 Button 按钮、Label 文本、Canvas 绘图和 ProgressBar 进度条等组件。有关各个组件的具体使用知识，请读者参阅 Toga 的官方文档。

例如，下面的实例文件 toga02.py 演示了使用 Toga 组件创建一个温度转换器的过程。

源码路径：daima\9\09-78\toga02.py

```
def build(app):
    c_box = toga.Box()
    f_box = toga.Box()
```

```
        box = toga.Box()

        c_input = toga.TextInput(readonly=True)
        f_input = toga.TextInput()

        c_label = toga.Label('摄氏度', style=Pack(text_align=LEFT))
        f_label = toga.Label('华氏温度', style=Pack(text_align=LEFT))
        join_label = toga.Label('相当于', style=Pack(text_align=RIGHT))

        def calculate(widget):
            try:
                c_input.value = (float(f_input.value) - 32.0) * 5.0 / 9.0
            except:
                c_input.value = '???'

        button = toga.Button('转换', on_press=calculate)

        f_box.add(f_input)
        f_box.add(f_label)

        c_box.add(join_label)
        c_box.add(c_input)
        c_box.add(c_label)

        box.add(f_box)
        box.add(c_box)
        box.add(button)

        box.style.update(direction=COLUMN, padding_top=10)
        f_box.style.update(direction=ROW, padding=5)
        c_box.style.update(direction=ROW, padding=5)

        c_input.style.update(flex=1)
        f_input.style.update(flex=1, padding_left=160)
        c_label.style.update(width=100, padding_left=10)
        f_label.style.update(width=100, padding_left=10)
        join_label.style.update(width=150, padding_right=10)

        button.style.update(padding=15, flex=1)

        return box

def main():
    return toga.App('温度转换器', '书创文化传播', startup=build)
```

通过上述代码设置了一个垂直堆叠的组件 Box；在这个 Box 组件中，我们放置了 2 个水平 Box 组件和 1 个按钮组件。因为在水平 Box 组件上没有宽度样式，所以会尝试将它们包含的小部件安装到可用空间中。TextInput 组件将被拉伸以适应可用的水平空间，边距和填充项确保里面的组件将垂直和水平对齐。执行效果如图 9-77 所示。

图 9-77　执行效果

实例 09-79：使用组件 ScrollContainer 实现滚动功能

在库 toga 中，实现布局功能的组件有 Box、ScrollContainer、SplitContainer 和 OptionContainer，

第 9 章 图形化界面开发实战

这些布局组件的用法都十分简单。例如，ScrollContainer（滚动容器组件）类似于 HTML 中的 iframe 或滚动 div 元素，包含了一个自身滚动选择对象。例如，下面的实例文件 toga03.py 演示了使用组件 ScrollContainer 实现滚动功能的过程。

源码路径：daima\9\09-79\toga03.py

```python
class ScrollContainerApp(toga.App):
    def startup(self):
        self.main_window = toga.MainWindow(self.name)
        box = toga.Box()
        box.style.direction = COLUMN

        for x in range(100):
            label_text = '文本 %d' % (x)
            box.add(toga.Label(label_text, style=Pack(text_align=LEFT)))

        scroller = toga.ScrollContainer()
        scroller.content = box

        self.main_window.content = scroller
        self.main_window.show()
```

通过上述代码创建了一个包含 100 行文本信息的滚动界面，执行效果如图 9-78 所示。

图 9-78 执行效果

实例 09-80：使用绘图组件

在创建 GUI 程序的过程中，较常用的功能是绘制和操作线条、形状、文本和其他图形。在库 Toga 中，可以使用绘图组件 Canvas 实现绘图功能。例如，下面的实例文件 toga04.py 演示了使用绘图组件绘制图形的过程。需要注意，本实例只能在 GTK 环境下执行。

源码路径：daima\9\09-80\toga04.py

```python
class ExampleCanvasApp(toga.App):
    def startup(self):
        #设置主窗口
        self.main_window = toga.MainWindow(title=self.name, size=(148, 200))
        canvas = toga.Canvas(style=Pack(flex=1))
        box = toga.Box(children=[canvas])
        #在主窗口中添加内容
        self.main_window.content = box
        #显示主窗口
        self.main_window.show()
        with canvas.stroke():
            with canvas.closed_path(50, 50):
                canvas.line_to(100, 100)
```

9.6　wxPython 实战

wxPython 是 wxWidgets（这是用 C++编写的）的 Python 封装，是一个流行的跨平台 GUI 工具包，由 Robin Dunn 和 Harri Pasanen 开发。wxPython 是一个 Python 扩展模块。

实例 09-81：开发第一个 wxPython 程序

可以通过如下命令安装 wxPython。

```
pip install -U wxPython
```

例如，下面的实例文件 wx01.py 演示了开发第一个 wxPython 程序的过程。

源码路径：daima\9\09-81\wx01.py

```python
import wx

app = wx.App()
window = wx.Frame(None, title="书创文化传播", size=(400, 300))
panel = wx.Panel(window)
label = wx.StaticText(panel, label="Hello World", pos=(100, 100))
window.Show(True)
app.MainLoop()
```

执行效果如图 9-79 所示。

图 9-79　执行效果

在上述代码中，使用 wx.Frame 类实现了一个窗体效果，类 Frame 拥有一个不带参数的默认构造函数，也有一个拥有参数的重载构造函数如下。

```
wx.Frame (parent, id, title, pos, size, style, name)
```

各个参数的具体说明如下。

- parent：窗口的父类，如果为 None，则对象显示在顶层窗口。
- id：窗口标识，设置为-1 可自动生成标识符。
- title：标题栏的标题。
- pos：帧（frame）的开始位置。
- size：窗口的尺寸。
- style：窗口的外观及样式风格。
- name：对象的内部名称。

实例 09-82：使用 StaticText 组件在窗体中显示文本

在 wxPython 中，wx.StaticText 类对象实现了只读文本组件功能。类 wx.StaticText 的构造函数如下。

Wx.StaticText(parent, id, label, position, size, style)

StaticText 组件中的预定义样式枚举器如表 9-1 所示。

表 9-1　　　　　　　　　　　　　　预定义样式枚举器

名　称	说　明
wx.ALIGN_LEFT	控制标签的大小及对齐方式
wx.ALIGN_RIGHT	
wx.ALIGN_CENTER	
wx.ST_NO_AUTORESIZE	防止标签自动调整大小
wx.ST_ELLIPSIZE_START	如果文本的长度大于标签尺寸，设置省略号（…）显示的位置（开始、中间或结尾）
wx.ST_ELLIPSIZE_MIDDLE	
wx.ST_ELLIPSIZE_END	

为了设置 StaticText 中的字体，首先需要创建一个字体对象：

wx.Font(pointsize, fontfamily, fontstyle, fontweight)

参数 fontfamily 有如下 3 个取值。

- ❑ wx.FONTSTYLE_NORMAL：普通字体。
- ❑ wx.FONTSTYLE_ITALIC：斜体字体。
- ❑ wx.FONTSTYLE_SLANT：斜体字体，且采用 Times New Roman 风格。

参数 fontweight 有如下 3 个取值。

- ❑ wx.FONTWEIGHT_NORMAL：普通字体。
- ❑ wx.FONTWEIGHT_LIGHT：高亮字体。
- ❑ wx.FONTWEIGHT_BOLD：粗体。

例如，下面的实例文件 wx02.py 演示了使用 StaticText 组件在窗体中显示文本的过程。

源码路径：daima\9\09-82\wx02.py

```python
class Mywin(wx.Frame):
    def __init__(self, parent, title):
        super(Mywin, self).__init__(parent, title=title, size=(600, 200))
        panel = wx.Panel(self)
        box = wx.BoxSizer(wx.VERTICAL)
        lbl = wx.StaticText(panel, -1, style=wx.ALIGN_CENTER)

        txt1 = "Python GUI development"
        txt2 = "using wxPython"
        txt3 = " Python port of wxWidget "
        txt = txt1 + "\n" + txt2 + "\n" + txt3

        font = wx.Font(18, wx.ROMAN, wx.ITALIC, wx.NORMAL)
        lbl.SetFont(font)
        lbl.SetLabel(txt)

        box.Add(lbl, 0, wx.ALIGN_CENTER)
        lblwrap = wx.StaticText(panel, -1, style=wx.ALIGN_RIGHT)
        txt = txt1 + txt2 + txt3

        lblwrap.SetLabel(txt)
        lblwrap.Wrap(200)
        box.Add(lblwrap, 0, wx.ALIGN_LEFT)

        lbl1 = wx.StaticText(panel, -1, style=wx.ALIGN_LEFT | wx.ST_ELLIPSIZE_MIDDLE)
        lbl1.SetLabel(txt)
        lbl1.SetForegroundColour((255, 0, 0))
        lbl1.SetBackgroundColour((0, 0, 0))

        font = self.GetFont()
        lbl1.SetFont(font)

        box.Add(lbl1, 0, wx.ALIGN_LEFT)
        panel.SetSizer(box)
```

```
            self.Centre()
            self.Show()
```
在上述代码中，3 个 StaticText 对象被放置在一个垂直的盒子大小测定器（BoxSizer）中。第 1 个标签的对象中心对准多行文本。第 2 个标签的文本设置为 200 像素。第 3 个标签显示省略号在文本的中间。执行效果如图 9-80 所示。

图 9-80　执行效果

实例 09-83：创建 4 种不同样式的文本框

在 wxPython 中，类 wx.TextCtrl 实现文本框组件功能，用于显示文本和编辑文本。TextCtrl 中的文本可以是单行的、多行的或密码字段。类 TextCtrl 的构造函数形式如下。

```
wx.TextCtrl(parent, id, value, pos, size, style)
```

例如，下面的实例文件 wx03.py 演示了使用 TextCtrl 组件创建 4 种不同样式的文本框的过程。

源码路径：daima\9\09-83\wx03.py

```python
class Mywin(wx.Frame):
    def __init__(self, parent, title):
        super(Mywin, self).__init__(parent, title=title, size=(350, 250))

        panel = wx.Panel(self)
        vbox = wx.BoxSizer(wx.VERTICAL)

        hbox1 = wx.BoxSizer(wx.HORIZONTAL)
        l1 = wx.StaticText(panel, -1, "文本域")

        hbox1.Add(l1, 1, wx.EXPAND | wx.ALIGN_LEFT | wx.ALL, 5)
        self.t1 = wx.TextCtrl(panel)

        hbox1.Add(self.t1, 1, wx.EXPAND | wx.ALIGN_LEFT | wx.ALL, 5)
        self.t1.Bind(wx.EVT_TEXT, self.OnKeyTyped)
        vbox.Add(hbox1)

        hbox2 = wx.BoxSizer(wx.HORIZONTAL)
        l2 = wx.StaticText(panel, -1, "密码文本")

        hbox2.Add(l2, 1, wx.ALIGN_LEFT | wx.ALL, 5)
        self.t2 = wx.TextCtrl(panel, style=wx.TE_PASSWORD)
        self.t2.SetMaxLength(5)

        hbox2.Add(self.t2, 1, wx.EXPAND | wx.ALIGN_LEFT | wx.ALL, 5)
        vbox.Add(hbox2)
        self.t2.Bind(wx.EVT_TEXT_MAXLEN, self.OnMaxLen)

        hbox3 = wx.BoxSizer(wx.HORIZONTAL)
        l3 = wx.StaticText(panel, -1, "多行文本")

        hbox3.Add(l3, 1, wx.EXPAND | wx.ALIGN_LEFT | wx.ALL, 5)
        self.t3 = wx.TextCtrl(panel, size=(200, 100), style=wx.TE_MULTILINE)

        hbox3.Add(self.t3, 1, wx.EXPAND | wx.ALIGN_LEFT | wx.ALL, 5)
        vbox.Add(hbox3)
        self.t3.Bind(wx.EVT_TEXT_ENTER, self.OnEnterPressed)

        hbox4 = wx.BoxSizer(wx.HORIZONTAL)
        l4 = wx.StaticText(panel, -1, "只读取文本")
```

```
                hbox4.Add(l4, 1, wx.EXPAND | wx.ALIGN_LEFT | wx.ALL, 5)
                self.t4 = wx.TextCtrl(panel, value="只读文本", style=wx.TE_READONLY | wx.TE_CENTER)

                hbox4.Add(self.t4, 1, wx.EXPAND | wx.ALIGN_LEFT | wx.ALL, 5)
                vbox.Add(hbox4)
                panel.SetSizer(vbox)
                self.Centre()
                self.Show()
                self.Fit()

        def OnKeyTyped(self, event):
                print(event.GetString())

        def OnEnterPressed(self, event):
                print("Enter pressed")

        def OnMaxLen(self, event):
                print("Maximum length reached")
```

在上述代码中，不同属性的 wx.TextCtrl 类的 4 个对象放置在窗体中。其中，第 1 个是普通的文本框，第 2 个是一个密码字段文本框，第 3 个是多行文本框，最后一个是不可编辑的文本框。当第 1 个文本框 EVT_TEXT 绑定器触发 OnKeyTyped()函数时，可以处理每个按键事件。设置第 2 个文本框的最大长度为 5，一旦用户试图输入超过 500 个字符，EVT_TEXT_MAXLEN 绑定器会触发 OnMaxLen()函数。因为设置了 EVT_TEXT_ENTER 绑定器，所以多行文本框会响应按 Enter 键事件。执行效果如图 9-81 所示。

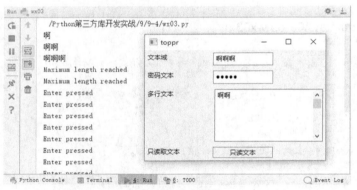

图 9-81　执行效果

实例 09-84：使用 RadioButton 组件

在 wxPython 中，类 wx.RadioButton 实现单选按钮组件功能，RadioButton 组件包含一个圆形按钮和一个文本标签。为了创建一组互斥选择的单选按钮，首先将 wx.RadioButton 对象的样式参数设置为 wx.RB_GROUP，其他单选按钮会被添加到一组中。例如，下面的实例文件 wx04.py 演示了使用 RadioBox 组件的过程。

源码路径：daima\9\09-84\wx04.py

```
class Example(wx.Frame):

        def __init__(self, parent, title):
                super(Example, self).__init__(parent, title=title, size=(300, 200))
                self.InitUI()

        def InitUI(self):
                pnl = wx.Panel(self)
                self.rb1 = wx.RadioButton(pnl, 11, label='Value A',
                                        pos=(10, 10), style=wx.RB_GROUP)
                self.rb2 = wx.RadioButton(pnl, 22, label='Value B', pos=(10, 40))
                self.rb3 = wx.RadioButton(pnl, 33, label='Value C', pos=(10, 70))
```

```
            self.Bind(wx.EVT_RADIOBUTTON, self.OnRadiogroup)
            lblList = ['Value X', 'Value Y', 'Value Z']
            self.rbox = wx.RadioBox(pnl, label='RadioBox', pos=(80, 10), choices=lblList,
                                   majorDimension=1, style=wx.RA_SPECIFY_ROWS)
            self.rbox.Bind(wx.EVT_RADIOBOX, self.onRadioBox)
            self.Centre()
            self.Show(True)

        def OnRadiogroup(self, e):
            rb = e.GetEventObject()
            print(rb.GetLabel(), ' is clicked from Radio Group')

        def onRadioBox(self, e):
            print(self.rbox.GetStringSelection(), ' is clicked from Radio Box')
```

- 通过指定 wx.RB_GROUP 样式对 3 个单选按钮进行分组并将其放置在面板上。
- 在 RadioBox 中从 lblList 对象读出标签按钮。
- 设置两个事件绑定器，一个单选按钮组和其他单选按钮的 RadioBox 被声明。
- 通过相应的事件处理函数确定被选择的按钮，并在控制台输出消息。

执行效果如图 9-82 所示。

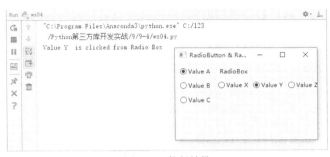

图 9-82　执行效果

> **注意**：wxPython 还包含很多其他重要的组件。限于本书篇幅，有关 wxPython 的其他重要知识，如界面布局、绘图、对话框和拖放处理等功能，建议读者参阅其官方资料。

9.7　GUI 高级实战

到目前为止，Python 开发中的常用 GUI 模块介绍完毕。本节将通过几个综合实例的实现过程，讲解 GUI 和其他技术融合使用的方法。

实例 09-85：实现 tkinter+ SQLite3 图书馆系统

下面的实例演示了实现 tkinter+SQLite3 图书馆系统的过程。实例文件 backend.py 的功能是创建各种常用的 SQL 语句以实现对 SQLite3 数据库的操作，具体来说包括如下 SQL 操作。

- 创建数据库表 book，实现函数是 __init__()。
- 向数据库中插入新的图书数据，实现函数是 add_command()。
- 查看数据库中的所有图书信息，实现函数是 view_command()。
- 搜索数据库中的某本图书的信息，实现函数是 search_command()。
- 删除数据库中的某本图书信息，实现函数是 delete_command()。
- 修改数据库中的某本图书信息，实现函数是 update_command()。

实例文件 frontend-bookstore.py 的功能是使用 tkinter 创建一个可视化界面，通过文本框组件显示图书信息，通过按钮组件调用执行要处理的 SQL 语句。实例文件 frontend-bookstore.py 的具体实现代码如下。

源码路径：daima\9\09-85\frontend-bookstore.py

```python
from tkinter import *
from backend import Database

database=Database("books.db")

class Window(object):

    def __init__(self,window):

        self.window = window

        self.window.wm_title("图书馆系统")

        l1=Label(window,text="书名")
        l1.grid(row=0,column=0)

        l2=Label(window,text="作者")
        l2.grid(row=0,column=2)

        l3=Label(window,text="出版时间")
        l3.grid(row=1,column=0)

        l4=Label(window,text="ISBN")
        l4.grid(row=1,column=2)

        self.title_text=StringVar()
        self.e1=Entry(window,textvariable=self.title_text)
        self.e1.grid(row=0,column=1)

        self.author_text=StringVar()
        self.e2=Entry(window,textvariable=self.author_text)
        self.e2.grid(row=0,column=3)

        self.year_text=StringVar()
        self.e3=Entry(window,textvariable=self.year_text)
        self.e3.grid(row=1,column=1)

        self.isbn_text=StringVar()
        self.e4=Entry(window,textvariable=self.isbn_text)
        self.e4.grid(row=1,column=3)

        self.list1=Listbox(window, height=6,width=35)
        self.list1.grid(row=2,column=0,rowspan=6,columnspan=2)

        sb1=Scrollbar(window)
        sb1.grid(row=2,column=2,rowspan=6)

        self.list1.configure(yscrollcommand=sb1.set)
        sb1.configure(command=self.list1.yview)

        self.list1.bind('<<ListboxSelect>>',self.get_selected_row)

        b1=Button(window,text="所有图书", width=12,command=self.view_command)
        b1.grid(row=2,column=3)

        b2=Button(window,text="搜索图书", width=12,command=self.search_command)
        b2.grid(row=3,column=3)

        b3=Button(window,text="添加图书", width=12,command=self.add_command)
        b3.grid(row=4,column=3)

        b4=Button(window,text="修改选定图书", width=12,command=self.update_command)
        b4.grid(row=5,column=3)

        b5=Button(window,text="删除选定图书", width=12,command=self.delete_command)
        b5.grid(row=6,column=3)

        b6=Button(window,text="关闭", width=12,command=window.destroy)
        b6.grid(row=7,column=3)

    def get_selected_row(self,event):
```

```
            if len(self.list1.curselection())>0:
                index=self.list1.curselection()[0]
                self.selected_tuple=self.list1.get(index)
                self.e1.delete(0,END)
                self.e1.insert(END,self.selected_tuple[1])
                self.e2.delete(0,END)
                self.e2.insert(END,self.selected_tuple[2])
                self.e3.delete(0,END)
                self.e3.insert(END,self.selected_tuple[3])
                self.e4.delete(0,END)
                self.e4.insert(END,self.selected_tuple[4])

    def view_command(self):
        self.list1.delete(0,END)
        for row in database.view():
            self.list1.insert(END,row)

    def search_command(self):
        self.list1.delete(0,END)
        for row in database.search(self.title_text.get(),self.author_text.get(),self.year_text.get(),self.isbn_text.get()):
            self.list1.insert(END,row)

    def add_command(self):
        database.insert(self.title_text.get(),self.author_text.get(),self.year_text.get(),self.isbn_text.get())
        self.list1.delete(0,END)
        self.list1.insert(END,(self.title_text.get(),self.author_text.get(),self.year_text.get(),self.isbn_text.get()))

    def delete_command(self):
        database.delete(self.selected_tuple[0])

    def update_command(self):
        database.update(self.selected_tuple[0],self.title_text.get(),self.author_text.get(),self.year_text.get(),self.isbn_text.get())

window=Tk()
Window(window)
window.mainloop()
```

在上述代码中，通过组件 Label 显示图书信息，通过组件 Button 实现了 6 个操作控制按钮，单击每个按钮会调用对应的函数实现数据处理。例如，单击"所有图书"按钮后会显示数据库中的所有图书信息，如图 9-83 所示。

图 9-83　显示数据库中的所有图书信息

实例 09-86：实现 tkinter + SQLite3 多线程计时器系统

本实例不仅实现了 tkinter + SQLite3 多线程计时器系统功能，而且实现了一个本系统使用的日志记录功能，日志记录功能是通过 SQLite3 数据库实现的。本实例的实现文件是 sql.py，其具体实现流程如下。

源码路径:daima\9\09-86\sql.py

(1) 导入需要的模块,对应实现代码如下。

```python
import threading
import time
import datetime
import sqlite3
import os
import functools
import tkinter as tk
from tkinter import ttk
from tkinter import messagebox as msg
```

(2) 定义多线程统计类 CountingThread,首先定义初始化函数 __init__(),在此创建一个框架来容纳项目所需要的内容组件,包括 start 按钮、Pause 按钮、剩余时间。在默认情况下,暂停按钮是禁用的,这是因为我们不能暂停计时器,直到计时器开始。

(3) 定义函数 run(),功能是通过参数 start_time 和 end_time 判断计时器的开始和结束,跟踪线程是否应该继续执行它的循环,循环是通过调用函数 main_loop()实现的。

(4) 编写函数 main_loop(),功能是获取剩余的时间,并适当地更新计时器的时间。通过 now()获取当前的时间,并检查它是否仍然低于 end_time。如果是则计算两者的差额,然后调用 divmod()得到时、分、秒格式的时间字符串,并再次检查当前时间是否为 force_quit,以防止超过之前的时间。对应实现代码如下。

```python
def main_loop(self):
    now = datetime.datetime.now()
    if now < self.end_time:
        time_difference = self.end_time - now
        mins, secs = divmod(time_difference.seconds, 60)
        time_string = "{:02d}:{:02d}".format(mins, secs)
        if not self.force_quit:
            self.master.update_time_remaining(time_string)
```

(5) 编写函数 LogWindow(),实现本项目的日志系统界面效果。这里需要查询数据库 SQLite3 记录的信息,并将获取的信息显示在窗口中。

(6) 定义计时器类 Timer,编写函数 Timer()初始化计时器功能,设置指定大小的窗体,并在里面布局各个计时器组件。

(7) 定义函数 setup_worker(),这是一个单独的线程,包含了对计时器实例的调用和引用,其功能是调用一次启动一个定时器,设置了计时时间是 25min,这需要创建一个新的实例。因为线程只能执行一次,所以我们不能将其用 __init__()初始化。对应实现代码如下。

```python
def setup_worker(self):
    now = datetime.datetime.now()
    in_25_mins = now + datetime.timedelta(minutes=25)
    #in_25_mins = now + datetime.timedelta(seconds=3)
    worker = CountingThread(self, now, in_25_mins)
    self.worker = worker
```

(8) 定义函数 start(),功能是启动计时器。如果当前没有计时任务,则创建一个计时任务;如果当前处于暂停状态,则切换到单击 start_button 按钮功能,并设置时间标签是 25min。

(9) 定义函数 pause(),单击 Pause 按钮后实现暂停计时功能。

(10) 定义函数 finish_early(),如果提前完成计时功能(时间未到直接单击 Finish 按钮),则只需要交换 Start 按钮和 Finish 按钮,并设置完成变量 end_now 的值为 True。对应实现代码如下。

```python
def finish_early(self):
    self.start_button.configure(text="Start", command=self.start)
    self.task_finished_early = True
    self.worker.end_now = True
```

(11) 定义函数 finish(),单击 Finish 按钮后实现完成计时功能。

(12) 定义函数 update_time_remaining(),功能是在屏幕中更新显示剩余时间。对应实现代码如下。

```
def update_time_remaining(self, time_string):
    self.time_remaining_var.set(time_string)
    self.update_idletasks()
```

（13）编写函数 add_new_task()，实现新增计时器功能，对应实现代码如下。

```
def add_new_task(self):
    task_name = self.task_name_entry.get()
    self.task_started_time = datetime.datetime.now()
    add_task_sql = "INSERT INTO pymodoros VALUES (?, 0, ?)"
    self.runQuery(add_task_sql, (task_name, self.task_started_time))
```

（14）编写函数 mark_finished_task()，用于标记当前计时器是完全完成的计时器。完全完成任务与未完全完成任务是相互对立的一个概念。当第一次启动任务时，会在数据库中添加一个条目，其中包含任务的名称和日期/时间。这个计时器任务开始（通过 add_new_task()）被标记为没有完全完成（25min），一旦单击 Finish 按钮，这个计时器任务就属于完全完成任务。

（15）定义函数 safe_destroy()，功能是如果用户要启动计时器，则关闭窗口后将留下一个执行线程。在这种情况下，当线程不能到达计时器实例时，线程就会抛出异常并退出。最好确保始终不离开应用程序，仍然保留活动线程，它会连接系统资源，让用户通过某种任务管理器关闭它们。对应实现代码如下。

```
def safe_destroy(self):
    if hasattr(self, "worker"):
        self.worker.force_quit = True
        self.after(100, self.safe_destroy)
    else:
        self.destroy()
```

（16）编写函数 runQuery()，功能是执行指定的 SQL 语句。

执行效果如图 9-84 所示。

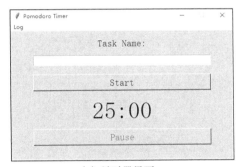

(a) 计时器界面　　　　　　　　　　　　(b) 日志界面

图 9-84　执行效果